DAVID SHEFF

GAME OVER

David Sheff's articles have appeared in *Playboy, Rolling Stone, The Observer*, and *Foreign Literature* (in Russia), among other publications, and on National Public Radio's *All Things Considered.* His book *The Playboy Interviews with John Lennon and Yoko Ono* was a Literary Guild Selection. Sheff lives in Northern California with his wife, Karen Barbour, and son, Nicolas.

DAVID SHEFF

Acclaim for **DAVID SHEFF's**

GAME OVER

"An intriguing portrait of what it takes to succeed in today's competitive computer industry."   —*Washington Post Book World*

"A fascinating story of the birth and growth of video games . . . Sheff has done a fine job explaining the high-tech computer world story of complicated games that children can play—and the people who are bringing this new world to us, whether we want it or not."   —*Kansas City Star*

"A fascinating look at the Japanese way of doing business . . . At times, *Game Over* reaches the pitch of a Cold War spy novel."   —*L.A. Daily News*

"Writing with the playful pluck of Mario, the little protagonist of the *Super Mario Bros.* games, Sheff . . . unfolds an engrossing tale of how Kyoto-based Nintendo, once a small playing-card company, transformed the U.S. toy and computer businesses."   —*People*

"Vivid portraits of the principal players and shrewd insights into Japanese and American corporate culture make this an unusually enjoyable business history."   —*Entertainment Weekly*

"My advice is simple. Read Sheff's book."   —*Wichita Eagle*

"David Sheff lays bare the corporate machine beneath the high-technology 'fun' of Japan's most profitable company. . . . Sheff painstakingly documents the history of Nintendo and its relentless rise to dominance of the global toy industry."   —*Maclean's*

"*Game Over* is a . . . fascinating look at the Nintendo phenomenon, filled with insiders' insights and plausible predictions about the future. . . . Sheff's book is an absorbing read, even for non-Nintendo junkies."   —*Birmingham News*

ALSO BY **DAVID SHEFF**

The *Playboy* Interviews with John Lennon and Yoko Ono

GAME OVER

How Nintendo
Conquered the World

This book is dedicated to Karen Barbour, who insists that Donatello, Rafael, Leonardo, and Michelangelo are painters, not Teenage Mutant Ninja Turtles, and to my son, Nicholas, who introduced me to Nintendo, but now prefers reading.

FIRST GAMEPRESS EDITION, JANUARY 1999

ISBN 0-966 9617-0-6

Portions of this work were originally published in *Focus, Men's Life* and *Rolling Stone.*

Grateful acknowledgment is made to the *San Francisco Examiner's Image* magazine for permission to reprint an excerpt from "Condemned to Be Mario" by Scott Rosenberg. Reprinted by permission of the *San Francisco Examiner's Image* magazine.

Originally published: 1st ed. New York: Random House, c1993. With new afterward. Includes bibliographical references and index.

1. Nintendo Kabushiki Kaisha. 2. Electronic games industry.
3. Nintendo video games. I. Title

CIP

Book design by Oksana Kushnir

Cover design by Michael Bouse

Photography courtesy of Nintendo of America

Manufactured in the United States of America
10 9 8 7 6 5 4 3 2 1

ACKNOWLEDGMENTS

Brief portions of this book appeared in *Rolling Stone, Playboy, Men's Life,* and *San Francisco Focus.* Thanks to sources named and unnamed, the hundreds of interview subjects. Thanks particularly to the following: At NCL, Hiroshi Yamauchi, Hiroshi Imanishi, Sigeru Miyamoto, Masayuki Uemura, Genyo Takeda, Gunpei Yokoi, Reiko Wakimoto, and Yasuhiro Minagawa. At NOA, Minoru Arakawa, Howard Lincoln, Peter Main, Al Stone, Phil Rogers, Gail Tilden, Don James, John Sakaley, Toshiko Watson, Sandy Hatcher, Sherrie Mennie, Tony Harman, Blaine Phelps, and numerous others, particularly Bill White. Yoko Arakawa was especially gracious and forthcoming. At Sega, Al Nilsen. At Atari Games, Hide Nakajima, Dennis Wood, Dan Van Elderen, and Barry Kane. At Electronic Arts, Trip Hawkins, Bing Gordon, Larry Probst, Danny Brooks, and particularly Holly Hartz, who sets the standard for public relations in this industry.

At other Nintendo licensees, Henk Rogers, Sheila Boughten, Greg Fischbach, Bruce Lowry, Gilman Louie, Les Crane, Bob Lloyd, Allyne Mills, Joe Morici, Kathleen Watson, Kathy Prall, and the many others. At Hill & Knowlton, Jeff Fox, Karen Peck, Don Varyu, and especially Lynn Gray, for support above and beyond the line of duty when she initially championed this book. At Golin/Harris, Alison Holt and Susan Iannetta, and at Manning, Selvage and Lee, Charlene Gigliotti.

Analysts, including David Leibowitz (American Securities), Manny Gerard and Sean McGowan (Gerard Klauer Mattison & Co.), Robert F. Kleiber (Piper, Jaffray & Hopwood), and Andrew J. Kessler (Morgan Stanley), periodically contributed their expertise.

I'd also like to thank Robert M. Callagy, Vladimir Pokhilko, Vadim Gerasimov, Howard Phillips, Nolan Bushnell, Robert Stein, Ron Judy, Suzuki Eiichi, Miyuki Grace, Jim Mackonochie, Steve Arnold, Elliot Luber, Deborah Brown, Phil Adam, David Ellis, Ben Myron, Mark Smotrof, Les Inanchy (Sony), Greg Zachary of *The Wall Street Journal*, Casey Corr, Tim Healy and Tom Farrey of the *Seattle Times*, Rich Karlgaard at *Forbes ASAP*, Jaron Lanier (J.P.L.), Sharon Fitzpatrick (The Learning Co.), Lynn Hale and Sue Sesserman (Lucasfilm and LucasArts), Marty Taucher (Microsoft), Linda Goetz, and Jenifer Van Horn. In Japan: Keisuke Ono, Yukio Miyazaki, Tsunekazu Ishihara, Yoshio Ito, Koh Shimizu (Sony), and Nishi Saimaru.

Special appreciation goes to Alexey Pajitnov, the creator of "Tetris." Also to many sources who spoke under the condition of anonymity. At Random House, I would like to acknowledge the contributions of Deborah Aiges, Carol Schneider, Lesley Oelsner, Mitchell Ivers, Gail Blackhall, Lawrence LaRose, Becky Simpson, Brian Hudgins, and designer Oksana Kushnir, Amy Edelman for pushing deadlines, and Veronica Windholz, Ed Cohen, and Sybil Pincus for tireless copyediting.

Special thanks to my editor for his insights and his devotion to books, to Binky Urban, my agent, for her counsel and commitment, and to Barry Golson for assigning the original Nintendo article from which this book grew. My thanks as well to Arthur

Kretchmer and Steve Randall, my editors at *Playboy* magazine; to Mike Moritz and Fred Bernstein for their insights and advice. Thanks also to Amy Rennert, who assigned the article on Nolan Bushnell for *San Francisco Focus;* to Don and Nancy Barbour, as impeccable a research department as anyone could ask for, and to the rest of my family, Joan, Sumner, Debbie, Mark, and Jenny; Steve, Susan, and Don, and my extended family of friends, including Armistead, Terry, Peggy, Susan, Buddy, Nick, and Doug.

CONTENTS

GAME OVER

GAME OVER: PRESS START TO CONTINUE

GAME OVER

Games are popular art, collective, social *reactions* to the main drive or action of any culture. [They] . . . are extensions of social man and of the body politic. . . .

As extensions of the popular response to the workaday stress, games become faithful models of a culture. They incorporate both the action and the reaction of whole populations in a single dynamic image. . . . The games of a people reveal a great deal about them.

—MARSHALL MCLUHAN
Understanding Media:
The Extensions of Man

A NEW LEADER
OF THE CLUB

Most people think video games are kids' stuff, and it is true that in "Super Mario Bros. 3," mushrooms give super strength, enemies have names such as Morton Koopa, Jr., and a pudgy, suspendered hero jumps on the heads of Little Goombas. Yet behind "Super Mario Bros. 3," a video game played on the Nintendo Entertainment System (NES), is a business that is very grown-up indeed. In America alone, revenues for that one game have topped $500 million; in the field of entertainment, only the movie *ET* has grossed more.

In the video-game market, where shooting and mass destruction were the norm, the first "Super Mario Bros." game created a revolution in 1985 by introducing elements not often associated with computer terminals and controllers: wit and humor. Mario, the main character, made an unlikely hero—a plumber who can wisely choose to avoid enemies as well as to confront them. In this whimsical world, bright green and red mushrooms make Mario grow taller and more powerful. There are bomb-hurling mice, waltzing cacti, and turtles who can use their shells as missiles. Surprises that give players more time and extra lives lurk in the most unlikely places. Children, who loved the characters and became ensnared in the maze of the game, which was replete with Pavlovian rewards and punishments and carefully programmed increases in challenge, were captivated.

When "Super Mario Bros. 2" was released, the beloved characters from the original game trekked through new cartoon scenery. This time they confronted foes not with cannon or lasers but with turnips, carrots, and pumpkins. Thus equipped, players headed into uncharted waters, where perseverance, wit, luck, and interminable hours of practice counted for everything. "Super Mario 2," like its predecessor, was a great equalizer. The game gave kids the sort of power they couldn't get anywhere else. It was safe for them to make mistakes while playing, because there was always another chance. The things that ordinarily made kids popular at school were not important when they were playing. Also, they had found an arena in which they could beat the pants off their parents, not to mention confound them with an incomprehensible vernacular ("I'm in the second world of the Sub-Con, but I can't get past the miniboss").

Months before it appeared on the market, there were rumors about the next "Super Mario Bros." sequel, but no one saw it until, in the winter of 1989, a movie hit the nation's theaters. *The Wizard* was less a piece of art than a one-hundred-minute advertisement for Nintendo that millions of families paid to see (it grossed $14 million). The excitement in movie theaters was palpable when kids realized they were glimpsing the latest Mario game, complete with new bells and whistles: Mario could don a raccoon disguise and—best—could fly.

Kids spread the word on playgrounds and in schools. Legions of parents were strong-armed by eight-year-olds. The pressure was enormous to be among the first to own "Super Mario Bros. 3."

Some parents remained oblivious and others refused to bend to the pressure, but many millions succumbed. "Super Mario Bros. 3" would sell more copies than any video game in history—7 million in the United States and 4 million in Japan. By record-industry standards, "SMB3" went platinum eleven times. Michael Jackson is one of the few artists to have accomplished that feat.

The money earned from its video games and the NES system that played them transformed Nintendo into one of the world's most profitable companies. By 1991 Nintendo had supplanted Toyota as Japan's most successful company, based on the indices of growth potential, profitability, penetration of foreign and domestic markets, and stock performance. Nintendo made more for its shareholders and paid higher dividends between 1988 and 1992 than almost any other company traded on the Tokyo Stock Exchange.

Nintendo's profits per employee were consistently greater than those of any other Japanese company (excluding finance, stock, and insurance companies). Fujitsu, with profits similar to Nintendo's, had 50,000 employees. Nintendo had 850. Nintendo, in 1991, earned about $1.5 million per employee. Internationally, Nintendo employed some 5,000 people. That year Sony, with 50,000 employees, earned $400 million less than Nintendo. By 1992, the company was consistently earning pre-tax profits of more than a billion dollars a year.

The multitentacled video-game business swelled to consume larger and larger segments of the entertainment and consumer-electronics industries as well as the toy industry.

In the entertainment business, Nintendo had become a force that could not be ignored. In early 1992, the company profited more than *all the American movie studios* combined and the three television networks combined.

The consumer-electronics industry watched as the Nintendo Entertainment System, in just five short years, was brought into more than a third of the households in the United States and Japan. Although twice as many homes had VCRs, the movie-playing machines were made by various companies, while one company alone made all the Nintendo machines. Moreover, the VCR companies sold just the machines, not the videotapes that played on them. Nintendo, on the other hand, was making hefty profits from an ever-expanding list of games in addition to the machines to play them on. Consumer-electronics giants like Sony and Matsushita Electric Industrial finally woke up to the fact that by the turn of the century, consumer-hardware companies would be archaic if they had no involvement in software. In a game of catch-up, Sony bought Columbia Pictures and Matsushita purchased MCA, the American movie-and-entertainment giant—attempts to wrest some participation in the entertainment software market.

Nintendo had completely blindsided the American computer industry, too. The founders of the personal-computer revolution had predicted in the early eighties that computers would soon be commonplace in most homes, like toasters. Yet a decade after the personal computer was launched, only 24 million American homes had them—almost 10 million fewer than had Nintendo systems. Worldwide, there were about an equal number of Nintendo systems and PCs, some 50 to 60 million. As with VCRs, the PCs were manufactured by dozens of companies; less than 10 percent were made by the number-one PC company, IBM. With the exception of a growing number of illegal pirate versions coming out of Hong Kong and Taiwan, just *one* company manufactured and sold all the Nintendo systems. The huge Japanese computer company NEC and a video-arcade-game company, Sega, tried to compete with Nintendo, but in spite of investments in the hundreds of millions of dollars, they shared less than 10 to 15 percent of the market through 1991. Companies such as Apple and IBM looked over their shoulders and saw Nintendo on their heels. When Apple Computer president Michael Spindler was asked in March 1991 which computer company Apple feared most in the 1990s, he answered, "Nintendo."

The computer industry understood why Nintendo had a jump on them: Nintendo had predicated its entire strategy on the control of both hardware and software. In 1991, Apple and IBM announced an alliance to take on Microsoft, the software giant. In 1992, IBM also announced an alliance with Time Warner. Like the consumer-electronics companies, the computer-hardware giants realized that to remain competitive, they needed access to and control of software. The software edge would become particularly important in the looming battle for shares in the next technological revolution—multimedia and networking. That industry, which combines computer power with home-entertainment systems, integrating television, video recorders, CD sound systems, and the telephone, was judged by the *Los Angeles Times* to be worth a mind-boggling $3.5 trillion annually by the next century. The question was which company would become the Maytag of home computers in the potentially mammoth industry.

Did Nintendo have the foresight and wherewithal—and sheer temerity—to be that ambitious? Evidence that it did was hidden in the belly of the NES unit.

On the bottom of the innocuous, gray game-playing machine was a panel. When it was removed, a computer cable connector was revealed. A two-way doorway to the main processor, this port allowed the Nintendo system to work as a terminal that could be connected to a modem, a keyboard, or auxiliary storage devices. The Nintendo system was rolled into living rooms by children who welcomed (and worshiped) it as a game, while inside it lurked the potential to be transformed into the integral component of the largest electronic network in the country. With a phone line plugged into it, the Nintendo system could be used to shop, call up movie reviews, buy pork bellies, do research, make airline reservations, and order a pizza.

The machine's greater possibilities were first tapped in Japan when Nintendo announced the Family Computer Communications Network System. A similar network planned for the United States had the potential to dwarf the Prodigy network (a joint venture of IBM and Sears Roebuck), the most-used network in the country, which had only 1.3 million subscribers by January 1992.

Nintendo's success had an enormous impact on numerous industries worldwide. Besides the competing hardware companies, more than a hundred companies that made video games rode the Nintendo wave. In 1992 they had worldwide sales of 170 million cartridges, at an average cost of $40 each—almost $7 billion worth.

Once it became clear that the huge video-game market showed no signs of disappearing, all kinds of companies began lining up to become licensed producers of Nintendo-compatible games. Companies such as Electronic Arts and Software Toolworks, which made only games on floppy disks for computers, had tried to hold out, but Nintendo became too big to ignore. They signed up, as did some entertainment companies, including Lucasfilm and Disney (through another company called Capcom). Companies that were already cleaning up with coin-operated video games set up in arcades, malls, bowling alleys, and pizza parlors entered the Nintendo business too, adapting their hit arcade titles to the home system.

Other companies affected by Nintendo's success were outside the video-game industry. Since computer chips were used in both the Nintendo hardware and software (the game cartridges contained special chips), Nintendo used more of certain kinds of semiconductors than any other company in Japan. Nintendo products accounted for more than 3 percent of total Japanese semiconductor production in 1991.

Quietly, Nintendo sailed past stalwart American corporations such as IBM, Disney, and Apple Computer, not only in profitability but also in impact on American culture. In the last part of the twentieth century, leaps in technology ushered in a new era in which children and a substantial part of the culture as a whole would be more influenced by interactive electronic media—in their simplest form, video games—than by television, which had defined the previous generation. The signs of the first Nintendo generation appeared as early as 1989 and 1990. A study by Nielsen Media Research, the company that monitors television viewing, showed that within a particular age group more kids were playing Nintendo than were watching the major children's TV network, Nickelodeon, at certain times on certain days of the week. Kids already spent more time in electronic environments (TV, radio, records) than they did in school or talking with friends or parents. Some of them were spending an additional two hours a day on Nintendo.

Even during the hours when kids weren't playing video games, they were being showered with the culture of Nintendo. Television cartoon shows based on Nintendo games and characters were watched by more kids than any other TV programs. Other cartoon shows (including *The Simpsons, Teenage Mutant Ninja Turtles, Chip 'N Dale Rescue Rangers,* and *Duck Tales*) *became* Nintendo games. A record of Nintendo songs and a feature film were developed;

there were Nintendo magazines, books, videos, cereals, note-books, drinking mugs, T-shirts, board games, puzzles, dolls, wall-paper, and bed sheets. Nintendo infiltrated every conceivable market until the question was not whether Nintendo's invasion would succeed but what the invaders would leave in their wake.

What, asked parents, teachers, and sociologists, were the long-term effects of so much game playing on kids' self-images, rela-tionships, and social skills? How did Nintendo affect learning? Did the games encourage violence? Did they empower kids or make them passive? Did the impact vary among different age groups and genders?

Some saw video games as insidious hypnotizers and mind de-stroyers; others viewed them as training tools for the cybernetic world of the future. One proponent claimed that children who excelled at one game, "Tetris," scored higher on intelligence tests.

Besides the attempts to figure out the effect of Nintendo's inva-sion, there was also a great deal of intellectualizing about *why* Nintendo had become so pervasive. In an essay in the *San Fran-cisco Examiner's Image* magazine in September 1991, "Con-demned to Be Mario: The Video-game Plumber as Existential Hero," Scott Rosenberg wrote, "Mario is a character, a dumpy fellow with a big mustache; but he is also a stand-in, your iconic representation in the video universe. . . . If millions of children and adults have melded with Mario . . . it may not be simply a matter of our shortening attention spans, our craving for novelty or our susceptibility to expensive ad campaigns. It may be that in Mario's fate—stuck in a world not of his own choosing, charged with a nearly impossible mission, doomed to perish sooner or later, yet free while he lives to grow, learn, slay demons and stop to smell the Fire Flowers—people are catching a crude, bright, hypnotic reflection of their own lives." Or maybe it *was* just the ad campaign.

One thing was inarguable, however. Nintendo had successfully entered the collective consciousness. "Q" ratings, which indicate the popularity of politicians, movie stars, and other public figures based on controlled surveys, showed that in 1990 the Nintendo mascot, Super Mario, was more recognized by American children than Mickey Mouse. How significant was the news? Uncle Walt

Disney and Mickey Mouse were as American as—well, nothing was more American than Walt and Mickey. The idea that Mario had become more popular than Mickey was to some a travesty that signaled the next phase of the Japanese invasion. Japan had already captured America's wallets. The country's minds—beginning with children's minds—were next.

Nintendo had become Japan's biggest cultural export. Indeed, whereas the rest of the world devoured Japanese hardware—cars, Walkmans, TVs—Japanese "software," such as movies, books, art, and music, had little impact outside Japan. The exception was video games. The most widely known Japanese cultural ambassador was Mario and with him came a new set of values.

Generations of children had been imbued with Mickey's message: *We play fair and we work hard and we're in harmony. . . . M-I-C . . . See you real soon. K-E-Y . . . Why? Because we like you . . .* Mario imparted other values: *Kill or be killed. Time is running out. You are on your own.* Donald Katz, in a February 1990 *Esquire* magazine article, observed that the lesson from Mario is "there's always somebody bigger and more powerful than you are [and] . . . even if you kill the bad guys and save the girl—eventually you will die."

Oh no! Not again! At the end of summer 1991, the children of America heard that the sequel to "Super Mario Bros. 3" was on its way. For their parents, this was even more terrible news than the last time, because Nintendo had also introduced an entire new video-game system more powerful and of course more expensive than the original. "Super Mario Bros. 4" would play only on the new Super Nintendo System.

In Japan, kids swamped stores to get their Super systems as soon as it was released. Most of them went home empty-handed; the game sold out in three days. Stores illegally parceled them out, sometimes inflating the prices, other times forcing buyers to purchase additional products if they wanted a Super Nintendo System.

In the United States, in the midst of a recession, Nintendo was less confident that the $200 Super NES would sell. By 1991, some observers of the video-game industry, noting slower sales and gloomier projections, suggested (almost gleefully) that kids seemed to be cooling off to Nintendo—that this might be the beginning of Goliath's fall. But Nintendo had no intention of going gently. To push the new system, Nintendo packaged "Super Mario Bros. 4," which went under the name "Super Mario World," together with the Super Nintendo hardware, like the prize in a box of Frosted Flakes.

Soon the playgrounds were abuzz with news about "Super Mario World." Even those children who had grown bored with the original Nintendo system were excited, encouraged by the company's $25 million ad campaign. Parents, who had watched with relief as their children's fanaticism for Nintendo seemed to be fading, were greeted with a new wave of fervor. "Dad," the children enthused, "you won't *believe* what Mario can do now . . ." In October 1991, Nintendo issued a press release titled "Nintendo Loyalists Put Long-term Entertainment Value Ahead of Short-term Budget Cuts," in which it was reported that the Super NES was selling at a rate of twelve units every retail minute, or one every five seconds. In spite of the recession and all the gloom and doom, Nintendo projected that 1992 would be its biggest year yet, with sales in America topping $4.7 billion.

IN HEAVEN'S
HANDS

In the eastern part of Kyoto, the ancient Japanese capital city, near the famous Heian shrine, is a slumbering side street now called Higashi-ogi. In the fifteenth century, it was a pathway of hard-packed dirt that led to the Shogoin Gotenso, the summer castle of the emperor. Across from the castle stood the home of the emperor's special doctor.

Centuries later, this home was bought by Fusajiro Yamauchi. The Yamauchis would live there for generations, behind an immense gate held together with large bolts and rusted diamond-shaped brackets. The huge metal hinges that hold it up are serpents, attached to beams more than a cubic foot thick. The gate, which has survived for five hundred years, locks shut with a heavy cross-bar.

In the year Heisei 4 (1992), the gate is still flanked by a high fence that winds around the perimeter of the Yamauchi property. The fence is crowned with coils of razor wire, iron spikes with dagger points, and deadly-sharp bamboo spears. Their purpose is unambiguous.

Inside is a pathway made of flat stones. Immediately to the right is a two-story guardhouse the size of a child's treehouse. By the pathway across from the guardhouse are flagstones that have been set in the earth for centuries, all but buried by velvety moss. They lead to the center of a lush and tangled garden of small foot-bridges, a patchwork quilt of textures and greens: feather-leafed bonsai, sculpted, dome-shaped bushes, and gold-flecked field grasses. Amid the green are cabernet-red maples that seem aflame.

Hidden throughout the foliage are a lantern, a bronze crane, and a stone shrine. Overgrown pathways coil through the garden, which was once rigidly tended, corralled in, immaculate, until the current generation of Yamauchis let it grow wilder.

One path through the garden leads to a teahouse with sliding cedar-framed paper doors and floors of straw tatami mats. Generations of Yamauchis prepared the tea ceremony there. Now it is a storage closet filled with boxes and mattresses.

Behind the garden is the residence, a traditional home built in the style of a Japanese temple. Homes in Japan are measured not in square feet or meters but in tatami mats—the book-thick, rect-angular sections of sweet-smelling, woven straw. In past genera-tions, wealth was measured by the number of mats (each smaller than a twin bed) in a family's home. The average home has eight or ten tatamis; the Yamauchi home has 152.

The rooms are partitioned by shoji screens, and the walls that look onto the garden are floor-to-ceiling glass, outlined by dark, weathered wood. Newer beige stucco clay covers some walls, but the peaked roof is finished with aged gray-blue tiles in the pattern of a choppy sea. Around the periphery the tiles have circular faces carved with the symbol of balance and harmony, the intertwined teardrops, yin and yang.

Fusajiro Yamauchi was an artist and craftsman during the Meiji period, at the end of the time that Western civilization describes as the nineteenth century.

Yamauchi, reputed to be fair, good-humored, and skillful, made *karuta,* playing cards. The Portuguese and Dutch had brought card games to Japan as long as 350 years ago, but the cards Yamauchi made had more in common with ancient Japanese games that were played with intricately painted seashells. *Hanafuda,* or "flower cards," smaller and thicker than Western cards, came to replace seashells, but the elaborate, richly colored images on them remained.

Instead of number and picture cards, the forty-eight cards in a *hanafuda* deck were painted with a scheme of symbols: for example, a deer, the wind, a chrysanthemum, a boar, the moon. There were twelve suits, one for each month of the year. The pine and the crane, which in Japan symbolize long life and good fortune, represent January; the nightingale and plum-tree blossoms, February; cherry blossoms, March, and so on through the year to the winter-blooming paulownia, with its fragrant clusters of violet flowers, for December. Individual cards within the suits had symbols—rain with a poet, for example, a highly valued card, worth twenty points; or wisteria, more common, worth one.

The most popular *hanafuda* game, matching flowers, appeared to be a simple game of matching images in order to make packs, but it could be as complex as bridge, and it was taken as seriously.

Yamauchi founded Nintendo Koppai in Meiji 22 (1889), to produce and sell the handmade cards. The *kanji* characters he chose to make up the name of his new company—*nin-ten-do*—could be understood as "Leave luck to heaven," or "Deep in the mind we have to do whatever we have to do." The most common reading of it was "Work hard, but in the end it is in heaven's hands."

Yamauchi made the paper for the flower cards in the traditional way, from the bark of mulberry, or *mitsu-mata,* trees. He pounded the bark into a paste and added clay to give it more weight. Thin layers were dried and pulled and shaped. The craftsman regarded paper as a living creature with a will of its own. Yamauchi fought with it until, at the end, it succumbed to *his* will, its new form.

Several layers were pressed together until the rigid thickness of a hardback book cover was achieved. Yamauchi designed a woodblock printing system to emboss the outline of the individual cards on a large sheet of the paper. He laid a series of stencils over the paper and, using luminous inks made from flower petals and berries, filled in the drawings. Backgrounds were red. Grass was black.

The full moon was left unpainted, the straw color of the paper. The pigments bled slightly together when they met, so each card seemed hand-dyed.

Nintendo's *hanafuda* cards, called *Daitoryo* (or President), became the most popular cards in the Kyoto region. They were sold in Nintendo's shops in Kyoto and Osaka. Nintendo also made cards with different symbols that were sold in other regions. Kanto's, for instance, had swords, mountains, and human beings.

So long as *hanafuda* was solely a domestic amusement, Nintendo's business remained small and only modestly profitable, but business increased when the flower cards began to be used for gambling. In the absence of horse or dog racing or sports pools, the *yakuza*, Japan's equivalent of the Mafia, operated high-stakes games of *hanafuda* in casino-like parlors. Nintendo profited handsomely, since professional players would begin each new game with a fresh deck, discarding the old one. To keep up with the demand, Fusajiro Yamauchi trained apprentices to mass-produce the cards.

He expanded his business again in 1907, when Nintendo became the first Japanese company to manufacture Western-style playing cards, which were becoming popular in Japan. Yamauchi, who had been selling his cards only in Nintendo's shops, now needed greater distribution. He negotiated a deal with Japan Tobacco and Salt Public Corporation, the tobacco monopoly, and that company began selling Nintendo *karuta* in its cigarette shops throughout the country.

It was a profitable arrangement. By the time Fusajiro was ready to retire, Nintendo was by far the largest Japanese playing-card company.

Fusajiro Yamauchi had no son. If Nintendo was to remain in the family, Japanese tradition required that his daughter, Tei, marry a man who could take over for him. A marriage was arranged with a stern, hardworking student named Sekiryo Kaneda. He agreed to take the Yamauchi family name, as was custom if he wanted to enter his bride's family's business. In 1929, Sekiryo Yamauchi became Nintendo's second president.

Although Sekiryo and Tei's home life was not easy (she stoically tolerated her husband's philandering), they prospered. Sekiryo's outside business passion was real estate, and the Yamauchis came to own a sizable portion of eastern Kyoto as Sekiryo worked to transform Nintendo.

In 1933, he established a joint-venture corporation called Yamauchi-Nintendo and moved from the original, modest *hanafuda* shop in Kyoto to a ferroconcrete building he had constructed next door. In 1947, he created a distribution company called Marufuku to sell new varieties of modern, Western-style playing cards—pinochle and poker decks—with fancy backs. He built a sales force that called on small and large shops all over Japan. To produce the cards more quickly and efficiently (with paper bought from suppliers) he developed an assembly line of workers. Nintendo became an efficiently run business with a rigid, hierarchical management structure. Industrious managers, pressured to surpass the performance of their colleagues, were notoriously tough on subordinates.

Sekiryo and Tei were the second generation of Yamauchis to have daughters but no sons. Kimi, their eldest, married Shikanojo Inaba, from a respected family of craftsmen. Like his father-in-law before him, Inaba adopted his wife's surname and thereby became the heir apparent to Nintendo. It was assumed that Shikanojo Yamauchi would take over when Sekiryo retired.

In 1927, Shikanojo and Kimi had a child, Hiroshi, the first Yamauchi male to be born in three generations. Hiroshi was five when Shikanojo ran away, abandoning his wife and son. The young boy was told that his father was worthless and deceitful. Nothing else was ever said of him.

Disgraced, Kimi began divorce proceedings and moved in with her sister, leaving Hiroshi in the care of her parents. They reared him with the same iron fist with which they ran Nintendo. They attended to his education, his grooming, and his manners with equal severity. But Hiroshi rebelled, and the older he got, the more intractable he became.

Arrogant and impudent, Hiroshi disregarded his grandparents as he grew into a balefully handsome and debonair gentleman who carried his small body with conceited sturdiness, his head jutting forward. He wore his thick hair back and his eyebrows swept downward around his dark eyes. Hiroshi dressed in expensive, well-tailored clothes. He kept his fingernails long, manicured, and polished. He was sullen and bitter, yet he disguised his moods with levity and a dust-dry wit. His temperament was the legacy of an absent father and his grandparents' scorn.

Hiroshi saw his mother on occasion, but Kimi, who never remarried, had become more like an aunt. She worked at Nintendo, where she ran a subsidiary that was involved in sales.

Hiroshi Yamauchi never saw his father again. Shikanojo had brought shame and dishonor to the family, and when he returned, aged and ailing, desperate to see his only son, Hiroshi refused to speak to him.

One day when Hiroshi was in his late twenties, his hair already graying, he heard from a half sister he didn't know he had: their father had died of a stroke. She said Hiroshi should honor his father's memory by attending the funeral.

Yamauchi sat alone for a full day before deciding he would go.

At the funeral, Hiroshi met his four half sisters, his father's second wife, and an aunt he had never known. He was overwhelmed when his aunt told him he looked exactly like his father, and he wondered what else he had inherited from Shikanojo. He also wondered what price a son would have to pay for refusing to reconcile and forgive his father.

Hiroshi grieved for months after the funeral. He cried freely; the death changed him, and a part of him never seemed to heal. Throughout the rest of his life, he made regular visits to Shikanojo's grave.

Hiroshi was sent to a preparatory school in Kyoto in 1940, and Sekiryo planned on sending him to a university to study law or engineering. But war came, and all lives in Japan were put on hold. During the war, Hiroshi's grandmother, Tei, emerged as the leader of the Yamauchi family. The rest of the family would go underground when the air-raid sirens wailed through Kyoto nights, but Tei resolutely went about her business as if nothing was out of the ordinary.

Tei would not consider allowing Hiroshi to enter the military. When the war began, he was too young to fight, and by the time he could have been called up, the tide had already turned and the Yamauchis knew that Japan would lose. To keep him safely out of

————————————————

the war, Tei made Hiroshi stay in school, and he was given an assignment in a military factory.

Rice and other food were scarce; most people in the area survived on little but potatoes. Yamauchi, however, carried a precious rice lunch to work each day, from the stockpile in Tei's pantry. During his lunch break, Hiroshi noticed that a supervisor was hungrily eyeing his rice. Hiroshi shared it with the man and was rewarded, that afternoon, with time off. Hiroshi went out to a field and took a nap. After that, Yamauchi brought two lunch boxes to work each day, one for himself and one for the supervisor, and each day he was excused from work.

When the war ended in 1945, Yamauchi enrolled in Waseda University to study law. He also entered into a marriage arranged by his grandfather. His bride was Michiko Inaba (no relation to Shikanojo), a descendant of a very high-ranking samurai, loyal to the *daimyo* who lorded over Shikoku Island during the early Meiji period. This powerful and wealthy samurai had moved from Shikoku to Kyoto, where he married and took the Inaba name. He opened a small business, making delicate cloisonné pieces—Inaba cloisonné would become known throughout the world.

In Japan, when a marriage was formally arranged, the couple's parents would meet to discuss the match. However, because Hiroshi's grandparents were making the match, the couple's first date was a meeting of the two clans: the matchmaker was host to Sekiryo and Tei and Kimi Yamauchi as well as Michiko's parents and four grandparents. The couple was married soon after in a traditional ceremony.

When Hiroshi was twenty-one his grandfather had a stroke. Sekiryo asked that the young man be sent in to see him. His grandfather, propped up on pillows on his bed, spoke soberly to Hiroshi. As the first Yamauchi son since Fusajiro, Hiroshi would assume the position that was supposed to have been his father's. He would have to leave school and immediately come to work at Nintendo, as president.

Hiroshi, responding without emotion, said he would take over the company, but he insisted on several conditions. The main one was that he must be the only family member at Nintendo. This meant that his cousin had to be fired. Hiroshi wanted there never to be a question that he was in charge, the sole heir.

Weak and saddened, Sekiryo had the cousin fired, and, in 1949, Hiroshi Yamauchi was appointed the third president of Nintendo. The old man died soon thereafter, never sure whether his family and the business would survive. Since Sekiryo never saw the success Hiroshi would eventually have with Nintendo, as far as Hiroshi was concerned, to Sekiryo he remained an ill-mannered, disrespectful, and spoiled child. Hiroshi lived with the knowledge that he had betrayed and disappointed the two most important men in his life, his father and grandfather.

Young President Yamauchi was not welcomed by Nintendo's employees. They resented his youth and inexperience and worried about rumors that Yamauchi planned a clean sweep of longtime employees. True to expectations, Yamauchi fired every manager, one by one, left over from his grandfather's reign, in spite of their years of dedicated service to Nintendo. Not only did he sever what he considered dead weight, but also anyone who had a stake in Nintendo's conservative past. He wanted none of the old guard around who might question his authority.

He changed the name of the distribution company to Nintendo Karuta (Nintendo Playing Cards) in 1951 and established a new corporate headquarters on a lot he bought in town, off a small street called Takamatsu-cho. He consolidated all the Kyoto manufacturing there, modernizing the card-making process.

In an attempt to compete with the modern, fashionable cards that were being imported from the West, in 1953 Nintendo began manufacturing the first plastic-coated cards in Japan (all varieties of cards until that time had been made of uncoated paper). In 1959, Nintendo made its first licensing agreement—with an American company, Walt Disney. Playing cards backed with pictures of Mickey Mouse and other Disney characters expanded Nintendo's market to include young people and families. The Disney cards were advertised on television. To reach the new market, Yamauchi structured a new distribution system that would get the cards into larger department and toy stores. The results were instantaneous:

Nintendo's sales shot up. The company sold a record 600,000 packs that year.

Yamauchi was still discontented. He wanted the company to expand faster, but he encountered stumbling blocks. In spite of his efforts to improve the quality of Nintendo's Disney cards and other Western-style pinochle and poker decks, they were still inferior to the ones being imported from America. The company's bread and butter, *hanafuda,* was, by its nature, a modest business; the profits were constant, but there was little room for growth in that market.

Yamauchi dropped the word *karuta* from the company name, which now became NCL—Nintendo Company, Ltd.—as the young president planned to branch out into new businesses. To finance them, he took Nintendo public, listing the company on the second tier of the Osaka Stock Exchange and on the Kyoto Stock Exchange, and became Nintendo's chairman.

The first product launched by the new company was a line of individually portioned instant rice. Add water and—*presto!* It was a dismal failure. Yamauchi then opened a "love hotel," with rooms rented by the hour. The business was, for Yamauchi, a personal passion; it was said that he was one of his own best customers (his infidelities were well known—even by his wife, who ignored them).

A taxi company Yamauchi started, Daiya, thrived, although he grew tired of negotiating with powerful taxi-driver unions, which demanded high salaries and expensive benefits for their members. He soon folded that company and closed the doors of the love hotel. He planned more changes as he moved Nintendo again, this time to a larger building, a three-story structure of beige bricks with black door and window frames and bars on the windows.

Yamauchi had concluded that he wanted new businesses that could take advantage of one of Nintendo's strongest assets, its *karuta* distribution system, which reached into toy and department stores throughout Japan. Nintendo's roots were in entertainment, and there would be no more rice or taxis or love hotels. Yamauchi set Nintendo on a new course as an entertainment company.

Hiroshi Imanishi had the demeanor and build of a Rottweiler. He gave the appearance of a highly cultivated sophisticate who was still common and accessible. Clearly he was also fiercely bright. A recent law graduate from Doshisha University, he accepted an offer from Nintendo in spite of the fact that he found the president glum, reserved, and formal. Hiroshi Imanishi was, nonetheless, intrigued by Hiroshi Yamauchi. There was an attractive cocksureness and obdurate ambition in Yamauchi's speeches about the dramatic expansion he planned for Nintendo.

He revealed the details sparingly. It sometimes seemed as if Yamauchi was obsessed by details—"trifles," Imanishi felt—and the worry over minutiae was tiresome. Other times there were glimpses of a calculated, secret plan. It was a frustrating process, as there was minimum communication, only commands, and meetings often turned into lectures.

As the company geared up for a series of new ventures, Imanishi worked in many departments: administration, finance, planning. He eventually became the general affairs manager. But regardless of the title he bore, he oversaw the majority of his boss's projects. The task by 1969 was to create a department that would set Nintendo on its new course. Called simply Games, it was the company's first research-and-development office, set up in a warehouse in Uji, a Kyoto suburb.

Gunpei Yokoi had grown up in Kyoto, where his father was the director of a pharmaceutical company. He graduated from college with a degree in electronics and then made the rounds of Kyoto companies, filling out applications. The short, solid, and unpretentious man who wore gray-lensed glasses was hired by Nintendo to maintain the assembly-line machines that made the playing and *hanafuda* cards. Yokoi was the entire maintenance department for several months before Hiroshi Yamauchi called him into his office.

Yokoi sat down in a large chair opposite the chairman's desk and folded his hands on his lap. Imanishi was already there, drinking coffee. Yokoi listened intently as Yamauchi assigned him to a new project in the newly founded Games division. He was to work under Hiroshi Imanishi to create a department for *komuki*—engineering—and make something for Nintendo to sell for Christmas.

"What should I make?" Yokoi asked.

Yamauchi said, "Something great."

Yokoi was the first of many weekend tinkerers hired by Nintendo, the sort of hobbyists who made toys, radios, and other mechanical gadgets from spare parts. For his own amusement, he had recently invented a wooden latticework connected by bolts and with a vice-grip device on one end. When the two leaves of the handle were pushed together, the crisscrossed pieces extended and the grip on the end closed. The gadget had practical, but mostly whimsical, uses as a groping, clasping extension of the hand.

The day after their meeting, Yokoi demonstrated his contraption for Yamauchi and Imanishi. It brought a trace of a smile to the chairman's face. He gave the go-ahead and assigned Imanishi and Yokoi to begin production of Nintendo's first toy, the Ultra Hand. Advertised on television, 1.2 million units of the novelty were sold, at about 800 yen (roughly $6 in 1970) each. For Nintendo in those days, it was a resounding success.

From then on, Yokoi's job was to come up with inventions and show them to the chairman. Yamauchi, although he had no engineering background, had an uncanny sense about products. He gave Yokoi suggestions and challenged the young man to improve his designs. Yamauchi, Yokoi says, instinctively knew if a new idea would sell. The chairman never sought a second opinion. If he liked it, he instructed Imanishi to begin production.

Yokoi's inventions resulted in Nintendo's Ultra series of toys. The Ultra Machine was a pitching machine that lobbed a lighter version of a baseball (that could be batted safely indoors). Seven hundred thousand units sold each year for three years through 1973. Somewhat less successful was the Ultra Scope, a periscope-like device with a lens that automatically refocused so that kids could see around corners, over fences, or behind them. It allowed kids to spy on each other, on their neighbors, or, often, on their parents. "It was a time of great fun," Yokoi recalls. "I saw myself as a cartoonist who understood movements in the world and created abstractions of them."

In the evenings, Yokoi dabbled with wires and oscilloscopes and various other electronic components and applied some of his experiments to a device called the Love Tester. Yamauchi loved it. A boy and girl grabbed on to the handles of the tester and then joined their free hands. A meter read the current passing through them and determined, with scientific inaccuracy, how much "love" they had between them. The true objective of the device had nothing to do with science and everything to do with holding hands in a culture in which holding hands was still pretty risqué. Yokoi noted that in America or Europe a love tester would have had to involve kissing. In Japan, however, a device that inspired hand holding was tantalizing enough to become a hit.

Yokoi's R&D department grew with a steady addition of young engineers. Yamauchi pushed them with his emphatic praise and scorn. He never pretended to want to foster a sense of team play; rather, he openly pitted them against each other. While many Japanese companies grew because of loyalty to the group and the corporate good, Nintendo grew because of its engineers' (and its other employees') desire to please Yamauchi. "We lived for his praise," one engineer says.

The next major Nintendo product, pivotal to positioning the company for its future success, came to the company by chance. Masayuki Uemura, with thick, wiry hair pushed to one side and an irrepressible ear-to-ear grin, arrived at Nintendo one day and requested a meeting with Gunpei Yokoi.

Born in 1943 in austere and beautiful Nara, Uemura had moved to Kyoto with his family to escape the wartime bombing, Kyoto then being one of the safest of Japanese cities. His father had sold kimonos but struggled to make a living (he eventually opened a record shop in Osaka). When Uemura was a child, his family's straitened finances meant he had to use his imagination to invent toys and games for himself. Masayuki learned to make radio-controlled airplanes from parts he found in junk piles. His desire to create more sophisticated devices led him to work his way through an industrial college, where he trained as an electronics technician.

By the time he graduated, Uemura could do far more than build toys, and he found a good job with an electronics manufacturer, Sharp, selling optical semiconductors used in the solar cells that might be used to power a lighthouse, or a robot that measured the amount of rainfall on the top of mountains.

The head of Sharp's Kyoto office sent Uemura on a routine sales call to Nintendo to see if he could drum up some business for the solar cells. The visit coincided with Yamauchi's most recent mandate—for Yokoi and his engineers to investigate electronic toys beyond the Love Tester. In a meeting, Yokoi and Uemura concluded that the Sharp cell had some interesting applications for entertainment products. Shortly afterward, Yokoi hired Uemura away from Sharp.

The young engineer was happy to leave sales, but he was even more thrilled to join Nintendo and return to what he had done for fun as a child: make toys. "There was something different about Nintendo," Uemura found. "Here were these very serious men thinking about the content of play. Other companies were importing ideas from America and adapting them to the Japanese market, only making them cheaper and smaller. But Nintendo was interested in original ideas."

When Yokoi saw the Sharp solar-cell battery, he envisioned a unique way to put it to use. He and Uemura experimented. Large solar cells, about the size of a silver dollar, were being used to collect and convert light into electricity. But a much smaller cell could be used as a sensor to detect light. Yokoi's idea was to adapt the technology to a shooting game, using solar cells as targets.

Yokoi and Uemura worked on a light gun that could be produced cheaply enough for the consumer market. The "bullet" would be a thin beam of light. If the beam hit a tiny solar cell, the cell would either produce or cut off a charge, depending on the circuitry. For instance, the electricity to a magnet could be turned off so that a spring-loaded target could be let loose—and a plastic bottle of beer, pieced together as if it were a puzzle, could be made to explode. A lion could roar. A stack of toy barrels could be blown apart.

Packed with light-triggered targets (the lion, the beer bottle, and the like), more than 1 million Nintendo Beam Gun games were sold for between 4,000 and 5,000 yen ($30 or so) in the early 1970s. Nintendo Co., Ltd., by then listed in the first section of the Osaka Stock Exchange, grew rapidly.

Now the company needed more space to keep up with the demand for its new products, so Yamauchi expanded once again. To build the new headquarters, Yamauchi bought out neighbors and a vacant lot alongside the company's older cement building, which was retained as the *hanafuda* factory. (Yamauchi kept the business going essentially for nostalgia's sake: cards were representing a rapidly diminishing portion of Nintendo's business.) The old building, whose front lawn held a few scattered trees, was dwarfed by the new structures (eventually three were constructed), which were high-tech slick: three floors, industrial white, huge rectangular slabs. A crisp blue NINTENDO sign in both *kanji* and roman characters could be seen by passengers on commuter trains and from the nearby hillsides, even from the gardens of the nearby Tofuku-ji Temple. A guardhouse was built, and blue-suited guards from the Kansai security firm were posted, day and night, out front.

Beam Guns were still flying off store shelves when Yokoi suggested to Yamauchi that the same technology could be used in other ways. At the time, skeet shooting was a popular sport in Japan. On a whim, Yokoi had bought a rifle and went to a range to try it. When he returned, he told Yamauchi that the technology in the light gun could work in a system that would replicate the experience of shooting clay pigeons.

Yamauchi absorbed the information and came up with a commercial application for it. There had been a bowling boom in Japan in the 1960s. Alleys sprang up all over the country. Since then, enthusiasm for bowling had waned and many alleys were closed down. Yamauchi decided they could be acquired very advantageously and easily transformed into "shooting ranges." Simulated clay pigeons would appear at the end of a lane. A solar cell would detect when a hit was made and tally it on an electronic scoreboard. There was nothing like it anywhere. It would be the closest to real shooting most people would ever get, far more realistic than the amusement-park shooting ranges, where cork bullets sailed off course with the slightest wind.

The technology, however, was still a bit tricky. Yokoi, who was told to get the system up and running, encountered a series of problems, from the operation of moving (or apparently moving) solar collectors to the coordination of the timing of a shot with the sound of its report.

Another engineer who had joined Nintendo was assigned to work with Uemura. The young man, Genyo Takeda, had responded to a newspaper want ad for a toy designer. When he saw the ad, he says, "some inspiration entered me."

Takeda, who wore a polyester trench coat even indoors in midsummer, was a colorful addition to the R&D division. Born and raised in Osaka, the son of the president of a fabric-design company, he had graduated in 1970 from Shizuoka Governmental University on Honshu, where his studies often took a backseat to his involvement in the student movement. In school, Takeda studied semiconductors, but for fun he built miniature locomotives and airplanes. When Gunpei Yokoi interviewed him, he realized that Takeda would easily fit into his growing engineering team. He put Takeda to work with Uemura on Yamauchi's shooting-range project.

The press showed up to document the grand opening of the world's first Laser Clay Range in Kyoto in 1973. Just as a television news crew was getting ready, its cameras rolling, the light-shooting laser gun malfunctioned. Before anyone realized there was a problem, Takeda climbed into the box behind the targets and, keeping himself hidden, shot clay pigeons off manually, and then kicked the controller, which lit up the targets that were supposedly hit. As far as the television audience knew, the ranges were running smoothly. They were packed from the first day they opened their doors to the public.

Nintendo's Laser Clay Ranges became the hip spot for an evening's entertainment in many Japanese cities, and soon there were variations on the original theme. In 1974 came "Wild Gunman." A 16-mm movie projector showed actual film footage of a "homicidal maniac" on a screen at the end of an alley. If the player blasted the wild gunman before he fired, the sensor on the screen detected it and the player scored. The image-projection-system games were sold to a trading company, which exported them to America and Europe.

The laser-gun ranges were going full tilt, but Japan had been hit by the world's first oil shortage in 1973; the country's economy went into a tailspin and, as it began to affect discretionary spending, the shooting galleries were soon empty. Orders from abroad were canceled and outstanding bills went unpaid. Nintendo's investment in the venture had been so great that the company was suddenly on the brink of collapse. Yamauchi was more desperate than ever for a breakthrough product.

The idea came from a boyhood friend of Yamauchi. The man had become an executive with one of Japan's largest electronics conglomerates. Over dinner one night in 1975, the talk was all about the latest technological breakthroughs in the electronics industry, primarily the importance of semiconductors and microprocessors. Yamauchi's interest was piqued when he realized that these technologies, as part of a revolution in office and consumer products, were becoming cheap enough to utilize in entertainment products. He researched the first traces of an industry that was emerging in America. Companies such as Atari and Magnavox were selling devices that played electronic games on home television sets.

Yamauchi negotiated a license to manufacture and sell Magnavox's video-game system in Japan. The machine played variations on the first commercial American video game, "Pong." A beam of light was batted back and forth between paddles that the players controlled. With plastic overlays affixed to the front of a TV screen, the light could be a football, a tennis ball, a soccer ball, or one of several other kinds of simple games.

Nintendo's operation wasn't sophisticated enough to develop and manufacture the microprocessor-based circuit boards that were the heart of the game system, so Masayuki Uemura suggested an alliance with an electronics firm. Nintendo teamed with Mitsubishi to build the video-game system and, in 1977, Nintendo entered the home market in Japan with the dramatic unveiling of Color TV Game 6, which played six versions of light tennis. It was followed by a more powerful sequel, Color TV Game 15. A million units of each were sold. The engineering team also came up with systems that played a more complex game, called "Blockbuster," as well as a racing game. Half a million units of these were sold. Nintendo had quietly entered the world of audio-visual entertainment and consumer electronics.

The successful TV game systems allowed Nintendo to tread water a while longer, but they were neither novel enough nor versatile enough to be the revolutionary product Yamauchi was searching for. He kept the pressure on his engineers, directing them to explore new ways of making video games. "We must look in different directions," Yamauchi said. "Throw away all your old ideas in order to come up with something new."

The electronic-calculator market was then booming. There were so many types available that the prices were shrinking almost as fast as their size—credit-card-size calculators were selling for a thousand yen, under $10. What else could be done with that miniaturized technology? "The Nintendo way of adapting technology is not to look for the state of the art but to utilize mature technology that can be mass-produced cheaply," says Gunpei Yokoi. He sought to make something smaller, thinner, and lighter than anything ever seen—something that was also fun.

It turned out to be Game & Watch, a video game the size of a calculator, with a tiny digital clock in the corner. With his engineers, he chose components from Uemura's old company, Sharp, and developed the smallest computer games ever seen. They weren't the easiest things to play—the controls were tiny—but they were a novelty. Nintendo shipped them all over the world by the tens, and then hundreds, of thousands. Many of the Game & Watches in circulation were illegal bootlegs—Nintendo lost potential millions because of all the non-Nintendo units made in various Asian cities—but the company nevertheless made many millions from the phenomenal number of Game & Watches they sold.

While Yokoi was "thinking small," Yamauchi had other engineers thinking big. After "Wild Gunman" and his first taste of the arcade business, he wanted more of the 100-yen pieces (and quarters) that were pouring into video games from the pockets of teenagers around the world. Popular games such as "Space Invaders" were behind a boom in the coin-op video-game business. Yamauchi wanted Nintendo to become a major player in arcades, and so his team came up with games like "Hellfire," "Sheriff," "Sky Skipper," and a battle game called "Radarscope."

Yamauchi, meanwhile, held endless meetings at Nintendo's growing R&D center with a group of designers under Masayuki Uemura. Uemura's team was working on what was emerging as the most significant new venture for Nintendo. It was a video-game system, but one that was much more sophisticated than Color TV Game 6 and 15. In America, systems had been released that played many games on interchangeable cartridges. The technology insured that the system would never become "old and stale," as Yamauchi put it—as long as "new and interesting" games would be available for it.

As Yamauchi learned about the technology—knowledge gained from late-night talks with Uemura and other engineers—he realized that the machine under development could do far more than just play games. "He had no concept that he was building a computer, but he nonetheless had his first glimpse of the incredible potential of a home-computer system disguised as a toy," says Uemura. "He saw far more than he let on to us." In the short term, Yamauchi saw a system that could be the basis of a profoundly expanded company if kids loved it enough and if they wanted more and more games to play on it and if Nintendo was the only maker of all those games.

There were many obstacles, including a number of competitors in the field. By 1983, systems had already been released in Japan by companies from the United States (the Atari 2600 and Commodore Max Machine) and from Japan (Epoch's Cassette Vision, Bandai's Intellivision, Takara's Game Personal Computer M5, Tomy's Pyuta, and Casio's and Sharp's small game-playing computer, the MSX). Yamauchi told Uemura he must "develop something that other companies cannot copy for at least one year. It must be so much better that there will be no question which system the customers will want."

For Uemura, the greatest challenge was not the technology; price was the crucial factor. Yamauchi wanted the system in many, many homes, so it had to be cheap enough so that almost everyone could buy it. All the machines currently on the market, except for the Epoch system, were selling for 30,000 to 50,000 yen ($200 to $350). Yamauchi set a goal of 9,800 yen, less than $75. At the same time, the system had to do what other systems, whether Japanese or American, could do, but more and better.

Uemura examined the competitors' machines. They were impressive in some ways. Built mostly by engineers with expertise in office computers, they could generate still pictures and alphanumerics, as well as perform complex calculations. But game play, Uemura believed, had essentially different requirements. The movement of characters and backgrounds on the screen had to be far more active, more believable, than on the other systems; it would have to approach the quality of fast-speed animation. A 16-bit processor could have done it all with ease, but Nintendo would have to make do with a less powerful 8-bit processor if the price was going to be kept low.

For ideas, Uemura picked the brains of the engineers who worked on Nintendo's arcade games. The arcade games had bigger, more expensive processors, but Uemura was most interested in the *thinking* behind the games, not the hardware. For a coin-operated game to make money, players had to become immersed in it as soon as the first coin was inserted. Many senses had to be taken over almost instantly to make the game play "hot," to use Uemura's term. The entire consciousness of a player had to be captured. There seemed to be two keys to accomplishing this: fast action, or a combination of fast action and intellectual challenge. The headier stuff was up to the game designers, but fast action required complex and expensive circuitry.

Uemura spent eighteen-hour days with the arcade engineers trying to determine the essence of the key components to the circuitry in the best coin-operated games. Only that essence could be carried over to the central processing unit of the new system. Finally he chose a relatively standard microprocessor called a 6502, but the one low-cost chip couldn't power all the aspects of a complex video game. One chip could control the information required for character movement and the interaction between the machine and the player, but if it had to do more than that, it would bog down. A second chip was needed to control the television screen itself—to generate bright colors, process pictures, and move them at a very high speed.

Other companies' game machines used an integrated circuit made for old-model personal computers, Texas Instruments' T19918, which allowed six to eight colors. Nintendo's machine had to have more colors for the better graphics Uemura sought (it ended up with fifty-two colors). Uemura, together with his growing stable of engineers and programmers, slaved over calculations and experiments to determine the maximum number of sprites (similar to the dots of a television picture) that could be generated on the screen in almost no time. The first calculation gave a number that was unsatisfactory; more sprites were needed if game play was going to feel noticeably more realistic than on the competitors' machines. The circuitry was modified. More experiments were conducted to determine how big an object and how many objects could move at a time, and how many changing functions could be built into one semiconductor. "We had to accomplish this exactly," Uemura says. "It was the order from the president. So much was riding on these experiments."

Nintendo's engineers could take the design of the two key chips only so far. They had to get the expertise of an outside company.

To determine who that would be, Uemura and Yokoi met with semiconductor-company representatives. It had been determined that the two custom chips that were needed were the basic central processing unit (CPU) and the picture processing unit (PPU). A supplier had to be able to help design and then produce them both, Yamauchi insisted, for a rock-bottom price.

It was not easy to be a supplier for Nintendo. The company's demands were rigid and exacting. Designs would change overnight and suppliers would have to be ready to change specifications on demand. "The most important thing we looked for in a supplier was the brain to cope with us," Yokoi says. "Unfortunately, we found that most companies are not flexible. Most companies move too slowly. That was not acceptable."

Given the prices Yamauchi would pay, the only way a partnership with Nintendo could pay off was in volume of sales—an *enormous* number of chips would have to be sold. Many major companies wouldn't gamble on Nintendo. "Now they wish they had," Uemura says.

Uemura went to Ricoh, the electronics giant, with his preliminary circuit diagrams. The chips he needed, he said, had to cost no more than 2,000 yen, which is why they had been scaled down to perform only essential functions. It happened to be a slow time for Ricoh's semiconductor division, so the company was willing to work with Nintendo, but they regarded the 2,000-yen price point as absurd.

Yamauchi was informed of Ricoh's stand and he made a proposal. "Guarantee them a three-million-chip order within two years," he said. "They will give us the price then."

Others inside Nintendo thought Yamauchi's tack was preposterous; leading manufacturers in Japan were selling 20,000 to 30,000 video-game systems, and the most Nintendo had ever sold of the TV Game 6 and 15 was 1 million. There was no way they would ever use anywhere near 3 million chips.

Yamauchi demanded that the proposal be made, and as he expected, Ricoh agreed. When an agreement was signed by Uemura and approved by Yamauchi, the head of Ricoh's five-man team in charge of the Nintendo project remarked politely that he was looking forward to bringing the new Nintendo system home to his children. He accomplished more than that: by the end of 1986, Nintendo would become Ricoh's largest customer, representing 60 to 70 percent of the company's semiconductor division's sales.

Ricoh was not Nintendo's only supplier. The larger companies that worked with Nintendo included Sharp, Mitsumi, Fuji, and Hoshi. There were, eventually, thirty subcontractors. Contracts with Nintendo would become among the most lucrative in the semiconductor and electronics industries. Ricoh and Sharp would form divisions that did nothing but supply Nintendo.

Keeping Hiroshi Yamauchi and his company happy, suppliers found, was difficult: Yamauchi insisted on lower defect rates than any other customer. But it was also rewarding. A 1991–92 survey indicated that Nintendo was spending $1 billion on semiconductors each year.

As development of the new video-game system continued, the engineers brought some of their questions to Yamauchi. What had to be included in the game console? Since the system was actually a small computer, it could have all the extras that computers could have. Should there be a disk drive which could read and write information? Should it have a keyboard? Should it have a data port, through which information could be sent and received to the system? The system could have a modem that would hook up, via telephone lines, to other game players or a central Nintendo terminal. It could have large amounts of memory that would accommodate more complicated programs.

Yamauchi was looking far down the road when he cautiously answered the engineers' questions. Although he eliminated anything that would add too much cost, he built in future expansion that went far beyond video games.

In the Japanese edition of *The Japan That Can Say No,* the book's coauthor (with Shintaro Ishihara), Sony's founder and chairman, Akio Morita, slaps American corporate wrists for shortsightedness. "We Japanese plan and develop our business strategies ten years ahead of time," he wrote. When he asked an American businessman if U.S. companies plan so much as a week ahead of time, he was told, "No, ten minutes." Nowhere is the repercussion of that difference more obvious than in the game system Hiroshi Yamauchi created. It anticipated a future that would not be revealed for a decade but which had the potential to propel Nintendo into the forefront of electronics and entertainment companies.

Yamauchi instructed Uemura to leave off the frills. No keyboard —it might scare off customers. No modem or disk drive. The system would play games on cartridges, not disks. Floppy disks were threatening to computerphobes and, more important, they were copiable.

The system would have minimum memory, since memory was so expensive, but it would have more than its competitors'. Atari's system had 256 bytes of RAM (random access memory, the amount of instructions a central processing unit can refer to at one time); the Nintendo system would have 2,000 bytes of RAM. In addition, games for the new Nintendo machine could be far more complex than the most powerful Atari games; a Nintendo cartridge could contain thirty-two times more computer code than an Atari cartridge.

Yamauchi cut out all extraneous devices to save money, but he told the engineers to include, for a trivial added cost, circuitry and a connector that could send or receive an unmodified signal to the central processor. The connector could pave the way for expansion —the addition of anything from a modem to a keyboard. It was why the machine would later be called Yamauchi's Trojan Horse: It slipped into living rooms with nothing but a pair of controllers, innocently toylike, yet it included the capability to do far more than play games. Nintendo boasted about it much later. "In the initial stages of [the system's] development, we foresaw these possibilities," reads a 1989 corporate report. "As such, we built a data communications function into the system and provided it with a connection terminal for an adaptor." Uemura modestly says that the plan worked so well because "we were lucky." But Genyo Takeda, a friend of Uemura, says, "He was so much an amateur that when Yamauchi told him to make this thing, he didn't know that it could not be done."

There were more practical and aesthetic decisions along the way. Steve Jobs's obsession to design the perfect mouse for the original Macintosh was no greater than Yamauchi's attention to the details of the controllers. Should there be one or two or more buttons? Should the system's casing have round or square edges? What color should the system be—a computer-like gray or beige, or a more playful color? Should the system box look more like a computer or a toy? (The answers they came up with were: two buttons on the right controller plus a directional pad and, on the left controller, a simple microphone through which one could "talk" to the system; softer, less threatening, rounded edges; and red and white plastic, to make the unit as toylike as possible.)

The year 1983 was a significant one for Nintendo. Yamauchi had the new Uji plant expanded to increase the company's production capacity. His stock became listed in the most respected first tier of the Tokyo Stock Exchange. And he began selling the system his men, after months of marathon work, shrouded in secrecy, had created.

The system was selling for more than Yamauchi had planned (about $100), but it was still less than half the price of the competition. In May, he addressed the Shoshin-kai Group, a wholesalers' group. He conceded that his new video-game player was priced so low that wholesalers wouldn't make much profit on it. "But," he said, "I guarantee that it will sell a lot because of the great games." He implored them to back the machine in spite of the low margin. "Forgo the big profits on the hardware," he said, "because it is really just a tool to sell software. That is where we shall make our money." At the meeting, Yamauchi announced the name of his new system. Here, he said, was Japan's first Family Computer. He dubbed it, for short, the Famicom.

Pushed by a barrage of advertising, 500,000 Famicoms flew off the shelves in the first two months. Six months later, however, a catastrophe occurred before the Japanese New Year, the toy industry's busiest season. There were at first a few calls from retailers. Then a few more. Masayuki Uemura and Gunpei Yokoi were called into Yamauchi's office. They were told that certain games for the Famicom caused the system to freeze.

The engineers nervously returned to their labs and worked on replicating the malfunction. Finally, there it was, trouble with one of the integrated circuits that got locked when certain information traveled on certain pathways, like a multicar pile-up on a badly designed freeway.

They trudged back to Yamauchi's office and explained the problem and the required solution. The circuitry on the chip had to be corrected. They expected Yamauchi to go into an explosive tirade. This was extremely *expensive* news.

Yokoi suggested that the company could replace units when customers complained. Hiroshi Imanishi, who was working on the marketing of the new machine, said that the problem could be more severe than whatever number of units had the defective chips; it could cost more than the hundreds of thousands, maybe millions, it would cost to fix or replace machines. Imanishi said it would hurt customers' opinion of Nintendo. Worse, much of Yamauchi's window of a year—the year that it would take competitors to try to copy the Nintendo machine and get it out the door—would be lost. A delay would allow competitors to swoop in and capture the customers that Nintendo had worked so hard, and spent so much money, to win.

Yamauchi listened to the opinions of his staff but ignored them. "Recall them all," he said.

Systems in stores and warehouses were pulled off the shelves, returned to the plant in Uji, and retooled (the bad chips replaced). In the end, Nintendo lost millions of dollars by missing the prime sales season, but Yamauchi's gamble paid off.

After the first million Famicom systems had been sold, there was still no sign of a slowdown. Once several million families had a Famicom and desperately wanted games, Nintendo could sell all it could produce. Yamauchi saw how Nintendo's emphasis would conceivably switch from hardware, with its limited market, to software, whose market was without limits.

Desperate retailers called Nintendo, frantically demanding product. New games were anticipated with a fervor that shocked store owners, distributors, and parents. Kids camped out in front of department stores and toy shops to snap up copies before the games sold out. Nintendomania was beginning, and Yamauchi, raking in more money than he had ever seen before, couldn't feed the frenzy quickly enough.

The success of the Famicom was unprecedented. Eventually, the fourteen competing home video-game machine companies withdrew from the market. The MSX was put in its place as a personal computer, not a game machine. Sega, a small arcade-game company, released a competitor called the SG-1000 the same year Nintendo released the Famicom, but it fizzled. And in spite of updated systems released by Atari, Nintendo had no competition to speak of. What had begun as the Yamauchi family business was inconspicuously on its way to becoming one of the most successful enterprises in the history of Japan—or, indeed, the world.

I, MARIO

"What if you walk along and everything that you see is more than what you see—the person in the T-shirt and slacks is a warrior, the space that appears empty is a secret door to an alternate world? What if, on a crowded street, you look up and see something appear that should not, given what we know, be there? You either shake your head and dismiss it or you accept that there is much more to the world than we think. Perhaps it really is a doorway to another place. If you choose to go inside you might find many unexpected things."

—Sigeru Miyamoto

Yamauchi's Famicoms were selling as fast as Nintendo built them. The success brought with it an unexpected, although not unwelcome, problem. A video-game system, like any other computer, could be elegant and powerful, yet it was only as useful as the software it showcased. The Famicom could have been as powerful as a mainframe computer, but no one would have noticed if the games were ordinary. Now the problem was that there were not enough good games.

Yamauchi had wisely anticipated the importance of software and prepared for it. One of the instructions he had issued to Uemura was that the Famicom must "be appreciated by software engineers." It had to be easy to program and able to do the kinds of things that game designers dreamed of doing. Any company, given the time, could copy the Famicom hardware. The key to staying ahead was software. By the time a competitor came out with a game that was as good as a successful Nintendo game, Nintendo had to be releasing a game that left the others in the dust.

Nintendo would, Yamauchi decided, become a haven for video-game artists, for it was artists, not technicians, who made great games. "An ordinary man," Yamauchi said, "cannot develop good games no matter how hard he tries. A handful of people in this world can develop games that everybody wants. Those are the people we want at Nintendo." He was interested only in the one genius, as he put it, who would drive Nintendo. He wanted to turn Nintendo into the single place the hottest game designers wanted to be associated with. Since, in Japan, most employees stayed with one company for their entire career, it was generally impossible to seduce good designers from other companies. That meant that they would have to come to Nintendo on their own, fresh from college.

Yamauchi wanted to create a place where his geniuses would be encouraged and inspired. But how? He was used to badgering and cajoling, or simply demanding—and that was certainly not the same thing as inspiring people, nurturing them. His reputation for aloofness and cockiness had grown with Nintendo. He luxuriated in his position as the merciless Goliath of his industry. He was already infamous for squashing people—or companies—that crossed him. He made up his own rules as he went along and he refused to play politics (which enraged government officials, who were used to being catered to). But could he *inspire*? "Research and development is the most difficult department to control," he observed. "It is difficult to control artists because they do not want to compromise."

The chairman had no engineering background, but he discovered how to stimulate innovative design. Isolated from the rest of Japan's industrial hubs in Osaka and Nagoya, and from the financial capital of Tokyo, Yamauchi ignored the textbook corporate examples. He had hand-picked his three chief engineers—Yokoi, Uemura, and Takeda—a long time ago, and they had done good work for him. In order to push them (and to learn more about how the engineers and designers worked), in 1984 Yamauchi made himself the supervisor of all R&D at Nintendo, "the heart of this company." He supported them with significantly more staff and resources.

Yamauchi arrayed his chiefs directly below him, each of them in charge of his own group: R&D 1, 2, and 3. Within an R&D group were many teams, which were pitted against each other. The teams in the groups working mostly on hardware tried to outdo the others in the virtual miracles they came up with, and the software teams competed to make the greatest games that had ever been seen.

Yamauchi has never played a video game in his life and he had little interest in them. Still, he alone was the judge and jury when it came to deciding which games Nintendo would release. It was audacious, and he was either remarkably intuitive or terrifically lucky. Yamauchi was criticized for being ruthless when it came to many of his business practices—manipulating the market, *terrorizing* employees—but no one questioned his genius when it came to choosing Famicom games. A Nintendo manager criticized Yamauchi for his obstinacy but praised his instinct: "It's like a sense for the fashion business, knowing what will become hot and popular next season. He can read a few years in advance. He is so certain that he is right that he listens to no one."

His R&D groups competed among themselves for Yamauchi's attention and praise, but there was no doubt about their collective place in the company. They were his stars. While most companies directed input from market research and from their sales force to their R&D sections, Yamauchi in those days insisted that R&D was sacrosanct: no one told his creative people what to create. The marketing department saw games only when they were completed. "He believes the marketing people will only look at what's popular right now," Hiroshi Imanishi says. "And if we make the game based on what's popular right now, the game will not be new and fresh."

The personal attention their leader lavished on his inventors was a mixed blessing. A nod from Yamauchi could make a designer's day—or week, or month. Engineers were ecstatic when they came up with a game that delighted him. On the other hand, an admonishment could be devastating. "Months of work can be disposed of with a scowl," says an engineer who left Nintendo. The project is dead, instantly. His victims suggested that Yamauchi's judgments were capricious or the product of his moods, and that his callousness caused a great deal of frustration and anger. Engineers occasionally left, and others, exhausted and disappointed, were sent on sabbatical. They were told, "The company is making money; don't worry. Spend the time, relax. Come back fresh." Most commonly, designers whose work was rejected would only redouble their efforts, determined to have *their* game chosen the next time. Yamauchi's autocratic, often brutal system worked.

The R&D groups worked in spacious, private laboratories in the development wing of the main Nintendo building. In these white-walled, white-ceilinged rooms, rows of computer monitors were set up on tables. Their screens shone with blow-ups of circuit boards that looked like magnified city maps. Other screens, stacked as if in a television showroom, displayed details of game characters—the left cartoon hand of a boxer, for example. Still other screens were filled with column after column of fluorescent, sallow-yellow numerals.

Here and there were drafting tables, covered with schematic blueprints for games or scribbled calculations. Laser printers, networked to dozens of terminals, spewed book-length programs, and Xerox machines churned out copies of sectional drawings of game worlds.

In the design rooms, the men (no women) worked methodically as they competed to make products that would become *the* product. The goal was excellence—anything less would wind up on the scrap heap. Yamauchi believed that it was far better to put all his resources into the production of one or two hit games a year rather than several minor successes. When he released new games, he only had to manufacture, package, market, and advertise those few, but that meant that the stakes for each Nintendo game were extremely high. The games had to warrant all the costs of development (up to $1 million per game) and marketing (up to several million more).

The high stakes meant there wasn't always *wa* (harmony) within the company. Yet in spite of the competitiveness, the three chief designers respected each other and, when they were called on to do so, worked together well. Part of the reason the competition didn't turn them against one another was that Yamauchi parceled out his praise. On the other hand, if any one team had too much success, it could be expected to be slapped down. The result was that each team came to excel in different areas and at different moments. In the end it was difficult if not impossible to determine which of the three design teams contributed more to Nintendo's growth.

Takeda says Yokoi, his mentor, was "the sharpest designer." Besides all the work they had done for Yamauchi in the past, Yokoi's R&D 1 designed Game Boy, which would become another extraordinary Nintendo success. His team of thirty engineers were "a band of samurai," says a colleague outside of Japan. They operated quietly, with less recognition than the others. Their leader was *nazonoyona,* an enigma.

Yokoi was the oldest of the top engineers (though still only in his forties) and the most like traditional, old-school engineers at other companies throughout Japan. He wore simple short-sleeved shirts, and his hair was cut so that there was a neat, clean line around his ears and neck. He was dedicated to the company over everything else; he was a Nintendo man.

The games from R&D 1 would be some of Nintendo's best. One was the phenomenal game "Metroid." In the video-game world of macho stereotypes, the game's hero was a surprise. Samus, the warrior, on the quest to destroy the Mother Brain, went to battle with a nifty array of weapons and slick moves, dressed in a space suit and helmet. At the end of the game, after the Mother Brain died a screaming, light-spewing death, Samus could finally relax and take off his helmet. Long blond hair fell out. Samus, the great warrior, was a woman.

The greatest contribution of Uemura's team, R&D 2, was the Nintendo hardware itself. R&D 2 also came up with peripherals, including the Communications Adapter for the Nintendo Network. Sixty-five people worked with Uemura, whose face wore a constant expression of astonishment. He spoke in a raspy, hushed tone—Tom Waits after a few bourbons—as if what he had to say was clandestine and dangerous, which it sometimes was. (Yamauchi had tapped Uemura's team for a top-secret project that was kept under wraps for years.)

Takeda ran R&D 3, which would design games such as "Star Tropics." More significant, however, was some of the technical magic the team performed. R&D 3 came up with technology that allowed the other groups to make games that the original Famicom hardware could never have powered on its own. The first Famicom cartridges used what were called NROM chips (N for Nintendo and ROM for Read Only Memory). Unlike computer programs on floppy or hard disks, these programs were not changeable. A game program was reproduced onto an actual integrated microcircuit. Using a photographic process, the circuit was duplicated onto thin silicon wafers that were sandwiched together and attached to connectors. Through them the information—the game program—was transmitted to other components in the system. The amount of information in a game was limited by the size of the ROM.

Each game cartridge had two main chips, one for the program itself, which could be up to 256K (kilobytes) and one for the on-screen characters, which could be 64K. Programs for games and characters had to fit within those chips until R&D 3 designed new kinds of cartridges.

R&D 3 created a cartridge called UNROM, which allowed greater memory size and bank switching. A RAM (Random Access Memory) chip was a place to store information until it was needed by the computer's processor. Bank switching was a process for grabbing, from that stored information, whatever was needed whenever it was needed. A new game screen, complete with new kinds of enemies and waterfalls and creatures (and the programs to make them work) could be retrieved from RAM when the player arrived in that "room."

There were still severe limits on the cartridges, however. The amount of information that could be switched was scant and the process was slow. Takeda's gang tackled the limits with new kinds of chips called MMCs (Memory Map Controllers). They made the system do things that the Famicom's 8-bit processor could never have approached on its own. Years after the Famicom was introduced, games seemed to get more and more complex. It was as if the old Apple II were suddenly powering Hypercard. Takeda's chips, by taking on some of the Famicom's processing power, essentially added RAM and other specific powers to the machine.

The Famicom could do things it was never designed to do: images could scroll diagonally, objects could move quicker, and far more could happen at one time. The system itself still had only 2 kilobytes of RAM, but this was supplemented by the custom-designed sets of circuits with specialized functions in the MMCs. Some of the circuits, called Logic Gates, increased the speed and efficiency of the background computing that made everything happen. Others directed the program to specific locations in the memory, traffic cop style. They were smaller and cheaper than the chips in the UNROM, and they allowed larger program and character memory size. With the addition of the first MMC chip, the potential for more complex and sophisticated games had arrived. The first examples were "The Legend of Zelda," "Metroid," and "Kid Icarus," three breakthrough games, all huge sellers.

Subsequent MMC chips allowed the Famicom to do even more. With MMC3, the screen could split into two parts, each moving independently. With MMC5, there could be more images on the screen at a time. Unaided, the Famicom could project pictures of 960 tiny square pieces, called tiles, but only 290 could be unique, which is why there were so many walls full of bricks or other repeated patterns in early games. MMC5 made it possible for all 960 tiles to be different. It also processed math problems on its own, freeing up the main processor. Memory size for games with an MMC5 shot up to 8 megabytes, thirty-two times more than the original cartridges.

R&D 3 also figured out a way to include a battery backup system in cartridges that allowed some games to store information independently—to keep track of where a player had left off, or to track high scores. The new battery system could store the data for the life of the battery (about five years).

Takeda's group obsessed over the highly technical and the obscure. The fruits of their labors were dramatic—most of the best Nintendo games would not have been possible without them—but they were not always obvious. R&D 3, nicknamed "Rumania," was isolated from the other groups. Its motto was grand: "There are no limitations, no boundaries; since we are on our own, there is nothing we cannot do; when you start with nothing you can do everything." Their leader, with his quizzical glances under heavy arching eyebrows and his arcane, light-bulb brain, boasts, "We *have* to have more talented people because we are given unthinkable tasks."

Takeda's twenty-person staff was a band of *otaku*—computer hackers and nerds. They were the consummate eggheads and dweebs. "Becoming maniacs," Takeda said, "is the idea."

The three R&D groups were immersed in their respective projects one day when Yamauchi required the talents of a designer. The project was not important enough for him to pull one of the key members from their work, so he called in the apprentice in the planning department.

Sigeru Miyamoto remembers the maze of rooms in the paper-and-cedar home of his childhood. Sliding shoji screens opened up onto hallways, from which there seemed to be a medieval castle's supply of hidden rooms. The tiny home was in the countryside near Kyoto, in the town of Sonebe, where his parents and grandparents had been born before him. The surrounding landscape was Miyamoto's playground: he fished in the river, ran along the banks of sodden rice fields, and rolled down hillsides.

Across the sand-and-stone street from his home was a rice field. After the yearly harvest, when the field was dry, it became a park for baseball and other games. He played there with neighborhood children in the afternoons, and in the evenings he attended Noh plays, heroic dance dramas, or puppet shows, or he gathered with his family at one of the neighbors' homes for festive dinners.

The Miyamoto family had no television and no car. Every few months they traveled by train to Kyoto to shop and see movies: *Peter Pan, Snow White.* At home, Miyamoto lived in books, and he drew and painted and made elaborate puppets, which he presented in fanciful shows. After school, he often lit out into the countryside for adventure. He had to pass a neighbor's house where a bulldog lay in wait for him. The dog charged every time, barking and snapping, and Miyamoto froze. At the last second, the dog's chain reached its limit and jerked it back. Miyamoto stood just out of the reach of its salivating jaws.

Investigating hillsides and creek beds and small canyons, Miyamoto once discovered the opening of a cave. He returned to it several times before he worked up the courage to go in. Lugging a homemade lantern, he went deep inside until he came to a small hole that led to another cave. Breathing deeply, his heart pounding, he climbed through. He never forgot the exhilaration he felt at this discovery.

The family moved to Kyoto, where Miyamoto and his new friends had secret meetings in the family's attic at which codes and passwords were traded. There were dares to explore forbidden places—a neighbor's yard guarded by an Akita; another neighbor's basement, which held a treasure trove of trunks stuffed with ancient costumes.

Miyamoto wanted to be a performer, a puppeteer, or a painter when he grew up. He carried pads of paper and pencils and drew nature scenes in parks and along the river that divided the city. In school, while his teachers lectured, Miyamoto daydreamed. At night, he constructed plastic models and wood-and-metal contraptions until his father sent him to his room to study. Math and grammar were put aside for drawing.

Miyamoto took cartoon-making seriously. He drew a figure and then invented its life and personality. The figures wound up in intricately drawn flip books. At school he organized a cartoon club that met regularly and had yearly exhibitions.

In 1970, Miyamoto entered Kanazawa Munici College of Industrial Arts and Crafts. It took him five years to graduate because he only attended class about half the time. Instead of studying, he spent his time sketching in his notebooks and listening to records. He loved the Nitty Gritty Dirt Band, the Country Gentlemen, and David Grisman. He taught himself how to play the guitar—American bluegrass music, of all things. It wasn't easy to find a banjo player in Kyoto, but he did, and the duo performed at coffeehouses and parties. His friends were artists and musicians. They hunted in record shops for hard-to-find (in Kyoto!) Kentucky Colonels LPs and traveled to Tokyo to see Doc Watson perform live.

When he finally graduated, Miyamoto agonized over what kind

of job he should get. He had no interest in traditional business, and he knew he would never survive the monotony of a rigidly structured corporation.

Then a revelation came to young Miyamoto. He asked his father to contact an old friend, Hiroshi Yamauchi, who ran Nintendo. The elder Miyamoto asked Yamauchi to meet with his son, a recent graduate with a degree in industrial design, who was looking for a job. "We need engineers, not painters," Yamauchi said, but he agreed to a meeting as a favor to his friend.

Miyamoto was twenty-four in 1977, when he entered the office of the Nintendo chairman. He had shaggy hair, boyish freckles, and a cat-who-swallowed-the-canary smile. He dressed nicely, and he behaved in accordance with traditional etiquette, yet there was mischief and wonder in his eyes. Yamauchi liked the young man and asked him to return for another meeting, this time with some ideas for toys.

Miyamoto returned with a portfolio and a large sack from which he produced a recent invention. It was a clothes hanger designed for children. Nursery schools could have a row of them along the wall, he explained. Or parents could put them in children's rooms. Regular metal hangers, he told Yamauchi, were dangerous for children; the pointed hook could hurt them, even poke out an eye. His hanger, carved out of soft wood and covered with cheerful acrylic paint, was in the shape of an elephant's head. Clothes were hung on the ears and turned-up trunk. The elephant's neck fit snugly like a puzzle piece onto a knob that attached to a wall.

Miyamoto had other hangers as well: a bird and a chicken. Then he showed Yamauchi some drawings for more elaborate toys—a whimsical clock for an amusement park; a swing within a seesaw on which three children could play at once.

Yamauchi saw ingenuity and resourcefulness in the work, and he hired Miyamoto to be the company's first staff artist, even though the company had no specific need for one at the time. Miyamoto was assigned to be an apprentice in the planning department.

When Yamauchi called Miyamoto into his office in 1980 the young man looked down at his hands, his long fingers folded on the smooth table in front of him. He listened intently as Yamauchi told him that he was looking for a video game. Miyamoto had played many video games at college in Kanazawa. He loved them. In video games, cartoons came to life.

He boldly told the Nintendo chairman that he would enjoy creating a game. However, he said, the shoot-'em-up and tennis-like games that were in the arcades at that time were unimaginative, simply uninteresting to many people. He had always wondered why video games were not treated more like books or movies. Why couldn't they draw on the great stories: some of his favorite legends, fairy tales, and fiction—*King Kong, Jason and the Argonauts,* even *Macbeth*?

Nodding impatiently, Yamauchi rushed to the point: A Nintendo coin-operated video game called "Radarscope" was a disaster. There was no one else available to come up with a new game design. Miyamoto had to try to convert "Radarscope" to something that would sell. Yokoi would oversee the project, but Miyamoto was on his own.

After consulting with the R&D 1 chief, Miyamoto returned to his desk with the schematic drawings of "Radarscope," which he found simplistic and banal. Enemy planes approached and players had to shoot them down. Miyamoto threw it away. He asked questions of technicians about the kinds of movements characters could make, the possibilities for different-size characters, and the variations of action and reaction that were possible. Nintendo was negotiating with King Features for the rights to use the *Popeye the Sailor Man* comic as a video game, and Miyamoto was told he could work with those characters. The Popeye license from King Features fell through (although the license was later renegotiated and the Popeye game was made), so he tried other ideas.

He thought about *Beauty and the Beast,* but simplified the story. He came up with his own beast, a King Kong–like ape, a humorous bad guy, "nothing too evil or repulsive," Miyamoto recalls. The ape would be the pet of the main character, "a funny, hang-loose kind of guy" who was not especially nice to the gorilla. "It was humiliating! How miserable it was to belong to such a mean, small man!" says Miyamoto. At his first opportunity, the gorilla escaped and kidnapped the guy's beautiful girlfriend.

The gorilla didn't take the woman to hurt her—an important point in Miyamoto's mind—but to get back at the little man. The man, of course, then had to try to save the girl.

Miyamoto wanted the main character to be goofy and awkward. He chose an ordinary carpenter, neither handsome nor heroic. He wanted him to be Walter Mittyesque, someone anyone could relate to. On a large sketch pad he drew a nose. "Having a nose or not having a nose is completely different," he says. "Noses say a great deal." The nose Miyamoto created was a distinctive bulbous orb made even more noticeable because of the exaggerated bushy mustache beneath it. From one of his old notebooks filled with characters, he chose a pair of large, pathetic eyes.

The engineers had taught Miyamoto that it was important to distinguish the body so it would be visible on a video-game screen. Therefore he clothed his chubby character in bright-colored carpenter's overalls. In order to make the movement obvious in the simple animation of video games, it was important that characters' arms moved, so he drew stocky arms that swung back and forth. The engineers said it was difficult to accurately represent hair in a video game because of inertia: when a character fell, logically his hair would have to fly up. To avoid the problem, Miyamoto added a red cap. "Also," he adds, "I cannot come up with hairstyles so good."

Many of his ideas for the game were rejected by Yokoi; Miyamoto's characters had to do simpler things than he wanted them to. He ended up having the carpenter maneuvering up the unfinished foundation of a building in order to reach the gorilla, who had climbed to the top with the girl. To get there the little man ran up ramps, climbed ladders, rode conveyor belts, and jumped on elevators while trying to avoid the objects the gorilla hurled at him—cement tubs, barrels, and beams.

Miyamoto was nearly finished, but the game needed background music. He wrote it himself, on an electronic keyboard attached to a computer and stereo cassette deck. When the game was complete, Miyamoto had to name it. He consulted the company's export manager, and together they mulled over some possibilities. They decided that *kong* would be understood to suggest a gorilla. And since this fierce but cute kong was donkey-stubborn and wily (*don-*

key, according to their Japanese/English dictionary, was the translation of the Japanese word for stupid or goofy), they combined the words and named the game "Donkey Kong."

Later, when the American sales managers who would sell the game outside Japan heard the name, they looked at one another in disbelief, thinking Yamauchi had flipped. "Donkey *Hong?*" "Konkey *Dong?*" "Honkey *Dong?*" It made no sense. Games that were selling had titles that contained words such as *mutilation, destroy, assassinate, annihilate.* When they played "Donkey Kong," they were even more horrified. The salesmen were used to battle games with space invaders, and heroes shooting lasers at aliens. One hated "Donkey Kong" so much that he began looking for a new job.

Yamauchi heard all the feedback but ignored it. "Donkey Kong," released in 1981, became Nintendo's first super-smash hit.

When Yokoi later needed help with games for Game & Watch, Yamauchi told him to use Miyamoto, since his other designers were busy with their own projects. "I asked him to do creation and I would supervise," Yokoi says.

The computer chips that were affordable and tiny enough to fit into a Game & Watch could store few characters and even fewer movements, so Miyamoto was limited to telling simplistic stories. He adapted a simpler form of "Donkey Kong" for Game & Watch, and after the agreement for the Popeye license was hammered out, he made a mini "Popeye the Sailor Man" game. The latter game has Popeye attempting to save Olive Oyl from Brutus. When Popeye is weakened by too much of Brutus's abuse, he gains strength by downing cans of spinach. Millions of "Donkey Kong" and "Popeye" Game & Watches were sold.

In 1984, Miyamoto was again summoned to the chairman's office. Yamauchi explained that he needed more games, this time for the Famicom. Miyamoto was to head up a new division, R&D 4. The group, Joho Kaihatsu, or the entertainment division, had one assignment: to come up with the most imaginative video games ever.

The decision was one of the smartest Yamauchi would ever make. Miyamoto, it was soon apparent, had the same talent for video games as the Beatles had for popular music. It is impossible to calculate Miyamoto's value to Nintendo, and it is not unreasonable to question whether Nintendo would have succeeded without him.

After meeting with Yamauchi, Miyamoto returned to his desk. He took a pencil and began sketching the suspendered hero from "Donkey Kong," who had been given the name Mario. Someone had mentioned that Mario looked more like a plumber than a carpenter, so he made the new Mario into one. Since plumbers spend their time working on pipes, large, radiant-green sewer pipes became obstacles and doorways to secret worlds in his next game, "Super Mario Bros."

The brother Miyamoto created for Mario was Luigi, as tall and string-bean thin as Mario was short and fat. That attribute, as well as the color chosen for his overalls (green to Mario's red), was simply to distinguish the two characters on the fast-moving game screen.

"Super Mario Bros." and the sequels Miyamoto designed soon became the most loved video games ever. The "Mario" games were more interesting because there were always new worlds to conquer, each one more magnificent than the last. There are walking plants, fish that Dr. Seuss might have created, dragons, serpents, flying turtles, fire-spitting daisies, and angel wings upon which Mario and Luigi can hitch a ride.

Humor was subtly introduced into the adventure. Miyamoto's mind bent around corners; players' minds follow, delighted. Eventually they figure out that the princess has to ride atop a ladybug if she is going to get to the boss of one level in "Super Mario Bros. 2." (The ladybug looks up her skirt as they head there.) The miniboss of that world—the chief bad guy—spits out lethal eggs larger than his head. In one sequel to "Super Mario Bros.," players have to figure out how to get through a seemingly unreachable door. Mario has to remove some of the coins that are floating in front of the door and take them back to another room to trip a "switch block" that changes the coins into stones. The stones can then be used as steps up to the door. Kids spend hours compulsively trying to figure it out.

Adults enjoy Mario too. They respond, Miyamoto feels, because the games bring them back to their childhoods. "It is a trigger to again become primitive, primal, as a way of thinking and remembering," Miyamoto says. "An adult is a child who has more ethics and morals. That's all. When I am a child, creating, I am not creating a game. I am in the game. The game is not for children, it is for me. It is for the adult that still has a character of a child."

Miyamoto borrowed freely from folklore, literature, and pop culture—warp zones from *Star Trek*, empowering mushrooms from *Alice in Wonderland*—but his most captivating ideas came from his unique way of experiencing the world and from his memories. When Mario jumps up in space at certain locations, nothing ought to happen because nothing is there, but Mario finds secret, powerful mushrooms and invisible doorways to new worlds. "I exaggerate what I experience and what I see," Miyamoto says.

In the "Mario" games and in some of Miyamoto's other popular games, such as "The Legend of Zelda" and its sequel, part of the adventure is wandering into new places without a map. "When I was a child, I went hiking and found a lake," he says. "It was quite a surprise for me to stumble upon it. When I traveled around the country without a map, trying to find my way, stumbling on amazing things as I went, I realized how it felt to go on an adventure like this." In the games, it often is quite a surprise to come upon a lake amid a forest, a rocket ship hidden beneath the sands of a desert.

"When I went to the university at Kanazawa, it was a totally strange city for me," Miyamoto says. "I liked walking very much, and whenever I did, something would happen. I would pass through a tunnel and the scene was quite changed when I came out." Tunnels in his games are doorways to unexpected things. At the other end of a tunnel the fog may be so thick that it is impossible to see what is ahead. In order to explore the new place, the player must return through the tunnel to search for a hidden torch. Armed with the torch, the player is able to go back through the tunnel and face what is hidden in the fog. In "Super Mario Bros. 3" and "Super Mario World," Mario can fly. However, as in Miyamoto's (and many people's) dreams, he often cannot fly high enough or long enough before he comes crashing down to earth.

There are often great risks attached to exploring the worlds in Miyamoto's games. "I was living in an apartment in Kyoto, and nearby was a building that had a small manhole cover mounted in the wall," Miyamoto remembers. "I walked by it every day and I noticed it. I wondered, Why is a manhole on the wall? Where does it lead?" Miyamoto never found out, but in "Super Mario Bros.," when the player encounters a manhole, he can choose to do what Miyamoto never did: open it and go inside. To do so is worthwhile.

Miyamoto as a child had worked up the courage to go beyond the periphery of the forbidding cave he had discovered. "The spirit, the state of mind of a kid when he enters a cave alone must be realized in the game," he says. "Going in, he must feel the cold air around him. He must discover a branch off to one side and decide whether to explore it or not. Sometimes he loses his way." Not just the experiences but the *feelings* connected to those events were essential to make the game meaningful. "If you go to the cave now, as an adult, it might be silly, trivial, a small cave," Miyamoto says. "But as a child, in spite of being banned to go, you could not resist the temptation. It was not a small moment then."

In Sonebe, Miyamoto had once climbed a tree and gotten high enough to see far-off mountains before he realized that he was stuck; there was no way he could get down. Super Mario gets himself into similar fixes all the time. Once while fishing when he was a young boy, Miyamoto reeled in a bony, grotesque little fish with snapping jaws. Mario encounters the fish that Miyamoto as a child *imagined* he had hooked: a monstrous creature that would happily devour him.

The memory of being lost amid the maze of sliding doors in his family's home in Sonebe was re-created in the labyrinths of the "Zelda" games, while in the Mario series Miyamoto made safe places that felt like the haven of his parents' attic. The dog that had terrorized him when he was a child attacks Mario. "I am especially proud of the dastardly, repulsive characters," he says. Miyamoto's dream was to make games that created worlds in which game characters could be more like players' companions, seemingly independent. "Perhaps they can even be ourselves at other times in our lives," he explains obliquely.

Older and more sophisticated players often miss much of the magic in the games. Young children, who do more leisurely exploring, and quiet and thoughtful children, who are more contemplative, have a better chance of finding hidden secrets than the kids who blast through, charging toward the goal. "The players must be thinking, 'Well, I don't see anything here, but it can be, it's possible.' Then the player is curious enough to visit that place. When he finds something he never expected, he feels, 'Ah, I did it. I made it.' It's a great kind of satisfaction."

The most wondrous surprises are timed to occur at intervals that keep things hopping. It is worth going forward because something good is waiting around the next corner, or in the next world. Some of the secrets are so well hidden that it is a miracle kids find them at all. Each level of each game ends with a flagpole, but a secret whistle in "Super Mario 3" is hidden *beyond* and above the flagpole—in a place that seems to be outside the game, or at least outside the part of the game that can be seen on a television screen. It is as if Mario has to fly out of the television set for a while until he reaches the entrance to a secret room. Who would ever think of trying it? Those who do are amply rewarded. The whistle gives Mario the power to travel to any world in the game at any time.

Many of Miyamoto's subsequent games not only had the same characters and roughly the same goals, but built on the skills that were learned in the preceding games. There were many new lands and new tricks, and with them the sense of accomplishing new things, yet there was also the comfort of not having to learn a game from scratch.

At Nintendo, Miyamoto's stature increased. After being made the director of his first games, he earned the title of producer. It meant a great deal to him: he had the same title as his idol, George Lucas (*Raiders of the Lost Ark* was Miyamoto's favorite movie). Now, instead of working on one game at a time, he oversaw the production of several, each budgeted at more than $1 million. From six to twenty people worked on each game for a period of twelve to eighteen months.

Technology eventually progressed to make some of the production stages easier. Originally Miyamoto had to paint each character. The colors in the painting were given numbers and the numbers were inputted into a computer, dot by dot. He showed programmers not only how the character looked but how it moved and what special traits it had (a bee, when hit, lost its wings but continued to stalk Mario; boats made out of skulls sank into a fire pit). The characters and their movements were written, line by line, as instructions in a computer program.

Tools were developed to eliminate much of the tedious work. Diagrams and drawings were translated into computer graphics with technology called Character Generator Computer Aided Design (CGCAD). "Character banks" of images were stored along with the codes that described them. Movement, too, was now programmable from a bank of choices.

Miyamoto was a terrible manager of his division; he needed an assistant to keep everything and everyone organized. Nonetheless, he oversaw all aspects of the creation of the games. He wrote the scripts and then worked with editors, artists, and programmers. When a game was nearly completed, he spread out its blueprint across a room full of tables that had been pushed together. The blueprint was the map of a game's pathways, corridors, rooms, secret worlds, trapdoors, and myriad surprises. Miyamoto lived with it for days, traveling through the game in his mind. As he went along, he determined which points were too frustrating or too easy. He added mushrooms or a star to make Mario invincible. He made certain that the moments that gave the greatest delight—a dinosaur that hatched from an egg, a feather that let Mario fly—came at sufficiently frequent intervals.

When he had edited a game to his satisfaction, Miyamoto went back to his director and technicians and had them incorporate his revisions. They worked for many days and nights on the changes, testing idea after idea, until Miyamoto was happy with the pacing.

When the game was ready, it was scored. Music was just as important for a game as for a movie: the same world could seem scary or lighthearted, depending on the music.

Miyamoto worked with a professional, in-house composer, most often with a brilliant young musician named Koji Kondo, who wrote the music for all the "Super Mario" games. Kondo's music became so popular that recordings of his Nintendo music were successful CDs and records. (In Tokyo, a symphony performed Kondo's "Mario" music, and the Jamaican reggae singer

Shinehead borrowed the "Mario" theme for the chorus of a rap song.)

After the music was added and the final edit completed, Miyamoto's games were ready. Kids were waiting. Between 1985 and 1991 he produced eight "Mario" games. An astounding 60 to 70 million were sold—either individually or packaged with hardware as an incentive to buy Nintendo systems—making Miyamoto the most successful game designer in the world. One designer suggests it is because he is left-handed. Miyamoto shrugs: "I think it is nothing more than destiny."

As his games' popularity grew, Miyamoto became well known in Japan and beyond. Westerners who made the pilgrimage to Kyoto to meet him included Paul McCartney, who, during a Japanese tour, said he wanted to see Miyamoto, not Mount Fuji. As a fan of the Beatles, especially *Abbey Road,* Miyamoto was thrilled, although he was never quite able to fathom the attention he received.

Meanwhile, Miyamoto had met a woman named Yasuko, who worked in Nintendo's general administration department. They dated and soon married. He had been living in a nearby Nintendo dormitory, and he and Yasuko moved into a small house near Nintendo's office. From there he walked or rode a bicycle to work. Yasuko stopped working when the first of their two babies was born. The family would walk down the street in Kyoto, and his fans, who reverently call him Dr. Miyamoto, often stopped him to pay homage. Miyamoto didn't change much. Even when he was approaching forty and started cutting his hair shorter (although no one would ever call it neat), he remained unassuming and shy. His mind never stopped wandering to new places—places that were re-created in his newest games.

In spite of a string of hits made by Miyamoto and by the other R&D groups, Yamauchi still was unable to meet the demand for games. Retailers were turning away hordes of customers, which distressed them. Yamauchi himself feared that customers who couldn't get enough games would move on to other forms of entertainment, perhaps a competitor's video game system. How, he wondered, could he increase the number of games available?

Many companies, mostly producers of video-arcade or floppy-disk computer games, had approached him, but Yamauchi hadn't wanted to relinquish any control over the games. If games of poor quality were released, his customers would become disappointed with the Famicom. But the real reason he didn't want other companies to produce games for his machine was that they would make piles of money, and Yamauchi wanted it all for Nintendo.

4

INSIDE THE MOTHER BRAIN

In a moss-carpeted park in the center of Kyoto's business district, amid still-dormant cherry trees, a man in a dark suit sipped tea and wrote haiku. The business day seemed to have thawed away into a tranquil pool of deliberation.

Across the street, in lounges, men sat before tall bottles of beer and delicate cups filled with warmed sake. The frenetic day shaken off like a brittle cocoon, the men felt replenished, even as the poet in the park took up his pen. A line of carbon-black ink assaulted the white parchment before him.

Night fell and an electric day was born in the Las Vegas blinking of the pachinko parlors and the electric street lamps, the spotlights on billboards, and the neon announcements for Coca-Cola and Sony. The poet vanished, but many of the businessmen ducked into nearby karaoke bars, where pretty, young hostesses giggled and made small talk and poured the next drink. Men—by day stern and forbidding—took turns climbing up on stage, where they took hold of a microphone and sang love songs to prerecorded accompaniment.

Karaoke had become a favorite after-work ritual for many businessmen; they took to it nightly, religiously. An important manager of a high-technology company arrived to join a group of men and was introduced not by his position in the company but by his distinction as the number-one singer in his office.

Across the river that divided Kyoto, it was quieter. Light emanated from the ribbon of windows that wound around the Nintendo compound. Inside, no one sang. Hiroshi Yamauchi had no tolerance for karaoke.

At the entrance of the main building was a large waiting area that had all the intimacy of an airport terminal. There were rows of uncomfortable molded-plastic chairs and couches and Formica-topped end tables. Behind a marble-topped reception desk were women in powder-blue skirts and smocks, some with tiny pillbox hats. The walls were devoid of all decoration. A maze of hallways with shiny waxed floors lay beyond the waiting room. Behind one anonymous door was Hiroshi Yamauchi's office, called by one employee, "the realm of the Mother Brain." In the game "Metroid" the Mother Brain was the pulsing, laser-spewing creature that hurled bolts of crimson electricity and survived by sucking the universe of all of its energy.

Inside Nintendo's Mother Brain was a substantial wooden desk that faced a small coffee table with couches on either side. The carpet was industrial gray, speckled with beige. There was a small television on a shelf.

A little after nine in the evening, Yamauchi concluded his final meeting of the day. Emerging from a conference room, he padded down the hallway in rubber sandals, his tie loosened, and headed back to the seclusion of his office.

Employees filed out—a succession of men and women wearing Nintendo-blue (hospital-blue) smocks or jackets, or else white shirts with dark business suits. They headed to their cars or to the train or simply walked down the road to their nearby corporate living quarters.

Gunpei Yokoi and Hiroshi Imanishi were huddled together in a conference. Some of Sigeru Miyamoto's R&D team, in the corner of a huge room under parallel rows of fluorescent lights, were playing a test version of a new game, searching for an irksome bug that had been detected earlier that day. (A bug is a flaw in a program that causes malfunctions.) From a cubicle in one corner of a large open office, the tearful voice of a female Japanese pop singer crooned desperately to the man who had betrayed her.

There were no sounds or voices along the corridor that led to the Mother Brain. Inside, another man had joined Yamauchi. They greeted each other and sat on the couches on opposite sides of the low table. Before leaving for the day, Reiko Wakimoto, Yamauchi's secretary, delivered a silver tray upon which was a bottle of fine Scotch, two heavy crystal tumblers, and a small bucket filled with ice. She poured drinks for the two men before departing, bowing respectfully.

Yamauchi's hair had thinned, but he still combed it straight back. The silver was more pronounced, more distinguished. As he spoke, he rubbed his hands on the wooden arms of his chair. He sat with his head jutting forward, which made it seem out of proportion to his small frame. He talked through clenched teeth, his chin taut and drawn.

"Your move," he said.

Yamauchi always wore dark suits with plum or navy ties and yellow-tinted glasses that gave his face a pronounced pallor. Without the jacket and with his tie removed, he seemed frail, his body shrunken in the oversized armchair. He leaned his head back and narrowed his eyes.

The two men clasped their drinks—Yamauchi's companion shook his in a circling motion; ice skated around the glass—and stared at the square board resting on the table between them. The board, made of blond wood, was covered with a grid of thin black lines, nineteen vertical and nineteen horizontal. The 361 intersections on the board represented the world. Smooth white "stones" (made of clamshells) and black ones (made of slate) were positioned strategically on the board. They represented the two forces in conflict, both trying to control the game board—the universe.

The game they played, *go,* is a Japanese game sometimes compared to chess, although it is really the antithesis. The object in chess is to whittle away at one's opponent's forces until the playing field is desolate and the king is hunted down. *Go,* on the other hand, is about building and balance—balancing aggression and caution, influence and restraint, friendliness and disharmony. The rules are much simpler than chess, yet the game is more complex. David Weimer, a professor at the University of Rochester who teaches *go,* says that Western games such as chess take "the Clausewitzean view of conflict—go for the capital and destroy everything along the way." But in *go* "you have to be patient; early moves may not have full consequences until much later."

Go is a difficult game to learn to play and takes a lifetime to master. A neophyte *go* player is rated as a Q 10. As he progresses, he works his way up through the Q levels, eventually making first Q. That is followed by first *dan,* which is the equivalent of a black belt in judo or karate. A player then goes up the scale of *dan*— second, third, fourth *dan,* and so on until tenth *dan.* Hiroshi Yamauchi was sixth *dan,* a sixth-degree black belt.

Yamauchi's opponent was one of Nintendo's licensees—his company developed and sold approved Nintendo games. Licensees were in a precarious position, for Nintendo gave away little and one had to play by Yamauchi's rules.

Because of this, Yamauchi's opponent felt it was prudent to learn all he could about the Nintendo chairman. An astute man could learn volumes about an opponent by his *go* game. "Yamauchi's game is obvious and clear. Nothing is hidden," the man observed. "It is very forceful when it has to be, yet there is give and take. But when he becomes strong, he does not look back. He takes advantage of weakness. He knows far in advance what will happen and he never loses his composure."

When Yamauchi decided to allow outside companies to create games for the Famicom, he initiated a licensing program. To become a Nintendo licensee, a company had to agree to unprecedented restrictions. Companies that were "invited" to become licensees were appalled at the terms of the agreement, but Nintendo's position was immovable. No one was forced to become a licensee, Yamauchi noted, and in spite of the complaints, companies signed up, because millions of customers were clamoring for games. The vastness of the Famicom market was enough to silence the complaints, and many companies made fortunes. Nintendo, of course, made the biggest fortune of all.

Two companies, Namco, the reigning arcade-game company, and Hudson, a computer-software maker, became the first two licensees. Hudson released a game called "Roadrunner." Before that, Hudson had sold a maximum of 10,000 copies of any computer game. "Roadrunner" sold 1 million units and was responsible for the quadrupling of Hudson's annual profits in 1984. Namco sold 1.5 million copies of a game called "Xevious." A new Namco building was nicknamed the Xevious Building because the game had paid for its construction costs.

Another company, Taito, founded in the 1950s as a jukebox manufacturer, was a large pinball-machine and coin-operated video-game company. Taito had made the game behind a surge in interest in video arcades. In "Space Invaders," rows of aliens descended in formation, unremittingly, on the black-and-white TV screen of a large console. The player controlled a mobile cannon at the bottom of the screen that fired shots at the invaders, which came faster and faster until they were entirely destroyed or their opponent—the player—succumbed.

While most companies sold arcade-game machines to distributors or licensees, Taito, in Japan, also owned and operated more than 100,000 coin-op games in arcades, which meant there was no middleman participating in "Space Invaders" earnings. Taito raked in so much cash that the company was in a strong position to diversify, and its chairman signed an agreement with Hiroshi Yamauchi. Taito and the other initial licensees (Konami, Capcom, Bandai, Namco, and Hudson) had the right to produce their own cartridges for the Famicom, a right no future developers or producers would get for many years. They paid Nintendo a large royalty on every game cartridge they sold (about 20 percent).

Konami, which was based in Kobe, had been successful at selling computer games, dedicated hand-held games (plastic Walkman-size games that had been programmed to be "dedicated" to play one game only), and coin-op games, but it grew enormously as a result of its Nintendo license. In five years, its earnings shot up from $10 million in 1987 to $300 million in 1991; there was a 2,500 percent increase in sales between 1989 and 1991 alone.

After the six licensees had begun selling games, Hiroshi Yamauchi realized that he had not only given away his ability to control the quality of cartridges (some defective games had reached the market), but some potential profits as well, because he had allowed the companies to manufacture their own games. Henceforth Nintendo would be the sole manufacturer of games for the Famicom. The licensees would develop them and then place an order with Nintendo for a minimum of 10,000 cartridges. The terms were elegantly simple: Nintendo insisted on cash, in advance.

In the new contract, Nintendo was paid about 2,000 yen per cartridge by the licensees, about twice what it took to produce them. Whether a company ordered 10,000 or 500,000 cartridges, Nintendo profited handsomely, even if the games didn't sell.

Licensees might have operated with caution and placed small 10,000-piece orders, but those could be as risky as large orders. Companies were in business for hits, "grand slams," as one game maker called them. If a company cautiously ordered a small number of games and found it had a big seller on its hands, Nintendo could take its time filling the order. The game's popularity might pass by the time the games were back in the stores. Companies, particularly small licensees without deep cash reserves, had to risk perilous amounts of capital on large orders if they wanted to gamble on big successes. They shouldered all the risk while Nintendo collected obscene profits, which came with almost no additional investment. (Nintendo subcontracted licensees' orders to outside manufacturers.)

It was common for a game to sell 300,000 copies, and the number was often three or four times that many. At the low end of such sales figures, Nintendo collected $2.2 million. For a million-seller, Nintendo's take was over $7 million. It was easy money, risk-free, and the licensing agreements accounted for a growing proportion of Nintendo's profits as more companies signed on. In 1985 there were seventeen licensees. A year later there were thirty. By 1988 there were fifty.

* * *

Sitting across the table from Hiroshi Yamauchi, staring fixedly at the smooth stones on the *go* board, Henk Rogers broke into a wide smile. He had found a hole in Yamauchi's defenses and had placed a stone on the board in position for an attack.

Yamauchi's expression remained impassive. He looked up for a moment, eyeing Rogers, a generation younger and as different a man as imaginable from himself. Rogers had a wolfish, pointed beard and longish ebony hair parted in the middle and swept back over his forehead. Above coffee-colored eyes hung lavish cuneiform eyebrows.

Both men drank the Scotch from their glasses and examined the board in front of them. Rogers, as a three-*dan* player, three levels below Yamauchi, enjoyed a handicap of three. He thus got to place three stones at the start of the game, the equivalent of three free moves. This meant that Yamauchi could win only if Rogers made three mistakes.

Rogers now made his third mistake. Yamauchi had seen in his opponent a broad streak of recklessness, so the move came as no surprise. He took advantage of the error and added a stone that determined the remainder of the game. Rogers was helpless.

The younger man shrugged. "Good move," he whispered. There was nothing he could do to save himself.

Henk Rogers lived with his parents in Amsterdam for eleven years before his father's gem business took the family to New York City in 1964. After Henk graduated from high school, the family moved to Oahu, where he enrolled at the University of Hawaii. Most mornings, before class, he surfed on the island's northern beaches.

Rogers spent most of his time on the U.H. campus in the computer-science building, playing games on terminals connected to mainframes. Game playing led to programming. "For a gamer, programming is the ultimate game," he says.

After graduating, he found a job in California at a software company that had a contract with the U.S. military. After a summer, he quit. "I didn't want to spend my life finding better ways to kill people."

In the meantime, Rogers's family had moved to Japan, and he joined them in 1976. He lived in Yokohama, a Tokyo suburb, and studied Japanese. He had connections that could have landed him a job at one of the major Japanese computer companies, but he felt that, as a foreigner, it would be a dead end. "The fact is," he says, "if you're not Japanese, you're not going to be president of NEC. It's just not going to happen."

He taught English and then accepted an offer to work with his father. The gem business was thriving. The Rogerses bought rough stones and had them cut in Bangkok and Hong Kong. They sold the finished gems throughout Asia and in a shop near Tokyo. Henk worked in the family business for seven years. He also learned to play *go* from his father, a six-*dan* player.

Personal computers had proliferated by then. A gamer no longer needed access to a mainframe to play and create games. Messing around with a PC, Rogers created an electronic version of "Dungeons and Dragons," the popular role-playing game that was an obsession on high school and college campuses back in the United States. The game, "Black Onyx," was, he believed, his ticket to freedom. He planned to sell it for a small fortune.

Rogers took the game to a number of computer-software companies until he found one that was interested. He shook hands on a verbal agreement with the company's president. When it came time to collect his advance and sign the contract, however, the man tried to pay less than he had promised. Computer games in those days were often created by struggling college students or unemployed hackers who were ecstatic if someone wanted to publish their game; they commonly signed contracts for almost nothing. Rogers, however, refused to sign, even when the publisher threatened to keep Rogers from publishing his game elsewhere.

Rogers decided to market "Black Onyx" himself. He placed advertisements in computer magazines throughout Japan and, his wife manning the telephones, waited for the calls to pour in. There were three phone calls in three months.

The problem, he diagnosed, was that there was no understanding of role-playing games in Japan; "Dungeons and Dragons" was not popular there. The solution, he concluded, was to educate Japanese gamers.

He talked his way into the editorial offices of computer magazines and convinced editors and writers to try his game. He set it up on their computers as they watched over his shoulder. After calling up an array of bodies and heads on the computer screen, he told the players to choose ones that looked like them, and he typed in their names below the figures they had created. The characters, he explained, *were them*. The essence of a role-playing game was to accept that premise. The player was not watching the character; he *was* the character.

Rogers was nothing if not enthusiastic, and his enthusiasm was contagious. He guided the editors and writers into the first of his game's dungeons, where they were shown how to explore and fight. When they won a battle, they gained experience and strength enough to venture into the next dungeon. He left the offices with the editors and writers enraptured; they continued playing for weeks. The magazines reviewed "Black Onyx," showering it with raves. Rogers sold 100,000 copies in 1980.

By the time he released a sequel to "Black Onyx," Rogers had formed a company, Bullet-Proof Software, BPS. It was a frustrating, uphill struggle to get BPS games into software shops. Rogers spoke Japanese and was accessible and respectful, but he was a *gaijin*, a foreigner. It was a formidable barrier, but he learned to use it to his advantage. He allowed the arrogance he encountered to placate business adversaries and catch them off guard. "I walked through the wall as if it didn't exist," he says.

With the success of his second game, "Fire Crystal," Rogers expanded BPS. He couldn't create all the games himself, so he went out in search of games to license. The Japanese game companies were the ones at a disadvantage at the international trade shows where designers hawked their games. Rogers had connections around the globe, spoke several languages, and had a remarkable ease with businessmen, who found him a willful negotiator, as well as with young gamers, who realized that he was one of them. When he could get away with it, Rogers wore colorful Hawaiian shirts in place of drab business suits and bestowed bear hugs instead of handshakes. From the moment he opened his mouth it was evident that he loved games. It led to licenses from all over the world.

Rogers wanted to release a computer *go* game. Although there was a proliferation of chess programs, *go* didn't lend itself as well to programming. In chess there are a limited number of good responses to any move, whereas in *go* the number is astronomical.

Go programs frustrated good players. Human opponents learned from their mistakes, but computers made the same moves over and over again. Artificial intelligence—later technology—would give computers the ability to analyze a game and "learn" from its mistakes, but even the best *go* programs were not yet capable of highly sophisticated play.

Rogers decided to release a *go* program for novice players. If nothing else, the computer was a tireless, patient teacher. He searched for *go* programs around the world and finally came up with one he considered appropriate for beginners. It was written by a man who had won the world computer *go* championship in Beijing, and who happened to be from Rogers's native Holland. A deal was struck and BPS released the game, which sold modestly well.

BPS grew, but the computer-game business was shrinking. The largest number of people playing the games were not using computers anymore; they were using Nintendo's new Famicom video-game system. People who owned computers might buy one game a year. Famicom owners bought many.

Some computer games could be converted to run on the Famicom, which was designed to have better graphics and faster action than computers. The Famicom wasn't as powerful as computers, however, so many of the games had to be simplified. Yet the trade-off was worthwhile because of the size of the market. By 1988, there were ten million Famicom systems in Japanese homes.

Though many computer-game companies had become early Nintendo licensees, others couldn't afford to. For an entrepreneurial company as small as BPS, becoming a Nintendo licensee was almost impossible, although that didn't stop Rogers.

An ordinary attempt to reach Hiroshi Yamauchi would be unsuccessful, Rogers knew. The man was unavailable to all but his biggest customers and suppliers, and they saw him rarely. They most often met with Nintendo managers, often referred to as

"Yamauchi's generals," who operated like guard dogs, trained to menace and intimidate.

However, Rogers did learn something about Yamauchi that might give him an entree. Rogers could appeal to him as part of the elite circle of men who played *go*.

In a letter, written on the stationery of his American office, Rogers said that his company sold the best *go* computer program in the world and that he was interested in releasing it for the Famicom. He said he was in Japan for only a few days and he would make time to visit Nintendo if Yamauchi was available.

The day after Rogers messengered the letter from Yokohama to Kyoto, he was contacted by Yamauchi's office and invited to a meeting. Rogers rushed to Kyoto on the bullet train and caught a taxi to Nintendo headquarters. A guard directed him to the lobby, where a receptionist told him to have a seat. An electronic version of a Beethoven cantata signaled that the lunch break was over. Reiko Wakimoto met him and instructed him to follow her to the chairman's office.

Yamauchi, sitting behind his large desk, gave a quick nod when Rogers bowed respectfully. He did not rise when the young man approached him to shake hands but gestured toward one of the chairs opposite the desk. Wakimoto placed glasses of green tea in front of the men.

Yamauchi listened as Rogers talked not about the *go* program or BPS but video games in general. Rogers sat up in the low-backed chair and spoke passionately. He knew what was hot in arcades and he theorized why. He revealed a keen understanding of the young people who played the games. Yamauchi let on nothing, but was impressed by both the young man's insights and his enthusiasm.

Rogers finally repeated what he had said in his letter: he wanted to create a *go* game for the Famicom. To do so, he would need more than just a license to work with Nintendo, since his small company didn't have the capital to pay for cartridges. He asked Yamauchi to back him.

Without fanfare, Yamauchi said he would work with Rogers. He could spare no programmers, only cash. Rogers said cash would be fine.

Yamauchi asked Rogers how much money he needed. Rogers had calculated how much it would cost him to develop the game and added a small profit. He threw out the figure. Yamauchi nodded. "Good," he said. "Done." Yamauchi had been so quick to agree that Rogers wondered if he had asked for too little.

Before Rogers left the meeting, he challenged Yamauchi to a game of *go.* Yamauchi nodded his acceptance. The game would follow their next meeting, he said. He would schedule it at the conclusion of a workday.

Rogers needed a new version of *go* that was simpler than his computer version. He contacted a programmer in England who had made a *go* program for the Commodore 64 computer, which had a variation of the same central processing unit as the Famicom. Rogers bought the rights.

By the time it was ready, Yamauchi had decided that the market for a *go* video game was too small. The whole point of *go* was the serenity of play, the feel of the stones, the patience, and he felt it was incompatible with the Famicom—not many people would want to play the ancient game on what he still viewed primarily as a toy. Yamauchi told Rogers he could keep the money, but that he should come up with another idea for a Famicom game.

Henk Rogers had too much invested in the *go* game not to see it released, and he asked Yamauchi if BPS could publish it. He said he would pay back the advance if Yamauchi would front him the manufacturing costs.

Yamauchi liked Rogers enough to agree. Nintendo manufactured the cartridges and sent the first order to BPS in Yokohama. Rogers brought a cartridge with him on his next visit to Nintendo, inserted it into a Famicom, turned on the television monitor, and invited Yamauchi to sit down in front of it. Yamauchi had never played a Nintendo game before. He held the controller awkwardly and became frustrated as he tried to follow Rogers's instructions. He put down the controller and refused to try again.

Rogers's *go* game sold 150,000 copies—unspectacular for a Nintendo game, but a huge number for BPS. Rogers easily paid Yamauchi back the money he had been advanced and found himself in a particularly enviable position. Not only did his company have a license to work with Nintendo, but he had something even rarer: a coveted relationship with NCL's chairman.

BPS released other Famicom games, including "Super Black Onyx," a newer version of Rogers's game. BPS thrived, and the relationship Rogers had with Nintendo proved valuable to both companies. When Nintendo, years later, sent an emissary to the U.S.S.R. to negotiate with the Soviets for the rights to a brilliant game called "Tetris," Rogers was the man the company dispatched.

The Famicom's popularity grew as licensees released their games. Another small company that signed on was Enix, a start-up formed specifically to create Nintendo games. Founded with a capital investment of 5 million yen, the company attained the status of video-game giant through a game called "Dragon Quest" and its sequels. The original "Dragon Quest" was a combination of two PC games. It was developed by a team of game designers, programmers, composers, and a well-known illustrator. Because Enix had so much confidence in its game, it put all its start-up money on the line and placed an order for 760,000 cartridges.

"Dragon Quest" was released in February 1986. The Enix team panicked when the game hardly sold. It began advertising in *Shukan Shonen Jump,* a weekly boys' magazine with a circulation of 4.5 million copies. The magazine's editors agreed to publish an article about the role-playing game's lore and mythology. It sparked "Dragon Quest" sales; a groundswell followed. So many players called and wrote in with questions about the game that the magazine's editors decided to publish an ongoing series of articles about "Dragon Quest." Both Enix and the publisher benefited: Enix ordered more games from Nintendo—1.4 million—and the magazine's circulation skyrocketed. Because of "Dragon Quest" sales, Enix's management that year gave its employees a bonus equivalent to twelve months' salary.

A sequel, released the next year, sold 2.3 million cartridges, and "Dragon Quest 3" sold 3.4 million. The degree of anticipation for the games was unprecedented. On its first day in stores, 1.3 million copies of "Dragon Quest 4" sold out in an hour, despite a price tag of 11,050 ($75) yen, higher than for any other Nintendo game.

The readership of *Shukan Shonen Jump* shot up to 18 million, and circulation grew to 6 million. For both magazine and licensee, the tie-in was a marketing coup that would not be lost on Nintendo (which, meanwhile, profited from the sales of the "Dragon Quest" games and from the publicity). Seven other publishers launched magazines that provided game tips, profiles of designers, and glimpses of upcoming Famicom games.

By 1990, there were seventy licensees selling millions of copies of hundreds of games, almost all manufactured by Nintendo. In turn, the licensees' games helped sell more Famicom systems, so that almost every household in Japan with children had one. Certain licensees made fortunes. The "Dragon Quest" sequels grossed several hundred million dollars apiece.

During those days, the only companies that still complained about Nintendo's strict licensing agreement and control of the industry were, for the most part, the ones that couldn't get in. As long as their games were selling, companies were happy to give Nintendo its large cut of the fortunes they were making. There was a slightly manic sense in the industry that anything would sell.

But with the proliferation of licensees—over ninety in 1991—something had to give. Although Nintendo spent a year or more and upward of $1 million to develop each of its games, smaller companies couldn't afford that. They used their limited resources to buy cartridges from Nintendo and for marketing and packaging. Not much was left for development. The result was an increasing number of boring games.

To curtail this disastrous flood of the market, Hiroshi Yamauchi modified the agreements with third-party licensees to limit the number of games they could release each year. Companies, he figured, would spend more to develop better games, since more would be riding on each one. Licensees, however, were incensed. Who was Nintendo to tell them how many games they could release? There were rumblings of discontent—out of earshot of Nintendo's executives—about unfair restraint of trade.

Tensions among licensees grew because of the increased competition. For part of 1990, Enix's "Dragon Quest 4" accounted for 25 percent of the entire Famicom software business in Japan. Amid the success stories there was a growing number of disasters.

When the licensees fought one another, it played into Hiroshi

Yamauchi's hands. In the industry, he said, there was room for "one strong company and the rest weak," and he manipulated the industry to make certain that Nintendo remained the one that was strong.

The licensees were in fear that any criticism of Nintendo would get back to Yamauchi. "They feared him like a marionette fears the puppeteer," says one distributor. "If a company upset Nintendo, he could cut the strings."

The chips that were the heart of the Famicom cartridges were in short supply during the years when consumer demand was soaring (from 1988 through late 1989) and Nintendo was obliged to ration cartridges. The company claimed to do the allotting fairly and without bias, but licensees knew better. "Nintendo has succeeded by monopolistic practices and intimidation," said one company executive. "We *all* were intimidated. Like a god, Yamauchi wielded power."

Nintendo anticipated that renegade companies unwilling (or unable) to become licensees would figure out ways to manufacture Famicom games on their own. To stop them, Masayuki Uemura's engineers had incorporated circuitry inside the Famicom that would reject non-Nintendo games. Periodically they modified the code inside new Famicoms so that only Nintendo-approved games could play.

Nintendo also deterred companies from releasing unapproved games through its control of the distribution channels. It was almost impossible for an outsider, against Nintendo's wishes, to get distribution. Wholesalers refused to carry unauthorized products for fear of being cut off by Nintendo. No distributor would risk alienating Hiroshi Yamauchi. A tacit threat pervaded: Yamauchi would crush any company that opposed him.

A small Japanese software company called Hacker International didn't have the capital to become a licensee. Moreover, Hacker made video games that included nudity and sex. Hiroshi Yamauchi allowed brutal violence in his games but forbade pornographic content. He felt that "unclean, dirty" games would tarnish Nintendo's reputation.

Hacker's engineers dismantled Famicoms and figured out ways to make games that would work on them. When Nintendo changed the circuitry, Hacker's techies found ways to get around the changes. Further, the company circumvented Nintendo's lock on the distribution chain by selling its games by mail. It didn't pay royalties or manufacturing costs to Nintendo, so it could make a healthy profit on relatively small sales. It sold 30,000 to 50,000 copies of many of its games, hardly enough to threaten Nintendo. Nonetheless, Yamauchi decided to wage war on Hacker.

The magazines devoted to Nintendo games, Nintendo bibles for millions of kids, sold more than any other magazines targeted to young readers. They were published by independent companies, but they were in fact completely dependent on Nintendo. NCL provided much of the editorial content of the magazines in the form of tips from game designers (where to find the whistle in "Super Mario 3"; how to fight Ganon in "The Legend of Zelda"), so the publishers did whatever Nintendo asked. NCL was allowed to review articles before publication, and it dictated when the magazines could write about games. The magazines gave Nintendo editorial control because it was the source of the insider information for the games; without it, the magazines were sunk.

To reach avid Nintendo players, Hacker placed ads in *Family Computer* magazine, the largest publication devoted to Famicom games. A day after the first ad appeared, Hacker received notice that its ads would no longer run. Even ads that had been paid for in advance were canceled. In the subsequent issue of *Family Computer,* the magazine's editors issued an apology to Nintendo. Unusual as this was, it wasn't enough to appease Hiroshi Yamauchi. Five top managers of Takuma Shoten, the company that published *Family Computer,* rushed to Nintendo to personally apologize. Yamauchi gave them a brief audience, during which the five men, heads bowed, vowed that no such breach of Nintendo's trust would occur again.

The message was unambiguous: Hiroshi Yamauchi had a complete lock on Japan's multibillion-dollar video-game industry. It was felt by retailers, publishers, distributors, wholesalers, licensees, subcontractors, suppliers, and many others in businesses that were both integral and peripheral to Nintendo.

One of the original licensees, Namco, was run by Masaya

Nakamura, who had been lord of the industry for many years, well before anyone had heard of Hiroshi Yamauchi.

Notoriously vain, Nakamura stamped his feet and ranted and raved in order to get his way. In one deal he passed up a huge windfall because it would have appeared that he was concerned about money more than principle. No one was fooled: power was Nakamura's obsession. A slight man who wore large metal-rimmed glasses, he founded Nakamura Manufacturing Company, which made coin-operated kiddie rides, in 1955. He entered the game business in the 1970s and changed the company's name to Namco, soon becoming the number-one video-game company thanks to one phenomenally successful game: "Pac-Man." A joystick controlled "Pac-Man," a hungry yellow dot that raced through mazes, gobbling up whatever was in its path. Enemies that looked like gumdrops appeared from nowhere, intent on eating "Pac-Man" before it ate them.

Pacmania struck worldwide when Namco licensed the game to companies in the United States and Europe. In America, "Pac-Man" made the cover of both *Time* and *Mad* magazines. A song, "Pac-Man Fever," topped record charts, and a *Pac-Man* cartoon was a popular Saturday morning children's television program. "Pac-Man" (and "Ms. Pac-Man," "Baby Pac-Man," and "Super Pac-Man") brought Nakamura hundreds of millions of dollars. (Nakamura awarded the engineer who came up with the game a piddling $3,500. Disgusted, the man left the video-game business.)

Soon after Nintendo's Famicom system was released, Nakamura instructed a group of his managers to see what it would take to enter the Nintendo market. They inquired and were told that no outside companies could produce games for the system. When that restriction changed, however, Namco was the first company to be contacted.

A meeting was set up between the two bosses. Hiroshi Yamauchi formally greeted Nakamura, and they agreed to work together. Nakamura would profit by selling his games, including "Pac-Man," to Nintendo players. At the same time, it was significant for Yamauchi to have as his first licensee the industry's dominant company. Nakamura expected, and received, favorable terms—certainly more favorable than later Nintendo licensees.

In 1989, Namco's original five-year contract expired. Feeling that he and Yamauchi were equal, Masaya Nakamura expected that the renewal of the contract would be a mere formality. Yamauchi, however, used the opportunity to humble Nakamura. Yamauchi had decided that all the agreements with licensees were going to be identical and there would be no exceptions.

When this decision was communicated to Nakamura, he exploded. Namco was earning 40 percent of its sales from Nintendo games, but Nakamura would not acknowledge that Yamauchi was the stronger. "All of a sudden Mr. Yamauchi was king," a Nakamura associate says. "Mr. Nakamura did not want to observe the rule created by Yamauchi. It was a slap in the face. It was unpardonable."

Nakamura did what no other licensee dared. He spoke out against Nintendo.

It began with an interview he gave *Nippon Keisai Shimbun* (*Japan Economic Journal*). He chose his words carefully: "The game industry is still new. I want it to grow soundly," he said. "Nintendo is monopolizing the market, which is not good for the future of the industry. . . . Nintendo should consider itself the leader of the video-game industry and accept the responsibility that goes along with it." He said that there was no competition in the industry, and that companies kept silent because Nintendo was too strong; to question Hiroshi Yamauchi when their businesses depended on Nintendo was suicide.

Everyone in the industry in Japan wondered if this was exactly what Nakamura had just committed.

In an interview, Hiroshi Imanishi said, "Namco has profited generously from the privileges awarded it by being the first licensee," but these privileges would be "omitted" in the future. Nakamura, in response, announced that Namco was developing games for Nintendo's competitor, Sega, which had released a system called the Megadrive. It was a futile gesture; Nintendo, with 95 percent of the market, was invulnerable.

Nakamura filed a lawsuit in Kyoto District Judiciary charging Nintendo with monopolistic practices. Hiroshi Yamauchi, in *Zaikai* magazine, dismissed it. "Frankly, Namco is envious of us. . . . If they are not satisfied with Nintendo and the way we do business, they should create their own market. That is the advantage of the free market." Before long, Nakamura withdrew the suit.

"Mr. Nakamura suffered the anguish of the defeated king," his colleague says. "The biggest blow was to have to crawl back to Yamauchi—the defeated king accepting that he must now be a courtier."

Nakamura sullenly instructed his staff to accept Nintendo's contract; he could not afford to continue without a Nintendo license. Namco's surrender was felt throughout the industry. If Masaya Nakamura had not been able to stand up to Hiroshi Yamauchi, no one could. Yamauchi's dominance was no longer questioned.

On February 21, 1986, after two major delays, Nintendo had released a new product called the Disk System, the "new media of family computers," the company boasted. The system was a disk drive for the Famicom that attached to the hidden connector in the machine's belly. With it, the Famicom could run software on credit-card-size magnetic diskettes instead of the traditional bulky Famicom cartridges.

The Disk System was a large investment, 15,000 yen (more than $100), but Nintendo made it seductive. Games would be better, it promised, since disks had more memory capacity than cartridges. They also would be cheaper. Famicom cartridges originally cost about 2,500 yen. That figure doubled by 1985 to 5,000 yen, almost $40. "Dragon Quest 4" sold for more than $80. Disk System games would be only 2,600 yen, about $20.

The biggest advantage to the Disk System for consumers was that the diskettes could be reused. It protected customers who bought games they didn't like or ones they grew tired of. Machines called Disk Writers, sort of like jukeboxes, would be set up in toy and hobby shops throughout the country. Instead of a menu of songs, the Disk Writer would have a list of the latest games. A disk card would slip into the Disk Writer and the existing program would be replaced with a new selection. The fee for a rewrite would be a mere 500 yen.

In an expensive ad campaign, Nintendo announced that some of the best new games would be available only for the Disk System. It said there would be 10,000 Disk Writers in retail outlets in the first year. Half a million Disk Systems sold in three months, almost 2 million for all of 1986.

There was dissatisfaction with the system, however. Licensees hated it. They had to determine whether to sell games in cartridge or disk form—or both. Nintendo charged them an ample fee to convert their games to disk, and the returns were much lower than on cartridge games. Nintendo required the licensees to sign a new contract if they wanted to make Disk System games, too, and it included new restrictions. Nintendo not only determined which games could be released on disk but, most galling, retained half ownership of the copyrights to all Disk System games. Copyright to the licensees' games was one of the very few things Nintendo *didn't* have up to that point.

There were other problems. Semiconductor technology improved and prices dropped, so Nintendo disks actually had less storage capacity than cartridges. Retailers complained that Disk Writers took up too much space in their stores.

By 1990, 4.4 million Disk Systems had been sold, far less than Nintendo had projected. The company backed off on its promise (threat) to release games only on disk. "Super Mario Bros." was supposed to have been a disk game, but it came out on both disk and cartridge. The best games were available only on cartridges, so many Disk Systems fell into disuse. The system was not a resounding success, but the sale of 4 million pieces of hardware at over $100 each can hardly be described as a failure.

Yamauchi had loftier expectations for another new venture, launched in 1988. It would, he believed, reposition NCL. Nintendo would no longer be a toy company; it would grow to be a communications corporation, among the ranks of Japan's largest company, Nippon Telephone and Telegraph (NTT).

The plan hinged on something called the Family Computer Communications Network System. At its center was the Famicom and another hundred-dollar piece of equipment that connected to it, the Communications Adapter (a modem), which allowed the Famicom to hook up to a telephone line.

A special cartridge transformed the Famicom into a terminal that could "talk" to other terminals or to mainframe computers.

At its most basic, kids could play video games—such as Henk Rogers's *go* game—with other players throughout Japan without leaving home. As significant as this was, it allowed for extraordinary new uses beyond games. The network was an appliance of the future, Yamauchi believed—one day as pervasive as the telephone —with Nintendo technology at its heart.

The Family Computer Communications Network System, according to an NCL corporate report, would "link Nintendo households to create a communications network that provides users with new forms of recreation, and a new means of accessing information." In a speech to his employees, Hiroshi Yamauchi expounded on this vision. "From now on, our purpose is not only to develop new exciting entertainment software but to provide information that can be efficiently used in each household."

Yamauchi saw that the video-game business in Japan was huge but not unlimited, partly because there were only so many households (with kids) that were potential customers. Communications was a bigger industry, virtually limitless. Other companies had sought to hook up households via telephone lines and computers, including NTT, but none had what Nintendo had: computers sitting in one third of the country's homes.

If Yamauchi succeeded in connecting a small percentage of the Nintendo households, it would be—instantly—the largest such network in Japan. Games, Nintendo's stock in trade, were only the door in. The network would offer a spectrum of business and services that would all be provided and/or licensed by Nintendo. Like any computer network carrier, Nintendo could charge customers for on-line time (the time that they used the system) while also charging the information and service providers for access to its customer base.

In Nintendo's 1989 annual report, Yamauchi summed up his vision: "We believe that the arrival of the high-information age has brought about a new opportunity for people to consider what vital information really is, and what information they really want. By employing the Nintendo Family Entertainment System as a domestic communications terminal, utilizing regular telephone lines, and the establishment of a large-scale network which to this point has been inconceivable, we plan to provide a vital supply of information for the domestic lifestyle in the fields of entertainment, finance, securities and health management, to mention but a few. . . ."

An agreement was reached between NCL and Nomura Securities Co., Ltd., the largest Japanese brokerage firm. Families would be able to use the Famicom to buy, sell, trade, and monitor stocks and bonds. After Nomura, the Daiwa and Nikko brokerage houses signed up.

It was only the beginning. Nintendo could make commissions or fees on home banking, shopping, and airline ticketing done on the Network. NCL could charge for information such as movie reviews, news, and recipes.

The network would also be a pipeline into Nintendo homes—a direct way to advertise new games and other new products. The Super Mario Club was formed so that Nintendo distributors across Japan could access information about games (including reviews) on-line.

Yamauchi approved a multimillion-dollar budget for advertising, and he personally met with representatives of the brokerage firms and banks to convince them to work with Nintendo. In spite of the chairman's intense personal commitment, the network had a slow beginning. There were difficulties installing and maintaining it. Information arrived in garbled form, and phone lines were cut off. The technical problems were solvable, but there were roadblocks. One was convincing adults to take the Famicom seriously enough to use it for stock monitoring and banking. Beyond that, all home networks *together* were attracting only a limited audience. People either found that banking and dealing with stocks were just as easy to do in more traditional ways or they resisted the newer technology. There was another obstacle: families didn't want their telephone lines tied up for long periods of time.

Only 15,000 to 20,000 customers used the stock-brokering services. Fourteen thousand customers used the network for banking. Three thousand businesses signed up for the Super Mario Club. The total number of households with Communications Adapters was 130,000.

The low number disappointed Yamauchi, but he never admitted that he had made a mistake. "It is," he said in 1991, "just a matter of time." New services brought new customers: soon you could buy stamps from the postal service, get odds and bet on horse races, and even exercise (the Bridgestone Tire Company used the Famicom fitness program for its employees). "When the people are ready for it," Yamauchi continued, "we have the Network in place."

The network's slow start and the problems with the Disk System didn't impede Nintendo's continuing growth. Other companies bringing in sales as high hired hundreds of new employees, but Nintendo maintained a streamlined operation. There were 200 people in research and development, 350 in administration, 180 at the main factory, and another 130 at the plant in Uji. Thousands of people worked for Nintendo's subcontractors, but NCL grew without major capital expenditures.

Without large numbers of personnel and gigantic plants, Nintendo could shift gears more easily. "We're not building factories that are tied to any specific technology," an executive said. "That differentiates us from 98 percent of the companies in consumer marketing today. We're far more flexible and far more responsive."

Throughout the 1980s, the number-one Japanese corporations, according to an independent rating by the *Japan Economic Journal,* were, alternately, Toyota and Honda. They were the best run, they performed best on the stock market, and they made the most money per employee and the highest overall profits. For 1989, however, the magazine announced that the number-one company was Nintendo. The company's ascension to this pinnacle was no abberation; the early 1990s continued to see Nintendo's economic dominance in Japan—with no slowdown in sight. And Japan was only the beginning.

5

COMING TO AMERICA

Minoru Arakawa's respected Kyoto family had been in the textile business for four generations. The company imported quality silk from China and cotton from the West and sold them to producers of linens, kimonos, *yukata,* and Western-style garments. The business had grown steadily into one of the largest in Japan and the Arakawas had invested in prime real estate throughout the most expensive section of Kyoto.

Waichiro, Minoru Arakawa's father, managed Arakawa Textiles diligently. Though not the shrewdest of businessmen, he ran his thousand-employee company with steadiness and efficiency and was satisfied with modest but consistent profits—in the neighborhood of $5 million to $6 million each year. Waichiro Arakawa didn't believe in debt or rapid growth. Instead, he was concerned with the fine quality of Arakawa products and with maintaining his good relationships with his suppliers and customers.

The family was aristocratic and deeply grounded in tradition. The tea ceremony was important in the Arakawa home. Neighbors and friends came by for formal visits. Waichiro did make a few accommodations to Western influence—he often wore a suit and tie—but the ancient home was in most ways as it had been for the past hundred years.

Minoru's mother, Michi, who dressed in kimono tied with an *obi*, a wide silk sash, and wooden clogs called *geta*, was from a family even more highly regarded than her husband's. The former Michi Ishihara was a descendant of the eighth-century emperor Uda and, after him, the first mayor of Kyoto. Her father had been a prominent member of the Japanese Diet. The Ishiharas had amassed a great deal of land that supported many families in sharecropping arrangements. When Michi Ishihara and Waichiro Arakawa married, the combined land holdings of the two families equaled approximately a fifth of Kyoto.

Michi Arakawa was an artist who spent afternoons in the family garden or in her studio, where she sang while painting. Her paintings hung on the walls of the family home alongside Picassos, Cézannes, and Renoirs.

The Arakawas had high expectations for their three children, who were constantly reminded that their position—and their families' names—came with responsibility. The children were raised to be soft-spoken, conscientious, and impeccably mannered.

Tradition determined the paths of two of the Arakawa children. As the eldest son, Shoichi was to take over the family business. He joined Arakawa Textiles after he graduated from college. His sister, too, did as she was expected—she got married (to a professor of medicine).

There was, however, no preordained course for Minoru. Sometimes second sons followed their elder brothers into the family business, but there was no rule. Counseled by his parents to do what would make him happiest, Minoru anguished over a career choice. His father offered simplistic advice: "Be unselfish, do something for others." Looking back, Minoru laughs. "I don't think I am that good," he says.

He entered Kyoto University in 1964 and took general courses until his third year, when he settled on a major in civil engineering, in which he went on to earn a master's degree. He graduated high in his class, but with no idea what to do next.

Minoru's family wealth put him in a privileged position. "It is difficult when you do not have to work," he says. "You have to think. It sounds easy, but sometimes it is not. I struggled to know why we are here and how we should best spend our lives."

Arakawa decided to look for answers abroad, away from the protected, insular world of Japan. He was accepted into the graduate civil engineering program at MIT, and headed to Boston several months before the school year began. There he bought an old Volkswagen bus and set out on a journey through America, choosing local roads as he crisscrossed the country. Allotting himself five dollars a day for expenses, he camped in state parks, national forests, and parking lots. "Each state is a different country," he says. "The people were like nothing I knew."

The year of his adventure, 1971, was a turbulent one in the United States because of the antiwar movement. Arakawa watched from the sidelines. He had been at demonstrations that turned into skirmishes between radical students and police at Kyoto University in 1968, but he admits, "I was throwing stones without really understanding what was going on." This time, in America, Arakawa's eyes were open to larger issues: the war in Vietnam, economic inequities, and racism.

In the towns he passed through he talked, as best as he could with his stilted English, to people he met in cafés, bookstores, and parks. He was fascinated by all the ways people chose to live. By the time he got back to Boston with 15,000 more miles on the bus's odometer, his mind was spinning with insights and even more questions than he had started out with.

Years later, Arakawa's colleagues in business would find it ironic that such soul-searching preceded his heading up a company that would be accused of everything from unfair restraint of trade to a failure to hire minorities. In a sense, however, his later success *was* partly rooted in that journey through America. After having been raised amid the tended gardens and peaceful temples along Kyoto's narrow streets, in cloistered tradition where conformity was valued and expected, he was on his own, six thousand miles from home, in the middle of America, where the young men and women he met were driven by individualism and independence. He was excited by it and felt, in many ways, not unlike them. "You must find what is your own, what you are good at," he says. "If you can find that and choose a goal you believe in and work to conquer it, you get the most satisfaction. There was nothing grand in what I found for myself. Still, it is a valuable view of life: to set your own sights. Then to do what is necessary to get there."

Arakawa found a place to live in Cambridge with a Harvard student who became his best friend. He continued his study of civil engineering, although he still had not determined how he would use his expensive education. After a year and a half, he earned a second master's degree in 1972.

On campus one afternoon he met a group of young Japanese businessmen who were visiting the United States as representatives of a trading company. Over beers, the men boasted about their jobs. There was an explosion of employment in Japanese trading companies that were doing business all over the world. Their work, they said, was stimulating. It involved financing, accounting, law, engineering, design, and even politics and psychology. There was a lot of travel, and they were given a great deal of responsibility and freedom. All of that interested Arakawa, who decided to try to find work with a trading company.

Leaving Boston was bittersweet. Arakawa had made close friends in America, and yet he was eager to return to Japan.

He had interviews with several trading companies in Osaka and Tokyo and was offered a position with Marubeni, a company that developed hotels and office buildings around the world. He moved to Tokyo, near the company's headquarters, and began an apprenticeship. He leased an apartment, and the woman he was dating moved in with him.

Arakawa returned to Kyoto to see his family at Christmastime, bringing along his old roommate from Cambridge, who happened to be visiting Tokyo. The two men accompanied Minoru's parents to an annual society ball, where the elite of Kyoto high society turned out each season. Everyone knew everyone else in that small, select circle. Before they arrived, Minoru warned his friend that the ball would probably be a bore. He didn't count on meeting

Yoko Yamauchi, an elegant and exceptionally pretty woman. Her face was self-assured and her eyes calm. Her satiny black hair was pulled back. When she smiled, her left eyebrow arched magnificently. She said she didn't like dancing, but she danced a great deal that night.

As a child Yoko butted heads with her family. Her great-grandmother Tei, who ran the household, scolded Yoko for playing under the eaves of the old house and called her down from the high branches of the trees. Tei, who was opinionated and overbearing, made most of the decisions about Yoko's education and discipline.

Yoko's mother was invisible during those years. Michiko had had a series of miscarriages after Yoko was born and she was often ill in bed. It was seven years before she had other children, a daughter named Fujiko and then a son, Katsuhito. Once she had regained her health, Michiko was more involved with her children, and she and Yoko sat together and talked for hours at a time. Much of the talk was about Hiroshi Yamauchi.

Yamauchi terrified his children. They hated Nintendo, for they saw how it consumed him. The only attention he paid to his daughters and son was to exercise his strong will and issue edicts. He laid down the law at home, enforcing a strict curfew. Yoko had to be home at the dinner table at six, although Yamauchi himself was absent on many of those evenings.

In his late thirties, Yamauchi was suavely handsome, a cigarette always dangling from the corner of his mouth. Even after he sold his love hotel, he was a familiar face among the Kyoto demimonde. Michiko said nothing, but the children resented him bitterly.

In 1970, on her twentieth birthday, Yamauchi shocked Yoko when he announced that she was going out on the town with him. She dressed up and accompanied him to a cabaret, a *sikikake,* where five geishas attended them, serving drinks. The women obviously knew him very well. Hiroshi toasted Yoko's coming of age, but when it got late, he sent her home in a taxi. He didn't come home until dawn.

Yamauchi was especially tough on his daughters. Young women, he felt, could not be trusted. He saw what they did when they were out on their own late at night; many of the girls he met were Yoko's age. Likewise, his promiscuity, his temper, and his absences made Yoko wary of men. She decided that if she ever settled down, it would be with a man who had nothing in common with her father.

After Yoko's twentieth birthday, her parents considered arranging a marriage for her, as was still sometimes done in Japan. Michiko tried a modern approach. She and other parents conspired and planned dates for their children. Yoko agreed to go because it was a chance to get out of the house, but she had no intention of becoming involved with anyone.

During Yoko's senior year at college, where she was studying English history, a friend asked her to a society Christmas ball. Her father, contemptuous of Kyoto's old monied families, would probably have forbidden her to attend, but he was out of town. Michiko gave her consent.

The ballroom was festive with decorations and the music of a small orchestra. Yoko watched from a hallway, where she sipped from a cup of punch.

A man approached her and asked her to dance. She could tell he was tipsy and she wanted to refuse, but it would have been rude. It was difficult to maneuver in the tight dress, particularly in the high heels she had chosen.

Doing her best to keep up with the man as he guided her awkwardly around the dance floor, she tried to catch the eye of a friend of her mother's, who was dancing nearby. The woman glanced in her direction, caught Yoko's SOS, and cut in, dancing off with Yoko's partner and leaving Yoko in the arms of the young man she had been dancing with.

Mino Arakawa was dressed in a wide-lapelled tuxedo. He was tall and thin yet athletic. He wore his shiny black hair, which swept and rolled, wavelike, to one side, slightly longer than most. He and Yoko had never met, although they knew about one another. Yoko's school friend was Minoru's cousin; she had raved about him, saying he was handsome and very bright. It struck Yoko immediately that he was different from the Kyoto men she usually met. He was cosmopolitan, educated in America, sophisticated, and funny.

Minoru and Yoko talked and danced through the evening. They

laughed about the fact that they had almost become cousins; her aunt had been engaged to his uncle in an arranged marriage. The wedding never happened because, scandalously, his uncle had broken off the engagement. It was an unforgivable embarrassment to both families, who had gone so far as to exchange gifts, the traditional way of validating an engagement. The Arakawas' gift was a treasured family heirloom, a family crest intricately embroidered onto silk. Tei Yamauchi, furious because of the humiliating breach of etiquette, cut the Arakawa crest in half, returning it in pieces.

Following the Christmas ball, Mino traveled by train back and forth between Tokyo and Kyoto so he could see Yoko. They took walks together and met for lunch. Soon Mino informed his girlfriend in Tokyo that she had better move out; he was becoming seriously involved with someone else.

Yoko and Minoru enjoyed each other immensely, but they were apprehensive about their families' opinion of the match. The Yamauchis may have been almost as wealthy as the Arakawas, yet there was a wall between their families. It had already stopped one marriage; the mutilated Arakawa crest was a reminder.

Not only was there the question of the standing of the Yamauchis in Kyoto society, but Hiroshi Yamauchi had only disdain for people like the Arakawas. The upper crust of society was, he felt, conservative and pompous. The Arakawa family was in one of the most respected businesses, and their fortune had been amassed over many generations. Yamauchi, on the other hand, had become the president of his family's company when he was a brash twenty-one. He was beginning to make it a success, but the men who ran the established businesses of Kyoto shunned him. When Yamauchi was able to get into one of their clubs, the older men either ignored him or were openly contemptuous. His style betrayed the distinction between them: Yamauchi showed his coarse emotion to anyone within earshot—his rages were routine—while the Arakawas were always dignified, revealing nothing.

Yoko Yamauchi consulted her mother, who said she would do her best to help. Over the coming weeks, Michiko pleaded Yoko's case. She told Hiroshi that the Arakawas were known to be different from the other aristocratic families; they were a well-respected family and he should give the young Arakawa boy a chance.

Michiko had Yoko invite Minoru to dinner at the Yamauchi home. Because of the terrifying portrait Yoko had painted of her father, Arakawa felt as if he were preparing to visit Don Corleone.

Dressed in a conservative suit, he arrived at the Yamauchis'. After the introductions were made, he joined the family at the low dining table, where Michiko and Yoko served the meal. Hiroshi sat back in his chair and studied his daughter's suitor.

The evening wore on and Yamauchi fired questions at Minoru as if he were conducting a job interview. He had to be convinced that Minoru was not a heavy drinker or a playboy.

"You went to Harvard, eh?" Yamauchi asked. "That is a good school."

Mino politely explained that he had gone to MIT.

"I have never heard of it," Yamauchi said.

Yoko and Mino had to convince him that MIT was okay too.

After the meal, the family withdrew to the living room for tea. There, Yamauchi looked at Arakawa and said, without emotion, "If you are going to marry my daughter, you should marry quickly."

Yoko and Minoru exchanged glances, and the young man nodded politely. "Yes, sir," he said.

Hiroshi ribbed Arakawa, saying he was a good choice because a woman shouldn't marry a man who was too good-looking. "If you have a nice-looking man, the girls won't leave him alone," he told his daughter.

At the end of the evening, after Minoru had gone, Yamauchi let on that he had actually been impressed by the Arakawa family all along. He told his wife, "Arakawa is such a fine man that no son of his could be bad."

Soon afterward, in March, Minoru officially asked Yoko to marry him. She was in love with him but worried that she should wait. Ultimately there was something about Minoru that convinced her. His sense of humor and contemplative nature were part of it. She was comforted that he was a second son who would never have to enter a family business. Nervously, she finally told him that she would marry him, and a wedding was planned.

The wedding, in November, was grand. There was a ceremony under the massive, red-orange Heian Shrine in the park near the Yamauchi home. There were 350 guests, friends of both large families (although the Arakawas' guest list was longer; Mino alone invited about fifty friends). An opulent dinner and champagne toasts in the luxurious ballroom of the Miyako Hotel followed the ceremony.

The couple moved into a small house in the Ogikubo district of Tokyo, near Shinjuku, close to Marubeni's headquarters. For Yoko, it was thrilling to be in Tokyo, away from her parents. She and her new husband were happy newlyweds, romantic. She imagined that their marriage would be like that of Mino's parents, who were still obviously in love after many decades together, taking walks together in the evenings and treating each other with affection and respect.

Yoko loved the tender and romantic attention she received from Minoru and believed that she had chosen wisely—he was as different from Hiroshi Yamauchi as imaginable. "It was like living with a boyfriend," she says.

By the end of the idyllic first year, Arakawa's new job was taking more and more of his time. After a training period, he became involved in some of Marubeni's foreign ventures. During his second year he was placed in charge of some small projects that took him away from Japan for ten months out of the year. He flew off to Caracas, Toulouse, Düsseldorf, and Vancouver. He was excited by the travel and the experience he was gaining, but Yoko felt deserted. "No one prepared me for this," she complained to her mother.

When Mino was in Tokyo it was better, but there was little time for the couple to be alone. He worked long hours, and Yoko was often dragged to business-related dinners and cocktail parties. She found the obligations of a Japanese businessman's wife appalling. Wives, Yoko Arakawa found, had no identity; their place in the unofficial hierarchy depended on their husbands' position in the company. The first time she went to a wives' luncheon, Yoko sat down next to the assistant general manager's wife and was bluntly asked to move. "They all looked at me," she says. "They were thinking that Arakawa married the wrong person."

When her complaints to her mother filtered back to Yamauchi, he called from Kyoto, fuming, berating her for having chosen Arakawa in the first place. He said she should divorce him. Yoko considered it, particularly when she had their first child, a daughter, and Arakawa was thousands of miles away. It struck her how bad things had gotten when, one time, she went to Narita International Airport to pick him up after a long absence and he passed right by her—they didn't recognize each other.

Minoru came home one evening in 1977 and announced that the family would be moving to Vancouver for at least a year. His firm had challenged him to develop a large condominium complex there. With a $1 million development budget, he would be in charge of everything from negotiating for the land to working with architects to selling the units. He told Yoko that Vancouver was a great place to live. It was cold, foggy, and rustically beautiful. The move scared her, but Yoko thought they might have more time together if they left Tokyo and the competitive company headquarters and settled into a new place together. They talked excitedly about the move.

The next night, Arakawa came home from work and told Yoko he had some bad news. Things had changed: he was going to Canada, but alone. The company was not allowing wives to go. He assured her he would send for her if it seemed the work in Canada would last, and in the meantime, he would be back and forth between Japan and Vancouver all the time.

Yoko spent a sleepless night. In the morning, she dressed up and headed to Marubeni. She stormed into the office of her husband's boss. "If I don't go, our marriage is over," she said flatly. The man listened and then shook his head. He said emphatically that she should not go. She would be a drain on Arakawa; she would hate living outside Japan. But Yoko told him that she would file for divorce if Arakawa went without her; it would be the company's fault that their marriage had fallen apart. The manager argued some more, but he finally shrugged and capitulated. "But," he said, "I wouldn't bother bringing anything with you. You'll be back very soon."

Yoko and Minoru left Japan with only their daughter, Maki, then three years old, and two suitcases. They arrived in Vancouver and checked into a hotel next door to a Denny's restaurant, where they ate greasy hamburgers and ice cream sundaes. "If this is America, I have made a big mistake," she said after one of these meals. She had smoked occasionally at home, but in Vancouver she acquired a three-pack-a-day habit. Frequently, she considered using the credit card her father had given her for emergencies to buy tickets home for herself and her daughter.

Saving his budget for real estate and construction, Arakawa, insisting that he and Yoko make it on their own, with no financial help from their families, felt he couldn't afford furniture or an exorbitant rent, so he relied on a trick he had learned when he was at MIT; he sublet the furnished home of a University of British Columbia professor who was on sabbatical. He bought a Honda Civic and drove to the new office every morning. He worked fourteen or more hours a day while Yoko, stuck at home, tried to begin a life in Canada.

She had thought that she spoke adequate English but found she could barely understand most people she met. She studied the language by watching television and developed an accent decidedly reminiscent of Peter Falk's Columbo. She wanted to enroll in English classes, but $10 an hour, the going rate, was too expensive. She hired an elderly Canadian woman to baby-sit for her daughter for $1.50 an hour and asked the woman to correct her English when she said anything wrong.

Her English improved, but she complained to Mino that she was stuck at home. He surprised her one day with an old Chevy he had picked up for $700. He opened a bank account for her with a modest deposit and gave her a map of Vancouver. "You have money, transportation, and a map," he said. "You are on your way."

Yoko was still miserable. Her first driving expedition resulted in a confrontation with the police. The Canadians in their neighborhood were impatient and rude. She had no friends, and she never saw her husband. Other than the baby-sitter, she spent time only with her daughter.

A year went by. A second daughter, Masayo, was born. The professor whose home they had rented returned, so the Arakawas had to move. Mino didn't want to spend money on movers, so

Yoko did all the packing and the two of them did the lifting and carrying. It was the first of eight moves they would make, from one professor's house to another, before buying their own home in West Vancouver.

Arakawa's single goal at that time was to build the condos. Marubeni was having financial trouble, and his successful execution of the project could help. It would also establish him as an important asset to the company. He portioned out his budget and began the development on prime land he had bought near Vancouver. He directed the clearing of the land, assisted in the designing and engineering of the complex, and oversaw the construction. When the first units were ready to be shown, Arakawa himself, occasionally with Yoko's help, worked as the agent. In order to sell one condominium he spent forty-eight hours applying a fresh coat of peach-hued paint for a prospective buyer. The exhausting efforts paid off when Arakawa began to bring in a healthy profit on his company's investment.

At the same time, Arakawa was trying to extricate Marubeni from another Vancouver venture that was losing money. In a joint venture with a Canadian real estate company, a Marubeni subsidiary had built a high-rise suburban condominium complex called Central Park Place. The 434-unit project was just completed when the real estate market in that area was falling off. The condos weren't selling, so Arakawa met with the Canadian partners to figure out ways to minimize the loss. It was a grim business for Arakawa, although one good thing came out of the experience. He met a man who would become a good friend—Phil Rogers, a tall Englishman with thinning blond hair and attentive blue eyes, who worked for the Canadian developers.

Yoko ran up large phone bills talking to her mother in Japan. The family flew back to Kyoto for a visit after two years in 1979. After a family dinner one night, Yamauchi asked Mino to give up Marubeni and Vancouver. He wanted him to join Nintendo and open a manufacturing operation in Malaysia. Many companies in Japan's toy business had begun manufacturing their products in countries with cheaper labor. It was, Yamauchi told Arakawa, an opportunity he should not pass up.

Yoko was mortified. She wanted nothing to do with Nintendo. Her father was married to the business. He brought home all his anger and frustration when things were going badly. Yamauchi himself blamed Nintendo for his chronic stomachaches. Throughout her childhood, the family had waited at home for him in the evenings, fearful of his moods. "We were all shaken by him," she remembers. "We all suffered."

Although Mino worked just as hard as Yamauchi, Yoko wasn't under anyone's thumb. There was no way she would allow Mino to accept a job with Nintendo; Hiroshi Yamauchi would then be back in control of her life.

Arakawa too was suspicious of his father-in-law's motives. "He decided I was very much like him," he says. "Or maybe he wanted to test me." Arakawa also had little interest in moving to Malaysia, which was Siberia as far as he was concerned.

They left Kyoto without giving Yamauchi an answer, but Yoko never vacillated. In a conversation with her mother from Vancouver, she said that she was absolutely against her husband joining Nintendo. Michiko was so angry she refused to speak to Yoko for months.

Arakawa was still immersed in his Marubeni project, which was becoming a great success. He had built a total of 350 condominiums and they were selling quickly. Marubeni had lost money on Central Park Place, but Arakawa's new project was bringing in healthy profits. He turned down his father-in-law's offer, and Hiroshi Yamauchi scrapped the idea of overseas manufacturing. However, he didn't give up on his plan to get his son-in-law into the family business.

If Hiroshi Yamauchi had been much younger and had spoken English, he would have done it himself. It was his dream to go to the United States, to live abroad and succeed there.

Yamauchi wanted to enter the American market, and he needed someone to manage the operation. Yoko had kept Michiko apprised of her husband's progress, and she, in turn, reported to Hiroshi. Hiroshi Yamauchi was impressed, not only by Arakawa's managerial and organizational skills, but by his perseverance and dedication. Yamauchi's son, Katsuhito, was too young and inexperienced to be given much responsibility at Nintendo, and he had gone off to work for Dentsu, an advertising agency. That left only Arakawa, and the international experience he had gained through his job with Marubeni was invaluable. It was imperative that Yamauchi convince his son-in-law to run the American subsidiary.

Minoru and Yoko Arakawa visited Kyoto again in early 1980. One night, the family had retired to the living room after a simple dinner at the Yamauchi home. As they settled into some wicker chairs, Michiko served glasses of Scotch. Yoko stared out the window that overlooked the garden, where she had played in her youth. Now it looked like a painting behind glass.

Hiroshi Yamauchi ran one of his delicate hands through his hair. Ignoring his wife and daughter, he turned to Minoru and began speaking softly, hesitantly. But soon he was ardent and emphatic, focusing his stare directly on his son-in-law. It took two hours for Yamauchi to outline his plan.

Though he was expecting it, it nonetheless jarred the younger man when Yamauchi told Arakawa that the plan depended on him. Arakawa sipped his drink and looked over at his wife, whose jaw was firmly set.

None of Yoko's misgivings about her father, none of her distrust or her fears about losing Mino to him, had disappeared, and she was worried because she could see that Mino was intrigued.

Arakawa didn't doubt that Nintendo was doing well in Japan, but he wondered if Yamauchi was overestimating the company's potential for expansion. Now that Arakawa had helped strengthen Marubeni, it was difficult to consider leaving the company. On the other hand, Yamauchi insisted that Nintendo was the uncontested leader of an industry that was still in its infancy.

He looked over at Yoko, who had finally come to enjoy life in Vancouver. She would fight him if he decided to do it. He looked at his father-in-law, who was pouring more Scotch into his glass. Yamauchi tugged at his soft trousers and leaned forward in his chair.

The video-game business had potentials that no one had yet been able to exploit fully, he said. His substantial investments in research and development had paid off, as his engineers were learning to adapt proven, inexpensive semiconductor technologies for new kinds of products. "I don't see any limitations," he said.

The offer was straightforward: Arakawa would not have to leave North America; Yamauchi now required the experience Mino had gained working there. And, he continued, Arakawa would be on his own, president of an independent subsidiary, generously backed by NCL. If he replicated in America even a small portion of NCL's growth in Japan, Arakawa would be the president of a substantial company.

"Yoko and I will discuss it," Arakawa told his father-in-law as he and his wife wished Hiroshi and Michiko good night.

Yoko saw that Minoru was becoming excited. Her independence from her father was suddenly in jeopardy; she envisioned herself in the middle of the inevitable battles that would arise between her father and Minoru. She was afraid the tension between the two men would strain her marriage.

For Mino, the idea of starting a new company in an industry he knew nothing about was intriguing. "Yoko and I were both from rich families," he says. "We could have lived our lives without working, so money wasn't a motivation. When it isn't, something else must compel you." He tried to reassure Yoko, but she did not relent. "No matter how much you accomplish, it will be viewed as mediocre, because you will be thought of as the son-in-law," she said. She looked longingly at the view from their Vancouver home. She had made friends there. She had begun painting and taking art classes. Vancouver was a good community in which to raise her two daughters. Minoru, however, wanted to take the job and she finally gave in. "Okay," she said. "We will see."

New York, the center of American finance and commerce, seemed to be the logical base for Nintendo of America. Yoko gritted her teeth when the Arakawas left Vancouver in May 1980.

In order to lessen the blow, Minoru decided to make the trip east a family adventure. The day they left, the four of them driving south in a car packed with suitcases and toys (the small amount of furniture they had gathered was shipped by truck), Mount Saint Helens erupted.

Volcanic fallout filled the sky with an eerie orange dust, destroying the paint job on the car. The family retreated back into Canada and circled east, through the Canadian Rockies. Yoko thought this an inauspicious beginning.

They arrived in New Jersey, where they rented a home in Englewood Cliffs. They carefully budgeted money earned from the condominium project; it was a matter of great pride to them that neither had ever asked for money from their families. Yoko had never used the credit card her father had slipped her, and she had never accepted money from her mother in spite of the phone calls from Kyoto in which Michiko pleaded, "Can't I at least send you a little something? Your father would never know."

Yoko was Nintendo of America's first employee, helping Arakawa choose a location for an office in Manhattan. They rented a small suite in a high-rise at Twenty-fifth Street and Broadway, on the seventeenth floor. Mornings, the two of them drove into the city together, dropping the children at a baby-sitter's on the way.

Yoko supervised the opening of the office. She watched a truck pull up to the building's loading dock to deliver the office furniture and equipment they had ordered, all in cartons marked FRAGILE. Truckers flung the boxes onto the sidewalk.

When the construction workers arrived to renovate the place, she stood by speechlessly. One man who towered over her opened boxes of fixtures or screws, taking his time about installing them. This man would stop to wait for another to show up to connect fixtures to plumbing or electricity.

The workers arrived sometime between 8:00 and 9:00 A.M. At ten they took a coffee break that lasted up to an hour, and at noon they broke for lunch. They were finished for the day at three or so. After days of this, she mustered up the nerve to politely question one of the men about it. "Welcome to New York," he said.

Nintendo of America's first task was to break into the coin-op arcade business, which at $8 billion per year was, at that time, the largest entertainment industry in the United States—bigger than movies and television. Its customer base was decidedly narrow: mostly teenagers.

Without a presence in America, Nintendo Co., Ltd., had only a limited ability to cash in on the boom. The arcade games it made and sold in Japan were licensed through a trading company to

American companies. Nintendo saw only a fraction of the take made by the companies that sold their games directly in the market.

Minoru and Yoko spent many evenings at video arcades. They looked over players' shoulders until it made the young kids nervous. "What the fuck's your problem, mister?" one kid in a Kiss T-shirt barked at Minoru. Arakawa asked him, "Would you like a job?"

He watched kids stand in front of the machines, transfixed, their hands melded to controllers, their bony arms like umbilical cords joining human and machine. He asked the kids questions about what made a game good. Arakawa realized that the most successful games had something the players couldn't articulate. The words used to describe them were those usually reserved to describe forms of intimacy between people. It was as if the players and the game itself somehow merged.

The other phenomena that made successful games were more obvious. They had to have immediate impact and exciting noise and graphics; a player had to be captivated within the first thirty seconds. There could be no letup in the intensity for two minutes, the time that one quarter lasted. If the players weren't engrossed by then, they left the machine for good. If the game snared them, the string of quarters to follow could have bought dinner for a family of four.

Arakawa hired kids he met at video arcades. They worked in a run-down warehouse he rented in New Jersey. There were enormous rats, a loading elevator that worked once in a while, and loading-dock doors that always got jammed. There was no heating or air conditioning. In winter the place was damp and frigid, in summer a muggy, relentless oven.

Coin-operated arcade games were shipped by boat from Japan to the port of Elizabeth, New Jersey. From there they were transported to the warehouse.

Since there were so few employees—a warehouse manager and a few kids—everyone helped out. When Mr. A., as the Americans called him (Yoko hated the nickname because it reminded her of the TV character Mr. T), wasn't making calls on customers, he worked alongside the half-dozen young employees. He wore jeans just like the kids, and he put in as much effort as anyone.

One of Arakawa's first tasks was to hire a sales team; Nintendo of America needed an entrée to the large video-game operators and distributors around the country. He approached the pair who had been selling Nintendo's games in America for the past few years, Al Stone and Ron Judy.

Stone, who had attended Lowell High School in San Francisco and then the University of California at Berkeley, had once set his sights on a career in professional baseball, although he had more of a football-player's physique—the square head and wide shoulders of a linebacker. After playing some minor-league ball in Reno, he finished college at the University of Washington, graduating with a degree in finance and economics, and then tried selling sausages and representing a steamship line. Eventually he moved to Silicon Valley to work for Intel, the semiconductor giant.

Stone had met Ron Judy when they were both at the University of Washington. They lived in the same fraternity house and became business partners, running the *Business Week* franchise on campus. On one occasion they bought a huge shipment of cheap wine that a local company was going to dump and sold it to students.

Judy was a compact man with deep blue-gray eyes, eyebrows that seemed stenciled on, and a pencil-thin mustache. For fourteen summers, from grade school through college in Seattle, he had worked in his father's construction business. Most of his time was spent clearing land for interstate freeways. For a time he worked on the Alaska pipeline in fifty-degrees-below-zero weather, so that Chicago, where he headed later to finish his degree (at the University of Illinois), seemed mild by comparison.

After graduating in 1972, Judy moved to New York to consult for Chase Manhattan Bank on mergers and acquisitions. Then he moved to San Francisco to work for a small company that consulted with high-technology firms. He and Al Stone, then at Intel down the peninsula, got together sometimes after work to drink beer and mull over ideas. They agreed on two things: they were fed up with working for other people, and they wanted to make more money.

They formed their first business in Seattle, a trucking company they called Chase Express. They managed a fleet transporting containers from Seattle's piers when a friend of Judy's called from Hawaii with a proposition. He said it was a sure thing. Judy found himself at the air-freight dock of the airport, claiming a giant crate. He tried, unsuccessfully, to get it into the back of his station wagon. He finally gave up and tied it on the roof.

Judy unpacked the crate in his living room. Inside was something that looked like a cocktail table with a TV screen that faced upward, like a tabletop. It was, he had been promised, the latest thing in Japan, great for a lounge or pizza parlor, where customers stuffed in coins while they sucked down pizza and beer. A video game called "Space Fever" was played on the machine, and it was made by a company called Nintendo.

Judy convinced a local hotel owner to let him put the game in the cocktail lounge, agreeing to a sixty-forty split of the take. It did fairly well, although one game wasn't worth the effort. His friend sent along two more tabletop games from Hawaii, explaining that all the big Japanese companies had representatives in the United States except for Nintendo. A trading company in Japan was buying games from NCL and sending them to him in Hawaii, and they needed someone to bring them into the mainland.

When the games arrived from Hawaii, Judy arranged to have them set up in another Seattle bar. The games made, in Judy's words, "obscene profits." It was enough to convince him and Stone to enter the video-game business. They formed a new company, Far East Video, and set out to sell Nintendo machines throughout the United States. They bought machines from the trading company and tried to sell them to local operators or small distributors. In the process, they became masters at convincing airline-ticket agents to ignore baggage weight and size limitations so they could get their game machines on commercial flights. From Peoria to Phoenix, they begged taxi drivers to let them load the game machines into the trunks of their cabs. Video games were so hot that even the bad games were selling. They bought the machines for less than $1,000 each, and customers were happily buying five or ten of them at roughly $1,500 each. Distributors sold them to operators for as much as $2,500. It made their backaches worthwhile.

Mino Arakawa contacted Judy and Stone in late 1980 and asked them to see him on their next trip to New York. The meeting coincided with Hiroshi Yamauchi's first visit to his new subsidiary. Yoko worried that her father couldn't avoid meddling; she feared a blowup.

Although he timed the visit so that he could accompany Arakawa to an arcade-game industry trade show, the other reason Hiroshi came to New York was to check up on Yoko and his grandchildren. His wife had pressed him to go; she felt that Yoko was drifting away from them. She seemed distant on the telephone; she was working too hard; America was too far away.

Soon after Yamauchi arrived in New York, he began to see that Michiko's perception was correct; something *had* changed. Yoko, however, was hardly unhappy. She seemed content and confident in a way that surprised him. In all the years she smoked, Yoko had never lit up in front of her father. Despite his lifelong habit, Hiroshi was part of the generation that held the double standard that smoking was unladylike. It never stopped Yoko from smoking, only from letting him know about it.

Sitting in a Manhattan restaurant after dinner one night, Yoko turned to her father and asked nonchalantly, "Do you mind if I smoke?" He looked at her sharply, paused for one long moment, then pulled out his own pack and shook out a cigarette, offering it to her. She accepted it and leaned over to meet the match he held out for her. Nothing was said, but that moment was the start of a change in their relationship. Yamauchi saw for the first time that his daughter was a strong, independent woman. He also realized that he liked her very much.

"Dad," she said as they walked through midtown Manhattan, "I know you're worried about how things are going here." She took a drag of the cigarette and continued. "But hold back a little longer; let Mino do it his way."

Yamauchi said he would—for a while.

One afternoon, Al Stone and Ron Judy came in to meet with Arakawa and Yamauchi at the Manhattan office. Arakawa did the talking, although Stone and Judy noticed that he frequently looked to Yamauchi.

Arakawa said he intended to import Nintendo's arcade games directly to America and he hoped that Stone and Judy would work with his new company. From their perspective, it was a risk-free deal: they would remain independent and Arakawa would cover their expenses. Their per-game take would remain about the same, so they stood only to gain by the arrangement.

Stone and Judy told Arakawa and Yamauchi that they could sell every game Nintendo made if the games were better. Games such as "Space Fever" and "Sheriff" did fairly well, but what they all hoped for was a killer game from NCL—a "Space Invaders" or a "Pac-Man."

Yamauchi said nothing. He left New York repressing his intense urge to take over. There hadn't been even one conflagration with Arakawa.

Judy and Stone took their contract to an attorney, Howard Lincoln, of the Seattle law firm of Sax and MacIver. Lincoln had quite a reputation in Seattle. Someone said of him, "To call Howard Lincoln forthcoming would be an oxymoron—like jumbo shrimp." The thirty-eight-year-old attorney had the down-home affability of an old Kentucky colonel, but he was also capable of intimidating almost any adversary. And when he lost control, which was rarely, wise men and women got out of his path. He carried himself with effortless confidence and ease. When he spoke, he was jocular and warm; there was reassurance in his voice. He was dapper, almost preppy, and fox-sharp.

Lincoln had intense, circumspect brown eyes, small for his face, and the flushed cheeks of an outdoorsman (his fishing buddies had given him the nickname Cato, after the Green Hornet's sidekick, for his reckless pursuit of king salmon). Every chance he got, he headed into the mountains or, at least once a year, up to Alaska to fly-fish in high-country rivers. He relaxed then, but at other times his neck muscles were taut and he held his shoulders high, as if ready to protect his face in a fight. It was one of the few signs that betrayed the quick thinking beneath his composure.

Lincoln was born in Oakland, California, where his father was an executive with the Pullman Company, makers of Pullman railroad cars. The elder Lincoln was a reserved man, hard of hearing, who was well liked in his community and by his employees. His wife was an elegant woman, delicate and gracious.

Throughout World War II, the American government moved troops by the nation's railroads, which kept Lincoln's father particularly busy. Howard remembers being taken, at age four or five, to visit trains carrying wounded soldiers who had been shipped back to San Francisco.

Lincoln attended high school in the fifties. A skinny, short-haired boy, he dressed in short-sleeved shirts and cuffed pants. He was active in his local Baptist church and the Boy Scouts. In 1954 he was a model for a Norman Rockwell painting that appeared on the Boy Scout calendar for 1956. A scoutmaster stood over a well-tended campfire near a bunch of scouts in sleeping bags. Lincoln was the blond boy with the angelic smile.

With the rise of commercial air travel, the Pullman Company's fortunes inevitably declined. After forty-eight years with the company, Howard's father presided over its slide, but by then his son had entered college. At the University of California, Berkeley, he gravitated toward law. When he arrived in September 1957, fraternities still controlled campus life. The situation soon changed: the budding Free Speech Movement rocked the campus. Lincoln watched it from the distance of a student determined to go to Boalt Hall, where he succeeded in gaining admission in 1962. He graduated high in the same class as Rose Bird, who would later become a chief justice of the California Supreme Court.

The draft for the Vietnam War was in effect, and three days before he took the bar exam, Lincoln was called to report for a physical. He headed downtown to Oakland's Naval Reserve center, where he signed up to be a seaman recruit. Before he was called into service, he took a short-term job working for the Alameda County district attorney, Frank Coakley. (Earl Warren had served in that office before he went on to the Supreme Court.)

Lincoln worked on an embezzlement case in Coakley's department (a county assessor was accused of accepting a bribe), and then for an associate district attorney named Ed Meese. He aided Meese on several cases and enjoyed working with the ambitious and contentious future U.S. attorney general.

When Meese went off to work for Governor Ronald Reagan as his clemency secretary in 1966, Lincoln went into the navy with a commission as a judge advocate. He headed to Newport, Rhode Island, to take the officers' training course and learn the ropes of military adjudication. When he was offered duty in either New York City or Seattle, he headed west.

Stationed at Sand Point Naval Air Station on Lake Washington, headquarters for the Thirteenth Naval District, Lincoln tried court-martial cases. While hanging out at the bachelor officers' quarters, he met a striking woman, a naval lieutenant from Kansas, where she had worked as a schoolteacher. Bored, she joined the navy and received a commission as a line officer and served a tour of duty in a photographic intelligence unit. Afterward she was sent to Seattle to recruit Wave officer candidates.

She and Lincoln married six months after they met.

Grace Lincoln, blond, with large green eyes, and curvaceous, cut an impressive figure in a navy uniform. When Lincoln accompanied her to the base grocery store and they were both in uniform, men winked or flashed him the A-okay sign. He felt that he had to tell them, "No, it's not like that—we're *married.*"

Many courts-martial for desertion were tried throughout the Vietnam War. Lincoln was assigned, variously, either to defend or prosecute the accused. Isolated on a naval base, he had no contact with antiwar protests, while Grace, recruiting on college campuses with a group of navy pilots, was attacked physically and verbally. When her tour of duty was over, she was relieved to retire from the military.

When Lincoln followed her into civilian life in 1970, the couple had to decide where to settle down. He considered looking up Ed Meese, then in Sacramento with Governor Ronald Reagan, but in the end they chose to stay in Seattle. He sent his résumé to all the top Seattle firms and was hired by Sax and MacIver, where he built up a sizable practice, specializing in banking and corporation law.

Lincoln often worked with a CPA who did the books for Ron Judy and Al Stone. The CPA asked Lincoln to look at a contract between his clients and Nintendo of America. It confirmed that the pair would be the sole U.S. distributors for Nintendo in the coin-operated video-game business, in exchange for the payment of a set figure for each machine they sold.

The contract was, Lincoln told them, "completely screwy." When he asked who had put the document together, he was told it was NOA's president, Minoru Arakawa. He helped Stone and Judy fix the contract before they headed off to New York to set up Nintendo's distribution network.

Nintendo's new sales team, Judy and Stone, had good connections throughout the industry. The two had worked with numerous distributors and operators (who often owned machines in locations such as bowling alleys, bars, and restaurants). Occasionally they ran into some shady characters—one aggressive operator, wanting more liberal payment terms, hinted that he was well connected with the mob—but more often the business was straightforward. If buyers liked a game, they paid cash for consoles. That was it. There were none of the complications of the consumer business: almost no advertising, marketing, promotion, or royalties. The industry had slowed, however, and the new company's sales were weak. Ron Judy told Arakawa that they needed a big hit if he and Stone were going to stay with Nintendo. By 1980 they were barely keeping above water.

The game that was to turn things around was "Radarscope," an uncomplicated shooting game in which the player lined up an enemy fighter in a site and blasted away. Arakawa placed some samples in test locations around Seattle and the results were good. "Radarscope," he decided, was the game he would use to push his operation into the big time. He ordered three thousand units. By placing an order that large, Arakawa was committing almost all of NOA's resources.

It took almost four months for the ship, heavy with the game consoles, to travel from Japan to Elizabeth. Arakawa was already in a panic when it arrived, because the excitement for "Radarscope" seemed to be evaporating. The games in the test locations sat idle, and he had no idea if he would be able to sell them. He was learning the hard way about the fickle tastes of video-game devotees. A hot game could become passé overnight. Tried-and-true games such as "Pac-Man," some of the classic shoot-outs, and sports and car-racing games always brought in decent returns, but it was impossible to count on new games. By the time "Radarscope" arrived, it seemed old, and Arakawa's in-house gamers gave him feedback he didn't want to hear: the game was boring.

Ron Judy and Al Stone did their best to peddle "Radarscope," but operators who previously had indicated they might buy dozens took only two or three. Arakawa lowered the price, but he was still left with more than two thousand games. Stone and Judy were going broke. To keep going, Judy borrowed $50,000 from his aunt, her life savings. He thought, I've lost everything. I'll be in debt for the rest of my life. Arakawa was deeply worried. "I was thinking it was a big mistake to take this job," he says. He was loath to tell Yamauchi he was in trouble, but he had no choice. Nintendo of America would sink under the weight of all those "Radarscopes."

Yoko Arakawa's worst fears were being realized. She was smack in the middle of the clash she saw coming between her husband and father. Arakawa had postponed making the telephone call, but finally he called Yamauchi and told him that "Radarscope" wasn't selling and NCL's fledgling subsidiary was in trouble. Yamauchi snarled, telling Arakawa that he knew he had overstepped himself with that huge order. What did Mino want him to do? He never would have made that many games if Arakawa hadn't ordered them. . . .

All coin-operated video games looked essentially the same on the outside: a cabinet, joysticks, and a screen. What made the games unique was inside the cabinet, the PC board, or "mother board"—the game's processor, chips, and circuitry. Arakawa could have the "Radarscope" cabinets repainted, and he could have the "Radarscope" program chips removed. The problem was that he had nothing with which to replace those chips. Arakawa weakly told Yamauchi that he needed a new game quickly.

On the phone to Yoko, Yamauchi screamed about her husband's ineptitude. He said he felt he had made a serious mistake of the kind he had never made before: judging character. Arakawa, irritated by Yamauchi's I-told-you-so arrogance, bellyached about his father-in-law. He felt he should have listened to his wife and stayed away from Nintendo. Yoko, meanwhile, felt torn apart. "It was like I was on a ship that was going down in the middle of the ocean,"

Yoko says. "Two captains were shouting into my ears about what must be done, about how disastrous things were." When she called to get sympathy from her mother, Michiko was anything but helpful. "Be patient, be cheerful," she said. The advice incensed Yoko. "If all I needed was encouragement, I wouldn't be calling," Yoko snapped. Reluctantly, when summer arrived a few months later, she followed through with her plans to take the children to Kyoto for a visit, even though the tension was hardly bearable.

In America, while they waited for the new game chip, Arakawa pushed Nintendo's other games, and also decided to move his base of operations. It was a mistake to be in New York. The "Radarscope" disaster illustrated the repercussions of being so far from Japan. Shipping to North America's West Coast would cut out weeks, even months, of delay. The decision was personal, too. He and Yoko had loved living in Vancouver, and neither of them was comfortable in New York. The frenetic pace was exhausting, and all the driving between the midtown Manhattan office and the warehouse and service center in New Jersey was inefficient.

Arakawa had agents search for offices in California, Washington, and Oregon. The major urban center in the United States that was nearest to Vancouver was also the closest to Japan: Osaka harbor to Seattle took nine days by boat. Seattle was a thriving region of new industry, yet it was hardly exploited. Real estate was still affordable, and there were other pluses: as a lumber-producing area, there were companies that made fine-quality wood products, and they could make cabinets for arcade games. There was a high-quality labor pool because of Boeing and the many high-technology companies in the Seattle area. The city's population, of about 2 million, was enough to support good restaurants and arts and entertainment, yet not so big as to make it overwhelming, like New York.

To search for a place in Seattle, Arakawa contacted Phil Rogers, from his days with Marubeni. Rogers put him in touch with a broker who found a warehouse to lease in a suburb called Tukwila, not far from Seattle-Tacoma International Airport.

In the Segali Business Park Nintendo leased a 60,000-square-foot warehouse with three small offices built into a corner. Arakawa took one for his main office, and Judy and Stone moved into another. The main warehouse became the assembly and distribution site for the game machines. The service center was there too. All the games remaining in New Jersey, including the two thousand "Radarscopes," were shipped by rail to the Seattle warehouse.

Minoru called Yoko in Japan and asked her to come back to help with the move from New York. She flew to Seattle and they found a home to rent—an attractive four-bedroom place on a half-acre in Bellevue, across Lake Washington from Seattle. Then, in New York, she arranged movers to ship their furniture and clothes back west.

In Japan, Yamauchi made a quick check of his R&D groups and found that all the key engineers and programmers were too busy to be diverted to help with Arakawa's problems. The United States represented an infinitesimal portion of NCL's business, and he couldn't justify taking one of his top designers away from more important work. He told this to Arakawa, who was growing more and more desperate.

Arakawa pleaded with Yamauchi until his father-in-law finally agreed to put someone on the project. The chairman told Gunpei Yokoi to oversee the work of the young apprentice he had asked to come up with something. "But he knows nothing about video games," Yokoi said.

Yamauchi responded that there was no one else available.

The young man Yamauchi had chosen wasn't from any of the engineering groups; in fact, he wasn't even an engineer, but he had enthusiasm and some interesting ideas about the ways video games should be designed.

When Yamauchi so informed Arakawa, his son-in-law fumed. He needed a superior game to save the business and Yamauchi had put an inexperienced apprentice on the job! Why had Yamauchi seduced him into going to America if he was going to sabotage the operation? But there was nothing Mino could do, and he weakly asked his father-in-law, "What is this apprentice's name?"

Yamauchi answered, "Sigeru Miyamoto."

6

FOR A FISTFUL
OF QUARTERS

The Nintendo employees Arakawa took to Seattle with him from
New York waited for the new game. The team included a service
manager, a technician, and a secretary. To manage the warehouse
Mino hired a friend of Ron Judy and Al Stone.

Don James, a stocky man with large arms and a thick mustache,
grew up in Seattle and studied industrial design at the University
of Washington. After graduating, he went to work for Far East
Video. At Nintendo, his first job was to prepare the warehouse to
receive the two thousand "Radarscopes" from New Jersey.
Arakawa gave him permission to hire an assistant. The help-
wanted ad in the *Seattle Times* read: "Have Fun and Play Games
for a Living."

Twenty-year-old Howard Phillips, who looked remarkably like
Howdy Doody with a Charles Manson beard, applied. He had
wavy orange hair, freckles, cobalt-blue eyes, and a goofy grin. Phillips, though originally from Pittsburgh, had grown up in Seattle,

where his father worked for Boeing. Howard had attended grammar school with Bill Gates, who went on to found Microsoft. They had been in the same carpool.

Outside school, Phillips employed his ample imagination in inventions. He made Frisbee shooters out of plywood, with a large rubber band and a clothespin for a trigger. He also made an enormous catapult that launched large rocks at passing cars.

Phillips had started playing video games when the first home systems were released in America. "I was the smallest kid on the block, so I was very competitive," he says. "I wanted to show my stuff and present myself as an equal." Atari's "Pong" and Magnavox's Odyssey were black-and-white tennis games. "The games got boring pretty quickly, but it was compelling because you could actually control something on your TV."

Then Phillips created his own game with a motor from an old clock, a discarded lamp, and circular Band-Aids stuck onto a globe (for targets). Electronic toys and computer games were just a better way of playing, he felt. "You didn't have to spend all the time setting up your army men each time," he said. The computer did it for you.

After college, Phillips was working at the Seattle shipyards painting boats when he saw the ad in the *Times*. After meeting Don James and Arakawa, he started work immediately at Nintendo. The ad, he quickly learned, had been misleading. The job wasn't as much playing games as lugging, crating, uncrating, assembling, loading, and delivering them—on occasion, in his own '68 Buick Wildcat convertible. The machines weighed several hundred pounds each.

The shipment of "Radarscopes" arrived by the truckload from the train station after their cross-country journey. Phillips excitedly uncrated a game, plugged it in, and stood before the console. "It was like when a car dealer gets in a new car," he says. "I would test 'em out, drive 'em."

Arakawa watched him. "It was like he was sleepwalking," he says. Phillips was disappointed with "Radarscope." "It's hopeless," he said. He joined the other Nintendo employees who were waiting for the new game to arrive from Japan.

One day a courier delivered a package that had arrived by air from Kyoto. Don James signed for it and delivered the small box to Arakawa. He opened it and saw the board that contained the new game's program. As the service technician installed it in a console, Arakawa called in Judy and Stone. They watched as the power was turned on. The opening screen announced the game: "DONKEY KONG."

They looked at one another. Stone swore. He and Judy tried the game and concluded that it was a disaster. Two thousand "Donkey Kongs" were worse than two thousand "Radarscopes." Al Stone walked out. "It's over," he said.

Arakawa worriedly complained to Yamauchi, who was thoroughly unsympathetic. He implored Yamauchi at least to change the name, but Yamauchi refused. "It is a good game," he said.

Arakawa had no choice but to attempt to sell it. Judy reluctantly agreed to try, and he convinced Stone to cooperate. There was one promising sign: when the new flame-haired kid in the warehouse, Howard Phillips, test-drove "Donkey Kong," they had to pry him off it to get him back to work.

Before NCL could begin mass-producing the game boards, the American team had to provide an English text in place of the Japanese in the introductory storyline, which flashed on the screen at the start of a game.

The NOA staff gathered in a corner of the warehouse around a couple of card tables. They came up with a simple translation of Miyamoto's story and they had to name the characters. Arakawa christened the princess Pauline, after James's wife, Polly. They were trying to decide what to call the rotund, red-capped carpenter, when there was a knock on the door.

Arakawa answered it. Standing there was the owner of the warehouse. In front of everyone, he blasted Arakawa because the rent was late. Flustered, Arakawa promised that the money was forthcoming, and the man left.

The landlord's name was Mario Segali. "Mario," they decided. "Super Mario!"

The Spot Tavern, near Nintendo's office, was a small, darkly lit hangout that served greasy hamburgers, excellent french fries, and draft beer in tall glasses. In a back corner was a pinball machine.

Ron Judy convinced the tavern's owner to allow him to set up a "Donkey Kong" game. The next morning he wheeled a console in on a dolly, set it down next to the pinball machine, and plugged it in.

He didn't expect much when he returned that night to check the cash box. But there were 120 quarters inside, $30, a phenomenal amount.

Judy assumed it was an aberration. He checked around 10:00 P.M. the second day and found $35. The next day there was $36. The Spot's owner was happy to have more "Donkey Kongs" delivered. His bar was packed with people lined up to play Sigeru Miyamoto's first game.

The NOA team, especially Stone, hated to admit that Yamauchi had been right, but Arakawa was greatly relieved.

The components—the mother board, power supply, and unassembled cabinets—arrived from Japan. Arakawa, James, Phillips, Judy, Stone, Yoko Arakawa, and virtually all the other employees assembled consoles throughout long days and nights. Completed games were crated and loaded onto trucks, which headed out to cities throughout America. Nintendo had a hit on its hands.

Nintendo of America had its first company Christmas party at a restaurant near the warehouse. All of the dozen founding employees attended. The centerpiece of the party was a sculpture of Donkey Kong made out of fifty pounds of butter covered with shredded coconut. Afterward, the sculpture was dried out, shellacked, and hung from a rafter in the warehouse, from which it watched over the Nintendo crew until it turned green with mold.

"Donkey Kong's" popularity grew, and Taito, the Japanese company behind "Space Invaders," offered to buy all rights to the game for an exorbitant sum. Almost everyone associated with Arakawa counseled him to take the money.

Arakawa pondered the offer for days and called Yamauchi to discuss it with him. Yamauchi felt that a large advance in hand was better than profits they might never see, but he said he would back Arakawa in any decision. Arakawa told Taito the rights weren't for sale. In Kyoto, Yamauchi told Hiroshi Imanishi, "Arakawa insists, so okay. We will see if it is a mistake."

All two thousand "Donkey Kongs" sold. The NOA staff was astonished. Arakawa called Yamauchi and placed orders for thousands more.

Arakawa enlarged his staff, hiring salesmen and service technicians. Don James hired twenty-five people to assemble and test games. When they were ready, serial numbers were affixed and they were packaged and shipped. Fifty games went out each day. To increase production James began to purchase control panels, cabinets, graphics controllers, and monitors from local sources. His staff, which grew to include most of Nintendo's 125 employees, were building up to 250 game machines a day.

Sixty thousand more "Donkey Kongs" were sold, and Nintendo of America's second year ended with more than $100 million in sales. The Arakawas realized they might be in Seattle for some time, so they bought the house they had been renting in quiet, peaceful Bellevue. Yoko took up tennis, and she dragged Mino onto the golf course.

Judy and Stone were having a good 1981 too. Their accountant called Howard Lincoln to tell him they had to form corporations for their clients. Lincoln thought it was a joke. "They're half bankrupt, what are you talking about?" he asked. The accountant explained that a Nintendo video game was selling like mad. Stone and Judy were making a fortune.

Lincoln formed Ron Judy Inc. and Al Stone Inc. and, with the accountant, closed the companies' first fiscal year. Stone says they each made in excess of $1 million.

That same year, Atari and Mattel were battling for the home video-game market in America. Coleco, which began in 1932 making swimming pools, went up against them with a system called ColecoVision. All three companies wanted the exclusive rights to create a version of "Donkey Kong" that would play on their system. Atari and Mattel contacted Arakawa, but Coleco went directly to Yamauchi in Japan, who decided that Nintendo should work with the company with the most at stake. He told Arakawa a deal should be made with Coleco. "It is the hungriest company," he said.

Before he could sell the rights, Arakawa needed a lawyer to

trademark the "Donkey Kong" name and copyright the game, so Ron Judy introduced him to Howard Lincoln. Lincoln had never done this kind of work and had no idea where to begin. He told Arakawa, "No problem." It was, he says, "a typical lawyer response. You never tell a client that you don't know what you're doing."

Once "Donkey Kong" was legally protected, Lincoln helped put together the deal with Coleco in 1981. He used a legal form book and prepared an agreement. When Arakawa went over it, he asked why Nintendo had to warrant that they owned any of the rights to the game. Lincoln explained that it was just the way it was done. Arakawa asked again. "But why do *I* have to do it?"

Lincoln shrugged. "I guess you don't *have* to."

Lincoln had Nintendo disclaim any warrant about ownership, and he noted in the contract that Coleco assumed the entire risk. There was no practical reason for Arakawa to change this provision other than his instinctive way of doing business: Never give anything away that you don't have to. "I knew Nintendo owned the game, but I figured, What the hell, might as well be bold," Lincoln says. Taking his cue from his client, Lincoln skewed the rest of the contract to favor Nintendo, "mostly because we didn't know how things were normally done," he says.

The agreement between Nintendo and Coleco was to be signed before the end of the year, and Lincoln spent most of the night of Christmas Eve 1981 completing it. On Christmas Day he sent it to Arakawa for his signature. Arakawa hurried it to Japan for Yamauchi to sign with a representative of Coleco, who had been sent to Kyoto to nail down the deal.

When they met at NCL, the Coleco executive told Yamauchi that he would take the contract back to America to have the company's attorney check it over and return it signed. Yamauchi refused, saying, "You must sign it now, or we are going to go with someone else." The man froze. He had no choice. Nervously, he signed the agreement.

Meanwhile, Nintendo had outgrown the warehouse at Segali Business Park, and Arakawa wanted to move to a larger and better location. Another manager would have been more cautious. One successful video game in that volatile business did not necessarily mean there would be more. Arakawa, however, viewed "Donkey Kong" as the break he needed, just the beginning. With Phil Rogers he went looking at real estate east of Seattle, amid the rolling hills of Redmond and Bellevue where high-technology companies had sprung up the way redwoods and fir trees once did. Set between lakes Washington and Sammamish, they were small logging towns before the 1920s, when the countryside had already been scalped of timber. The communities dependent on logging became pleasant bedroom communities surrounded by the beautiful lakes, protected woods, and meadows. Redmond's chamber of commerce was proud of the community's designation as the Bicycle Capital of the Northwest before it was discovered by the high-technology industry. (Since then the chamber boasts that Redmond is the City of Opportunity.)

Microsoft, which sprawled out in dozens of beige buildings at the Ninth Avenue NW exit off mall-lined Route 508, called its Redmond plant a "campus," which fooled no one. Microsoft was hardly distinguishable from the other electronics and aviation companies in the area—from Data I/O and Rocket Research to Sundstrand Data Control and, along the road to the Seattle-Tacoma airport, Boeing.

Arakawa invested "Donkey Kong" profits in twenty-seven (and later, an additional thirty-three) acres of cleared and plowed Redmond land. He paid cash and the deal was closed in July 1982. By then he talked Phil Rogers into joining Nintendo full time. Rogers said he wanted to be a vice-president. When Arakawa balked, Rogers said, "Don't worry about the title; everybody is a vice-president in the United States."

Arakawa said he would talk to Howard Lincoln, who advised against it, and Rogers was persuaded to accept the title of director of real estate. His first job was to oversee the construction of the first of several Nintendo corporate buildings—three-layer rectangles, like high-tech sheet cakes. Eventually he also bought 125 acres in nearby North Bend for a larger new warehouse, as well as several small parcels in Seattle.

The first building was ready in November 1982. By the time the additional warehouse and production buildings were completed and the warehouse at Mario Segali's industrial park was vacated,

Rogers was Nintendo's director of product development and manufacturing. Six months later he became vice-president of operations.

Rogers worked with Don James to design a production system to build video games more efficiently. After "Donkey Kong" there were "Donkey Kong Jr.," "Popeye," "Punch-Out" (later introduced as "Mike Tyson's Punch-Out!" as a home video game when an endorsement deal with the heavyweight champ was negotiated), and then a new kind of arcade game designed by Nintendo Japan. Called the VS System ("VS" for *versus*), it was a console video-game system that had two monitors, side by side, with two play options: you versus the machine or you versus another player. The first VS game was a baseball game. Four kids held on to joysticks and controlled pitching, batting, and fielding.

Arakawa came up with another system that Genyo Takeda's R&D 3 developed for him in Japan. This new arcade machine, dubbed Play Choice, had ten games to choose from. The selection of games could be changed by removing a panel and replacing a circuit board. It was similar to a jukebox; the machine stayed the same, only the records changed. McDonald's restaurants founder Ray Kroc had once forbidden video games from the golden arches, but when he died Play Choice systems were set up in various Big Mac franchises in Florida.

Income from "Donkey Kong" poured in. There were tertiary licensing agreements of the "Donkey Kong" character (a cartoon show, pajamas, and the like), as well as spin-off versions of the game.

As the arcade games continued to sell well in early 1982, Arakawa asked Phil Rogers to accompany a friend of his, Peter Main—his former neighbor from Vancouver, who knew the restaurant business—on a visit to the corporate headquarters of Chuck E. Cheese Pizza Time Theater. Chuck E. Cheese was a California-based chain of restaurants started by Nolan Bushnell, the man who had invented the first commercial video game and the founder of Atari. Pizza Time Theater was a combination entertainment center and restaurant. Families waiting for their pizzas would be entertained by stage shows performed by robotic stuffed animals and by the latest video games, which conveniently lined the walls.

NCL was interested in acquiring the rights to open Chuck E. Cheeses in Japan. It was a logical move for Yamauchi, who saw that Chuck E. Cheeses would, like his Laser Clay ranges, be an entertainment center where he could place Nintendo's technology. He would make money selling pizza, too. Arakawa was interested in Chuck E. Cheese franchises as well. Besides the other benefits, Nintendo-owned arcades could serve as testing centers for new games.

Rogers and Main arrived at the company's headquarters in Sunnyvale and noticed that the parking lot was full of Porsches and Mercedeses. Above the receptionist's desk was the smiling face of Chuck E. Cheese, the company's mascot—a rat. Below Chuck was an electronic sign that announced, in gleaming red neon, the price at which the company's stock was currently being traded.

Rogers and Main spent the day meeting with various executives (founder Nolan Bushnell was nowhere to be seen). One, in charge of the restaurant operation, said she was working on improving the quality of the pizza crust. All the others they met seemed obsessed with robotics, global conquest, and the value of the company's ascending stock.

When the two men returned to Seattle, they warned Arakawa that the Chuck E. Cheese people didn't know what they were doing. Main said that under its current management Chuck E. Cheese didn't have much chance of surviving.

Over a series of meetings, Arakawa investigated franchising Chuck E. Cheeses that he would manage independent of the Sunnyvale organization. Yamauchi, meanwhile, concluded that the large Pizza Time Theaters were impractical for Japan, where square footage was extremely expensive.

Arakawa went ahead and secured the franchise rights to open a Chuck E. Cheese in British Columbia (the Seattle rights had been sold). Rogers found a location near Vancouver on a busy thoroughfare, and in 1983 Nintendo opened the doors to 20,000 square feet of furry robots, video games, and pizza-making equipment. Out front was a huge Chuck E. Cheese rat.

Nintendo's Chuck E. Cheese did exceedingly well, bringing in $3 million in gross sales and $700,000 in profits its first year. Phil Rogers was looking into opening a second Chuck E. Cheese just when it was announced that Nolan Bushnell had resigned from the company, which was on its way to bankruptcy.

No other Chuck E. Cheeses were opened by Nintendo, but Arakawa liked the restaurant business. Advised by Peter Main, he opened two other Vancouver restaurants, Salmon House on the Hill, and Horizons, both of which featured fresh fish, thick steaks, and great views. Yamauchi felt the restaurants were a distraction. He warned Arakawa: "You cannot look forward if you are looking left and right."

By 1982, Howard Lincoln found himself occupied almost exclusively with Nintendo-related matters. He was asked to draw up a new contract between Nintendo and Al Stone and Ron Judy and terminate their past agreement. Judy was becoming Nintendo's vice-president of marketing and Stone its vice-president of sales.

Next Lincoln was asked to work with Nintendo to do something about the extraordinary number of counterfeit "Donkey Kong" games on the market—as many as half the games sold were counterfeit.

Lincoln and his staff went after the counterfeiters and their customers, employing detectives and cooperating with police and customs officers throughout the country. One distributor in Texas who dealt in phony Nintendo games disappeared, but many others were brought to court. Organized-crime connections were never concretely proved in any of the cases, but one prosecutor insisted that the distribution of the illegal games never could have become such a large business without the mob.

When Lincoln's team of lawyers tracked down a source of bootleg games, they would get U.S. marshals to raid the offenders, court orders in hand. Many thousands of mother boards and cabinets were confiscated.

Nintendo litigated thirty-five copyright infringement suits against individuals and companies selling counterfeit "Donkey Kong" games, but in spite of all the efforts, the company lost at least $100 million in potential sales because of counterfeiters.

While planning legal strategies, Lincoln and Arakawa spent a great deal of time together, consulting at each other's offices or, on occasion, on airplanes, flying between Seattle and Japan and throughout America. In Seattle, Mino and Yoko frequently dined with Howard and Grace Lincoln. These evenings, though ostensibly social, inevitably turned to Nintendo business.

One late night in April 1982, Arakawa called Lincoln at home. There was an unusual urgency in his voice. Arakawa explained that his father-in-law had just called from Kyoto, where it was early morning. A telex had arrived at NCL from Sidney Sheinberg of MCA Universal, the huge entertainment conglomerate. The telex was short and direct. It said that Yamauchi had forty-eight hours to turn over all of Nintendo's profits from "Donkey Kong" and to destroy any unsold games. The game, MCA claimed, was an infringement on a Universal Studios copyright—for the movie *King Kong*.

The telex was ominous. Just as Nintendo of America was finally succeeding, here was a threat from one of the most litigious, hostile men in the entertainment business. Sheinberg, MCA's indomitable president, second in that organization only to chairman and CEO Lew Wasserman, was an attorney who was known to work, as one colleague put it, "like a python, strangling his prey before devouring it."

After a sleepless night, Arakawa and Lincoln headed to their respective offices and consulted on the telephone. There was every reason to believe that MCA, with its enormous clout and bank account, could crush a relatively small, foreign-owned company like NOA. Why, Arakawa wondered, hadn't anyone considered that "Donkey Kong" might infringe on the *King Kong* copyright?

Lincoln tried to reassure him. He explained that the threat from MCA wasn't unusual in business, and it certainly didn't mean that Nintendo was in the wrong. If nothing else, it was undeniable proof that Nintendo had made it into the big time.

After a week or so of frantic research, Arakawa and Lincoln prepared to leave for Los Angeles for a meeting with the MCA brass. The pair arrived in L.A. and headed for Universal's main office building, known as the "black tower" (appropriately forbidding, the two felt), in Universal City. Bob Hadl, an MCA attorney, escorted the men into an elaborate office crowded with antiques and expensive artwork. There was a sweeping view of the smog.

Also present were two attorneys from Coleco, MCA's in-house attorney, and an outside counsel. Coleco, which had invested a fortune in its impending launch of the ColecoVision system, which would include a home version of "Donkey Kong," had apparently received the same threatening letter.

Hadl and another MCA attorney explained their boss's position: Sheinberg planned to sue and immediately seek a preliminary injunction that would stop both Nintendo and Coleco from selling any games while the litigation continued. There was only one alternative: Nintendo and Coleco had to settle.

It was expected, but chilling nonetheless. Coleco had virtually bet the store on "Donkey Kong." For Nintendo, "Donkey Kong" *was* the store.

MCA's lawyer waited for a response. Lincoln spoke first. "If you own *King Kong* and it is infringed by 'Donkey Kong,' then we'll settle," he said. "But I'm not going to buy the goddamn Brooklyn Bridge. First you'll have to prove to me that you own *King Kong.*"

There was silence. The MCA team looked at him and Arakawa as if they were Martians, Lincoln recalls—"Like, come on, what are you smoking?"

There was lawyerly huffing and puffing and large bodies fidgeting in uncomfortable chairs until the MCA outside counsel spoke. "Of course we own it," the lawyer said.

He was back on the offensive, listing everything that MCA was going to go after: royalties, inventory, damages. . . . He suggested that the Nintendo and Coleco representatives discuss their course of action in privacy. Before the others adjourned to a private conference room, MCA's attorney told Lincoln that the only way to avoid a lawsuit was a settlement. "You don't have a chance in court," he said.

When the lawyers and Arakawa were alone, the Coleco team said they agreed; there was no alternative but to acquiesce. "What else can we do?" they said. "The gorillas look the same, the name is almost the same . . ." They attempted to convince Arakawa and Lincoln to settle, never letting on that they already had. Coleco, terrified of MCA and looking to appease Sid Sheinberg, who had dangled the possibility of investing in their company, had promised to try to convince Nintendo to settle too. "Sid Sheinberg had put the screws to Coleco's CEO, Arnold Greenberg," said Howard Lincoln. "He told him to get Nintendo to settle up."

MCA also went after other companies to which Nintendo had licensed "Donkey Kong." By then Atari was selling a "Donkey Kong" computer game, and MCA was also pressuring Atari's parent company, Warner Communications. Sheinberg personally threatened Warner CEO Steve Ross. "I told him . . . I didn't want to be in the position of suing him," Sheinberg says. Ralston Purina, which had licensed the "Donkey Kong" character for a breakfast cereal, responded to Sheinberg's threat by offering a $5,000 settlement. This incensed Sheinberg. "It's the most stupid thing I ever heard of," he said. "Throw them out of the building." Thereafter Ralston Purina refused to settle, as did Milton Bradley, which had a "Donkey Kong" board game.

In the meeting in the black tower, Coleco's lawyers never so much as intimated that they had cut a deal with MCA. All they said was, "There is no alternative to settling." Lincoln listened intently and never let on that he felt they were probably right, that Nintendo would have to settle (he calculated that it would take a minimum of $5–$7 million). Still, he refused to agree to anything. "We need more time," is all he told the Coleco team.

They pressed him, seemingly in panic. Lincoln told them, "You can do whatever you want, since there was no warrant about the rights you bought from us."

The Coleco lawyers glared at him. "We know that," one of them said. Arakawa glanced over at Lincoln. They shared a fleeting moment of schoolboy glee. The no-warrant clause in the contract Yamauchi had intimidated Coleco into signing was paying off. Even if they did have to settle, Nintendo was not responsible for Coleco's losses.

As the men left the room, Lincoln tugged Arakawa's arm. They slowed up and Lincoln whispered, "Something's going on." The reaction to his crack about the Brooklyn Bridge made him wonder if MCA actually *did* own *King Kong*. "Don't agree to anything," he said.

Back in Hadl's office, the Coleco attorneys excused themselves, saying they would meet alone with Hadl at another time. It made Lincoln suspicious. "Something was wrong. Something was going on between Universal and Coleco," he says. He decided he had better buy some time. "We have to sort things out," Lincoln told Hadl when he and Arakawa prepared to leave. Hadl told him he had better not take too long.

Lincoln and Hadl spoke by telephone throughout the next month. In June, the MCA attorney said, "You've had all the time you need. It's time to meet and sew things up." Lincoln said he was ready and he and Arakawa planned to return to Universal City.

The morning the pair left for the Seattle airport, Arakawa was running a fever, fighting off the flu. He asked Lincoln if he was really needed. Lincoln insisted. Under no conditions would he do this alone.

They flew back to Southern California and taxied to the black tower. Soon after they met with Hadl, they were told that if they were ready to "resolve things" they should head down "to meet with Mr. Sheinberg." Arakawa looked at Lincoln, who said nothing.

They accompanied the MCA lawyer through the corridors to the elevator.

Sheinberg had risen in the ranks of MCA to become one of the most powerful movie moguls in Hollywood. MCA was comprised of a record company, a television production company, a book publishing house (G. P. Putnam's Sons), a retailing division, theme parks, and massive real estate holdings. Sheinberg was savvy about future technologies, and he did his best to place MCA in a strategic spot by investing in them.

Before Nintendo had been threatened, a meeting was arranged between him, MCA chairman Wasserman, and Coleco's president. Arnold Greenberg had big plans for Coleco, but he needed volumes of cash. The three men were talking about the possible acquisition of—or at least an investment in—Coleco by MCA when Sheinberg casually mentioned that Coleco's association with Nintendo was a stumbling block to any arrangement they might make. "We believe some rights of ours are being violated," he said.

Greenberg smiled wryly and asked, "What took you so long?"

Later Sheinberg said he remembered the moment. The line, he said, was "that . . . famous movie line where the hero rides off into the sunset and then somebody always comes running after him, and the hero turns around and looks at the guy or the woman, always John Wayne or Gary Cooper or somebody, and says, 'What took you so long?' and you do a fadeout." The message was clear: Greenberg had come prepared to negotiate.

A deal was made. As leverage to convince Greenberg to settle with MCA, Sheinberg dangled the prospect of a partnership or some agreement that would bring a cash infusion into Coleco. Coleco could continue to sell "Donkey Kong" in exchange for a modest royalty. "We entered a covenant not to sue," Sheinberg said. "And we received from Coleco an agreement that they would pay us 3 percent of the net sales price [of all the "Donkey Kong" cartridges Coleco sold]." It turned out to be an impressive number of cartridges, 6 million, which translated into $4.6 million. (In subsequent negotiations, Sheinberg's interest in Coleco evaporated.)

The Coleco negotiation wasn't Sheinberg's first attempt to bring MCA into the video-game business. "There was a point in time when the video-game business looked like it was going to be the biggest business around," he said. He wanted MCA to be, as he put it, "a principal" in video games, as it was in records, movies, and videocassettes. He formed MCA Video Games and later bought a company called LJN, makers of toys and Nintendo games.

On that June day when Arakawa and Lincoln followed Hadl out of the black tower, they walked through the Universal lot, past actors in costumes and palm trees traveling by on a trailer. When they reached a long warehouse-like building, Hadl announced, "The executive dining room," as he grandly opened a side door.

Inside was one massive table, that could have seated dozens, set for four. They sat down and waited. Arakawa's fever worsened.

Sidney Sheinberg, tall and lanky, wearing large horn-rimmed glasses, arrived and settled into his chair. Arakawa was feeling so ill that he felt he might pass out when, after a round of introductions, an elaborate lunch was served. There was interminable small talk, and by the time Sheinberg announced how pleased he was that Nintendo had agreed to settle the case, Arakawa was ready to keel over.

He looked at Howard Lincoln, who had been charming

Sheinberg and Hadl with stories of video games and fishing, and watched the attorney transform. "It's real simple," he said, eyeballing the MCA chief. "We have done a lot of research on this thing, looked at it from top to bottom, and we feel that there is no infringement." Firmly, coldly, he said, "We have no intention of settling."

Arakawa had never seen Lincoln this icy, and he was shaken. He had no idea how Sheinberg would react.

His face reddening, Sheinberg pushed back his chair, placed his hands on the edge of the table, and boomed, "What is going on here?" He took a deep breath. "I understood that you were coming down here to talk about a settlement! For Christ's sake, you're wasting my time. What the hell is going on?"

He looked toward Hadl, who went white. Hadl threw a pleading glance at Lincoln, who continued. "It's simple," he said. "I wanted to tell you we aren't going to settle and I wanted to do it by looking you right in the eyes. That, Mr. Sheinberg, is what I'm doing."

Rising from his chair, Sheinberg said, "I've heard enough. You'll hear from our legal department." He glared at Lincoln. "You are making a major mistake," he fumed. "I view litigation as a profit center."

The three men rose, too, and walked behind him to the elevator. Hadl followed Sheinberg in as Arakawa and Lincoln hung back, watching. Before the doors closed, Lincoln called out—"Mr. Sheinberg?"

Sheinberg looked at him and grunted.

"Have a nice day," Lincoln said.

Sheinberg kept his promise and filed the lawsuit in New York State, so Lincoln flew East to meet with John Kirby of Mudge, Rose, Guthrie, Alexander and Ferdon, the well-known New York law firm. Kirby was one of the top litigators in the country, and had handled antitrust cases for PepsiCo and Warner-Lambert and franchise cases for General Foods. "When I initially met him, I wasn't all that impressed," Lincoln says. "He was kind of disheveled looking and out of sorts. But it didn't take long to figure out that this guy was one hell of a lawyer." Kirby had a reputation for vigorously defending his clients. Importantly for Nintendo, nothing intimidated him.

Lincoln and Kirby headed to Kyoto to meet Yamauchi and discuss the case with him and other NCL executives. It was Lincoln's first meeting with Yamauchi, and he had heard enough about the Nintendo chairman to be on guard. Inside the Mother Brain they inspected each other. "I have no liking for lawyers," Yamauchi said.

Lincoln restrained a smile. "We have more lawyers in Seattle, Washington, than you have in all of Japan," he said. "Few lawyers is the greatest business advantage Japan has over the United States."

"I could tell that he was a very strong person who was used to getting his way," Lincoln says. "There was no small talk. He didn't waste time."

Yamauchi wanted to know one thing from the attorneys: What must be done to guarantee that Nintendo would win the lawsuit? "We *must* win," he said.

After interviewing Nintendo employees, including Gunpei Yokoi and Sigeru Miyamoto, Lincoln and Kirby returned to the United States and worked with their respective staffs over the next ten months to prepare their case. Arakawa was elated when he heard that Lincoln's instincts had proved correct: there was serious doubt that Universal owned the rights to *King Kong*. The company had never protected the *King Kong* trademark, and past litigation seemed to confirm that *King Kong* was in the public domain. The best news was that Universal had once prevailed in litigation by asserting just that: that the name King Kong was in the public domain and could not be trademarked.

There was more. Nintendo discovered that Coleco had settled with MCA, and that Sheinberg was negotiating an investment in Coleco with Arnold Greenberg. This brought to light other motivations for the lawsuit. Sheinberg wanted a place in the video-game business, and was suing the companies that would be his major competitors.

As the materials for the trial were being assembled, Arakawa asked Yoko one night if she thought Howard Lincoln would leave his law firm and come to work exclusively for Nintendo. Arakawa had come to rely on Lincoln's legal skills, but there was more to it than that. He had never found anyone, outside of his family, with

whom he felt as close. He respected Lincoln, professionally and
also as a friend, but he wondered if he was being presumptuous,
too pushy, or too eager. Yoko replied that there was only one way
to find out.

On one of their many flights together, Arakawa finally approached Lincoln. He said he knew that Lincoln was bored with
his legal practice and he asked him to come work with him at
Nintendo.

Betraying no emotion, Lincoln told Arakawa he would think
about it. After a brief deliberative pause, he said that if he decided
to gamble on Nintendo, he wouldn't come aboard as corporate
counsel. He wanted to help run the company, to be involved in
every aspect of the business.

Arakawa immediately responded that this was exactly what he
had in mind. In a subsequent meeting that lasted no more than
fifteen minutes, Lincoln and Arakawa came to an agreement. Lincoln would come aboard as Nintendo's senior vice-president, and
would work alongside Arakawa in the number-two position. At the
conclusion of the meeting, the Nintendo president shook Lincoln's
hand. "I got you," he said.

Lincoln scheduled a meeting with his partners at the law firm on
December 7, at which he gave them thirty days' notice.

In fighting the "Donkey Kong" case, MCA and Nintendo presented evidence and testimony in the New York courtroom of U.S.
District Court Judge Robert W. Sweet. Howard Phillips was called
to court one day to play "Donkey Kong" because John Kirby
wanted to demonstrate that the game had nothing to do with *King
Kong*. Sigeru Miyamoto was deposed in Kyoto. He recounted how
he came up with "Donkey Kong," explaining that he had called the
character King Kong before naming him Donkey Kong because
"King Kong" in Japanese was a generic term for any menacing
ape.

Kirby also presented testimony and judgments from the past
trials that brought into question MCA Universal's ownership of
King Kong. He sought to prove that after its previous trials, MCA
knew it didn't have those rights, and that they had filed suit against
Nintendo in full knowledge of this fact.

A strong case developed so quickly that Nintendo's lawyers moved for summary dismissal of the suit, and Sweet granted it. In his written opinion, the judge described what appeared to be his favorite day of the trial. " 'Donkey Kong,' " the judge wrote, "was demonstrated by a game master and pertinent parts of the 1933 movie and the 1976 remake were reviewed, an altogether satisfying court day enhanced by the argument of highly skilled and forceful counsel and marred only by the submission of affidavits, depositions, and briefs."

There was nothing lighthearted about the judgment itself. Sweet concluded that Nintendo had not infringed on MCA Universal's rights because the company didn't own them. He also ruled that there was no infringement even if MCA had owned *King Kong,* since the game was completely different from the movie. The judge criticized MCA for bringing the lawsuit in spite of full recognition that it didn't own the rights, and this paved the way for Nintendo to be awarded damages.

MCA appealed the case all the way to the U.S. Supreme Court, and lost in each round. At one stage of the appeals, Howard Lincoln was called to testify. He pulled out the notes he had taken after his first meeting at MCA, when he said that Nintendo would not agree to buy the Brooklyn Bridge. He detailed the lunch meeting, during which Sid Sheinberg said that he viewed litigation "as a profit center."

When Sheinberg testified, he was asked to recall his lunch meeting with Arakawa and Lincoln. "I believe that one of them was a relative or somehow related to the Japanese person who I understood owned or controlled Nintendo," he said, dismissing Arakawa. Of Lincoln, Sheinberg said, "I believe he was a kind of a general counsel to Nintendo, but not one of great authority, as it appeared to me."

In his cross-examination of Sheinberg, John Kirby earned all the money Nintendo was paying him by reminding the judge why MCA sued in the first place. "I gather it's a pretty big company you are running out there, Mr. Sheinberg?" he asked.

Sheinberg nodded. "It's a big company with lots of legal involvements."

Kirby asked about MCA's revenues for the past year. When Sheinberg said they were in excess of $1 billion, Kirby asked how much of this was profit.

"I think our profits were around $135 million. . . ." Sheinberg responded. Total sales, Sheinberg said, were $1.6 billion.

Kirby continued. "What portion of your total operating income was contributed by your *litigation profit center*?" he asked.

Sheinberg eyed him. "By our litigation profit center?"

The point was not lost on the judge. Kirby continued, asking Sheinberg to confirm a report in *Business Week* that put his salary at $4,638,000.

Sheinberg's attorney jumped in. "This is amusing and all," he said, "but what real relevance is there to it? We'll stipulate that he's a well-paid executive. Where do you go with this line?"

Kirby said that the line of questioning was concluded. He added, "And I don't even think I need to make the argument about relevance in terms of an exemplary damages claim."

Sheinberg's arrogance had, in the opinion of several spectators, infuriated the judge. But what sank MCA was the evidence that Sheinberg had already instigated and won a lawsuit proving that *King Kong* was in the public domain. MCA's suits (and the threats of other suits) were seen as MCA's "litigation profit center" at work.

Nintendo was awarded $1.8 million. Months later, the arrival of a big, fat check from Universal was followed by a celebration at Arakawa's home. Yoko brought out caviar and champagne. Arakawa and Lincoln toasted each other and John Kirby, whom they would reward lavishly. Soon after in New York, they took the attorney, his wife, and some associates to dinner in the private dining room of an elegant Manhattan restaurant. After dinner they presented Kirby with a framed photograph of a thirty-thousand-dollar, twenty-seven-foot sailboat. The boat, Nintendo's way of saying thank you, had been christened *Donkey Kong* and included, they explained, "exclusive worldwide rights to use the name for sailboats."

Coleco, which had sold Nintendo out in its settlement with MCA, filed against MCA Universal to get back the royalties they had paid. Universal settled. Atari and the other companies MCA had shaken down were also paid back.

The experience did more than bolster Nintendo's bank accounts. "We learned," says Lincoln, "that we could handle ourselves in the big leagues. And we learned that the kind of arrogance we saw at MCA is lethal."

NOA, strong from "Donkey Kong's" enormous success, headed cautiously into the consumer market. NCL sent over Game & Watches, which had sold like 40 million hotcakes in Japan and Asia, where demand was never a problem. In fact, the demand was so great that all manner of counterfeiters started making them; the cheap imitations being made all over Asia were the problem. Perhaps as many illegal Game & Watches were sold as ones made by Nintendo.

To launch the business in America, Bruce Lowry, a vice-president at Pioneer, was hired to run Nintendo's new consumer division. Just the idea that they *had* a consumer division delighted Arakawa and Lincoln.

Entering the consumer and toy businesses meant learning a new industry. At the Redmond headquarters, in a conference room they had named "Donkey Kong," Lowry tried to educate his new colleagues about the rules of the toy business. There was much to do to prepare them for their first trade shows in January and February 1983. Don James worked on display booths while Lowry outlined the steps Nintendo ought to take.

There was no dissent when Lowry said that Nintendo needed an office in New York (they would rent space in the Toy Center building, at 200 Fifth Avenue). But when he explained the way billing was done in the toy business, the Nintendo executives balked.

Lowry said that toy companies expected all invoices to be dated December 10. But Nintendo sold coin-operated video games on net-thirty terms—all bills were due in thirty days. Simple.

Lowry explained that in the toy business, orders came in in, say, January or February for a product that was to be shipped in the summer. The toy stores had the winter season to sell it. Then, finally, they began to pay their bills on December 10, once much of the Christmas business was over.

Howard Lincoln jumped in. "Wait a second! Why the hell would anyone agree to that? We build the product and take all the risk and we have to finance this whole thing and then sometime in December we're going to get paid—maybe?"

Lowry said this was correct. That was how the toy business worked.

In 1983, Arakawa attended his first Consumer Electronics Show to drum up business for Game & Watch. At this crucial trade show for businesses from Walkmen to VCRs, a company's placement in the CES's convention-center display rooms says volumes about its stature in the industry. Nintendo was hidden in a tiny booth on a high floor in an out-of-the-way building. Buyers couldn't have found the company's modest display of Game & Watches even if they had been looking for it.

Things weren't much more promising at the Toy Fair the following month. The shows were barometers of what was to come; NOA lost millions on Game & Watch in the United States. The lessons, however, were valuable. The next time NOA entered the consumer business, the Nintendo team was prepared. They knew never to call their product a toy, whether or not it was. That way they didn't have to offer December 10 invoicing. They also didn't have to give mark-down money, which was another unsettling new concept for Howard Lincoln.

When it finally came time for a major chain to pay its Game & Watch bills to Nintendo—after December 10, 1983—Lincoln received a call from the chain's controller. The gentleman asked for mark-down money, and Lincoln didn't know what he was talking about. "You know, mark-down money," the man said. "We still have a lot of those Game & Watches, you know. We have to mark them down to sell them."

"We sold them to you and sent the invoice," Lincoln said. "We shipped them to you, so you owe us the money. If the product doesn't sell, that's not my fault. You took the risk. You owe us the money."

The man said, "You may have learned a lot of things in law school, but we're in the toy business. If the product doesn't sell, you have to give us mark-down money; you have to reduce the price so that we maintain our margin."

"You gotta be nuts," Lincoln said, but the man explained that the practice was standard. "We're long-term partners," he told the Nintendo manager. "If you don't want to give us mark-down money, fine, but don't have your salesmen call anymore." Nintendo gave the mark-down money.

Another lesson Nintendo learned from the Game & Watch disaster was how *not* to make television commercials. An agency Nintendo had hired came up with a creative television campaign it dubbed "the bored campaign." In one thirty-second spot a boy was sick in bed. The second after his mother tucked him in, he pulled a Game & Watch out from under his pillow. In another "bored" commercial, a young boy was bored to tears at a wedding. He said, "My parents brought me to this stupid wedding and I'm bored . . ." until he pulled out his Game & Watch.

Rather than allow the agency to produce the commercial with professional actors, Ron Judy and the marketing group decided to cast it. They thought they could make the commercials more believable by using nonactors. It was an expensive, though hysterically funny, mistake. They used Nintendo employees. The company's new credit manager, for instance, a kindly older woman who froze in front of the camera, played the mom. The resulting ads were so bad that television stations refused to air them.

Nintendo's Game & Watch business was dissolved in the summer of 1985. By then, although NCL was making an enormous amount of money on the Famicom in Japan, almost all of the revenues of its American subsidiary were still from coin-operated games. Hiroshi Yamauchi wanted to change this and he told Arakawa it was time to launch the Famicom in America. To determine if it was feasible, Arakawa undertook an investigation of the American home video-game business.

It was like surveying a car wreck. In the early 1980s, companies such as Atari, Mattel, and Coleco had been sharing a multibillion-dollar business. By the end of 1983, all that was left was the wreckage of a devastating crash. The industry had shrunk to an insignificant size, amounting to only a few hundred million dollars. Companies that had been raking in the cash were bankrupt. It seemed clear that the American market just wasn't interested in home-video games.

Arakawa talked to people who had worked in the home video-game industry. He met with manufacturers, wholesalers, and distributors, with buyers for department stores, discount stores, toy stores, electronics stores, and software companies, and with parents. From everyone he heard the same message: the last thing anybody wanted to hear about was a new video-game system. Everywhere he went, he heard one name over and over again: Atari.

7

REVERSAL
OF FORTUNE

In a conference room of the Sands Hotel in Las Vegas, in January 1991, a Mendelssohn concerto played in the background as Irving Gould, CEO of Commodore International, the $900 million computer company, introduced Nolan Bushnell. The most persistent figure in California's Silicon Valley history, now forty-five years old, unfolded his Ichabod Crane frame from his chair and lumbered to the podium.

Though Bushnell once showed up at a top-level meeting of the directors of the company he founded, Atari, in a Black Sabbath T-shirt and sneakers, on that morning in Las Vegas he wore a dark six-button suit, shiny plum tie, and spit-shined English brogues. His mane—the dark mass of *fusilli* hair and beard that framed his face—had grayed, but the figure at the podium was still commanding. He adjusted the microphone, placed his palms on the lectern, and began to evangelize.

Bushnell, it was soon apparent, was on a new mission, preaching the gospel of multimedia through something called CDTV, Commodore Dynamic Total Vision, the first available consumer product designed for multimedia. *Multimedia* was the latest buzzword in the consumer-electronics industry, predicted to be for the 1990s what personal computers were for the 1980s. It was a dramatic new technology that pulled together the wizardry of television, personal computers, compact disks, video disks, and video games. Key to it was the ability to interact with the programming that comes into the home.

CDTV was a small, innocent-looking black box, much like a VCR machine, that plugged into an ordinary TV. It had many of the same capabilities as a computer, but it was neither intimidating nor complicated. It was operated with a palm-size remote-control unit. A half-dozen industry giants, from Philips to Apple, were exploring multimedia, but Commodore, a relatively small company among high-tech giants, was the first to launch a consumer product based on the technology. For Bushnell, it was not just new technology but a cause. He explained that the key to the new system was the ability to control virtually unlimited information. When a single compact disk was plugged into CDTV, a television became a library, one that included more than books—also pictures, even moving pictures, and sounds. In a CDTV encyclopedia entry on Dr. Martin Luther King, Jr., the civil-rights leader's biography came to life in high-resolution graphics and full-blown stereo sound. Bushnell asked, "What is the biography of Martin Luther King without his oratory?" King's "I Have a Dream" speech was available at the click of the remote control.

"It's the combination of the audio record, video records, text, and the interactivity that control these items that make it happen," Bushnell explained. With a CDTV children's book, kids no longer would be passively fed television. They would influence the stories they were being told, choosing whether the hero would be a prince or a princess, or whether he or she would kill the dragon or befriend it. All forms of media—movies, books, music—were becoming more like video games, which is why Bushnell was convinced that CDTV would revolutionize education. "A video game can teach geography and history as a player moves through the course," he said. "A game can teach complex decision making and critical thinking."

Education, Bushnell said, was why he was so committed to CDTV. (It was silently understood that the other reason for his commitment was the high salary he was getting from Commodore.) The system could teach kids at their own pace and at a fraction of the cost of a computer with similar capabilities. Posters on the walls explained the vision of CDTV's promoters. DARLING, WOULD YOU ASK THE KIDS TO LOWER THE VOLUME ON THE ENCYCLOPEDIA? Another: NOW, TURN ON THE TV AND DO YOUR HOMEWORK.

Still, the future of CDTV was anything but guaranteed. Its retrieval system was pretty sluggish, and the software available for it was limited, but there was, nonetheless, an audible buzz in the room when it was shown off. "This is multimedia at its best," Bushnell said. "It will revolutionize our lives."

After applause, Bushnell stepped away from the microphone and headed to his seat next to Irving Gould. A reporter for a consumer-electronics trade journal in the audience whispered loudly, "Wasn't he the cat who was here a few years ago with something else he thought was going to revolutionize our lives?"

Nolan Bushnell had been trying to revolutionize our lives so often that it was difficult to take him completely seriously. His life had been a roller coaster of revolutionary visions and promises— and the millions of dollars that come with them—but also the disappointments.

In the early 1980s, he had had two fully decked-out Learjets to shuttle him between his Woodside, California, estate and his homes in Aspen, Georgetown, and Paris (where his palace's backyard opened onto a view of the Eiffel Tower). When he wasn't sipping Dom Perignon above the clouds, he was often racing his sailboat, in training for the Transpac, one of the oldest yacht races in the world, which he won in 1983. When he took to the roads, he chose from among a Rolls, two Mercedeses, and a Porsche.

At the time, there were few models for the kind of wealth and notoriety Bushnell had acquired so quickly and so young; he was the first of that generation of much-hyped super-successful high-tech entrepreneurs, the founder of one of the fastest-growing companies in history.

Bushnell, the son of a cement contractor, grew up in a bleak outpost near the Great Salt Lake. When he was a child, he was already obsessed with innovation. "I read science fiction," he says, "and I really wanted to live there, without all the limitations we have in our world." He spent much of his youth in the family's garage, trying to create the things he read about. He was only six years old when he built a control panel for a spaceship out of an orange crate. He was the youngest ham radio operator in Utah—but not the shortest; he was six-foot-four by the time he was in seventh grade. When he launched a "UFO" he had made—a hundred-watt light attached to an enormous kite—he convinced a fair percentage of the local population that the planet was under attack.

Another time he took a few shotgun shells, removed the shot, and, wearing a ski mask, drove a borrowed car up to a buddy of his, who was standing in the schoolyard among a group of their friends. Bushnell aimed and fired both barrels into his friend's chest. The boy smacked two handfuls of catsup against his shirt and fell to the ground.

In college, at Utah State and later the University of Utah, Bushnell studied engineering, economics, business, and philosophy. Once, when he had lost his tuition money in a poker game, he had to take work guessing people's weights and ages at an amusement park. Eventually he ran the arcade there. At college he often hung out in the school's computer lab playing a game called "Spacewar," one of the first computer games ever made.

Bushnell graduated with an engineering degree in 1968 and moved to California, where he worked briefly as an engineer in the computer-graphics division at Ampex. At home, meanwhile, in a laboratory he made in his daughter's bedroom (the little girl was exiled to the living-room couch), he created a simpler version of "Spacewar," which by then could be found on computers at most universities around the country.

The game had been invented in 1962 by an MIT graduate student named Steve Russell (who based it on a series of science fiction space operas called *Lensman*, written by "Doc" Smith).

Bushnell's version, "Computer Space," was made of integrated circuits connected to a nineteen-inch black-and-white television. Unlike a computer, it could do nothing but play the game, a primitive simulation of air combat between a spaceship and flying saucers. The key to Bushnell's invention was that since it didn't require a full-fledged computer, it could be produced relatively cheaply. He envisioned video games like his standing alongside pinball machines in arcades, pool halls, and bowling alleys.

With hopes of having his machine put into production, Bushnell left Ampex for a small pinball-machine company, which manufactured 1,500 of them. They never sold, and Bushnell, then twenty-seven, left the company. He had determined that "Computer Space," which required players to read a full page of directions before they could play, was too complex.

He set out to make a simpler game, and this time he would sell it himself. Kicking in $250 each, he and a friend formed a company they called Syzygy (from a word meaning the nearly straight-line configuration of three celestial bodies in a gravitational system— for example, the sun, the moon, and the earth). The name was already taken by another company, so Bushnell chose the Japanese word that was the equivalent of *check* in the game go: atari.

In his home laboratory, Bushnell built a new game, "the easiest one I could think of. People knew the rules immediately, and it could be played with one hand, so people could hold a beer in the other." A "ball"—really a squarish dot of light—was batted back and forth by two inch-long paddles that were projected on a screen. The paddles, on the far sides of the "court," could be moved up and down when players twisted knobs on the front of a crudely built cabinet. "I made it with my own two hands and a soldering iron," Bushnell says. He named it "Pong," after the sonar-like "pongs" that sounded each time the ball made contact with the paddle.

In the fall of 1972, Bushnell placed "Pong," the first commercial video-arcade game, with a coin box bolted to the outside, in Andy Capp's tavern, a popular Sunnyvale pool bar that holds a place in Silicon Valley lore rivaled only by the garage in which Steve Jobs and Steve Wozniak invented the Apple computer.

Set beside a pinball machine, "Pong" was an oddity, a dark wood cabinet that held a black-and-white TV screen on which cavorted a white blip like a shooting star in a black sky. One of the bar's patrons stood over the machine, examining it. "Avoid missing ball for high score," read the only line of instructions.

The young man reached into his pocket, extracted a quarter, and slipped it into a slot on the console as he called a friend over. The machine, announcing its name with its trademark bleat, "served" a ball automatically from one side of the screen. The players missed the first few serves until they got the hang of the controls, but two bucks' worth of quarters later, they were having lengthy volleys. A crowd had gathered around to watch. There was a long line in front of the machine for the entire day, and the next day too.

The "Pong" machine stopped working toward the end of the second day; the coin box was stuffed with so many quarters that the game had short-circuited. A new coin box (actually a casserole dish) was installed inside the machine. It took about a week to be filled to its capacity of about 1,200 quarters. Bushnell was ecstatic. His simple, monotonous game was bringing in $300 a week. The pinball machine that stood next to it was earning only about $30 or $40.

Lacking the money to do a major "Pong" production run himself, Bushnell approached the established amusement-game makers, companies like Bally's Midway. The pinball companies unceremoniously showed Bushnell the door. He was left with two alternatives: he could either finance the venture himself or forget it.

To get some quick cash, Bushnell accepted jobs consulting for electronics companies, and he talked his way into a $50,000 line of credit with a local bank. Employing a band of long-haired techies whom he paid next to nothing, he started an assembly line in an abandoned roller-skating rink. The rowdy gang assembled machines for twelve to sixteen hours a day as the Rolling Stones and Led Zeppelin blared from a staticky stereo system. Dan Van Elderen, a young engineering graduate who came on to assemble games, recalls that "there wasn't even a monitor business in those days. All the original 'Pong' games were built with Motorola TVs. We threw away the plastic case and the tuner and RF circuitry and used the raw tube and video drivers."

Bushnell met potential customers, mostly distributors who handled pinball machines and jukeboxes, and was able to sell all the machines his small staff could make, about ten a day. If he was going to make any real money, however, he had to expand; he needed more cash. Banks and investment bankers declined to put up money because of rumors that Atari was connected to the Mafia. Also, they were worried about an inherent flaw in the whole idea of video games: people would steal the TVs from the consoles.

Investors whom Bushnell fast-talked past these concerns were turned off when they visited the company's site. Employees in ripped jeans and worn sneakers (if they wore shoes at all) worked whenever they wanted to. Staff meetings were a rarity. The founder wore T-shirts or flower-print shirts with polka-dot ties. As Steve Jobs, one of Atari's early employees, remembers, "The smell of marijuana ran freely through the air-conditioning system. A few of the people there had beards so large that I never once saw their faces."

Bushnell was a consummate salesman, obnoxiously persistent. This and his immodest, even grandiose, vision—he projected sales of hundreds of millions of dollars—finally convinced one of the Valley's most astute and credible venture capitalists, Don Valentine, to back the company. The cash infusion allowed Atari to grow. Bushnell hired more staff and rallied his team with cheerleading, charm, and anything else that worked—including lying; he persuaded one employee to work double-time on some revised circuitry for "Pong" by telling him that General Electric was waiting anxiously for units, even though GE had refused to return his calls.

"We were all so young," he says. "I was in my twenties. My vice-presidents were in their twenties. Many of the people were teenagers." Atari had a fund for "unwanted pregnancies" and another to bail staffers out of jail. The average age of the staff was so low that when Bushnell decided to get group health insurance, the rates for extraordinary benefits were dirt-cheap—"until everyone started getting braces," he remembers. "Not for their kids, for *themselves*!"

A typical early staffer was Steve Jobs, who came on board when he was only seventeen. At the beginning of 1974, he had dropped out of Reed College and returned to his parents' Los Altos home. When he began looking for work, in a local newspaper he saw a help-wanted ad that read "Have Fun and Make Money." He visited Atari. "We've got this kid in the lobby," Bushnell's partners' secretary announced one day. "He's either a crackpot or he's got something."

Jobs, skinny, long-haired, with a Ho Chi Minh beard, filled out an application and listed all the things he had done—nothing relevant, as it turned out, except for some courses in engineering at Reed. "Don't call us, we'll call you," he was told.

The phone rang the next day. Jobs became Atari's fortieth employee. He was hired as a technician, earning $5 an hour. His first assignment was to help an engineer on a new game, "Video Basketball." Atari was trying to model its games after field sports, for "Pong" circuitry was easily adapted to such simulations.

Although Atari games were selling, Bushnell was in over his head and Atari's survival was continually precarious. "A lot of people don't understand that you can be successful and profitable and still not have enough cash; a growing company consumes *tremendous* amounts of cash," he says.

The company literally couldn't afford the payroll twice one month. Don Valentine's money had helped build up production, but the returns lagged. A big success that followed "Pong" bailed them out. It was the first video car-racing game that was controlled by a steering wheel attached to the cabinet. The game, "Gran Trak," gobbled up quarters even faster than "Pong."

A friend of Steve Jobs, Steve Wozniak, an engineer at Hewlett Packard, was a "Gran Trak" addict. Most evenings after work he headed to a pub, where he put great quantities of quarters, money he could not afford, into "Gran Trak." Jobs began to sneak him into Atari's production facility at night, where he could play the game for free. In exchange for the free-game time, Woz, a whiz with computers, helped out whenever Jobs hit a stumbling block with some particularly tricky circuitry.

Bushnell found Jobs tactless on occasion, but he liked the headstrong young man and took him under his wing. Jobs was working in order to earn enough money to travel to India. Bushnell helped him by paying Jobs's expenses to Europe in exchange for a service call in West Germany. Some games Atari had shipped there were causing local TV interference.

Jobs flew to Europe, adjusted the "Pong" games, and continued on to New Delhi, where he met a guru who shaved his head. He stayed in India for six months before returning to Palo Alto and his old job, and then was assigned to work on one of Atari's oddest games. "Gotcha" had been dreamed up in a brainstorming session during which a young technician joked that the joystick used to control most arcade games was a phallic symbol. The boy suggested that Atari try out a "female" game; "Gotcha" was the result. On the game's console were two rounded mounds made of rubber that were squeezed to control game play. Insiders called it "the boob game."

Dan Van Elderen soon moved from assembling games to designing them. He worked on a game called "Tank," in which two players blasted each other's tanks as they sped through an obstacle-filled maze. It was the first game that used ROM chips to store graphics data (there was still no microprocessor in video games).

Other engineers refined a version of "Tank" that was a favorite of Bushnell's. As in the original game, each player controlled a tank that tried to seek out and destroy the other. When a successful hit was made, the on-screen tank exploded and the player controlling the disabled vehicle got an electric shock. "We did it so it wasn't lethal or anything," Bushnell notes. "But all of a sudden it was *real.* Certain people really liked it." The company's legal department, however, was not among them, and the game never made it out the door.

Steve Wozniak came over to Atari to help Jobs build another "Pong"-based game for Bushnell called "Breakout." A paddle hit a ball against a wall of bricks that disappeared, one by one, when hit, until there were none left. Bushnell liked the game, but the circuitry required too many expensive computer chips. He offered Jobs a bonus of $100 for every chip he was able to eliminate. Jobs made himself $5,000.

When they weren't working their day jobs, Jobs and Wozniak were busy on their own, in the Jobs family garage. They built a makeshift computer—a circuit board, really—which they called the Apple I. Some of the parts had been lifted from Atari. The Apple I didn't do much, but when Wozniak showed it off at a computer club meeting and the result was orders for fifty of the contraptions, it dawned on Jobs that there might actually be a market for personal computers, and he left Atari to found Apple.

Wozniak's interest was primarily technical; Jobs set about making the computer accessible to people. Together they added a keyboard and memory, and Wozniak developed the disk drive and added a video terminal. Jobs hired experts to design an efficient power supply and a fancy casing, and thus was born the Apple II—and with it an entire industry. Needing investors, Jobs went to Nolan Bushnell and asked him to become a partner in Apple Computer. Bushnell unwisely declined.

Jobs and Wozniak were only two of the computer and video-game industry executives who cut their teeth at Atari. A decade after Bushnell founded the company, there were Atari alums in high-level spots at Electronic Arts, Lucasfilm and LucasArts, Apple, Microsoft, and a number of other companies. "It's because we provided a place for creative people to be part of something completely new," Bushnell says. "These were people who wanted to create something intellectually stimulating and fun. They wanted to put their talent into making games, not bombs."

Atari, meanwhile, continued to grow. In 1973, after six thousand "Pong" games were sold for more than $1,000 each, Bally's Midway approached Bushnell with a huge offer to buy the rights to the game. Bushnell agreed, and Bally's sold about nine thousand more "Pongs." Atari now had eighty employees, "long-haired freaks, bikers and dropouts, hired not for their skills but on the basis of their good vibes," according to Scott Cohen in his book *Zap! The Rise and Fall of Atari. Fortune* reported that 100,000 "Pong"-type games were produced in 1974 alone. Although only a tenth of those were made by Atari ("Pong" was copied with abandon), the company earned $3.2 million that fiscal year. In the three years that followed, Atari sold $13 million worth of video games, including "Quadrapong," for four players, and "Puppy Pong," in a Formica doghouse.

In 1974, after Atari's success with "Gran Trak," Bushnell decided to make a "Pong" system for the home. The trick would be to compress a coin-operated game down to a few inexpensive components. Magnavox had been selling a home video-game system for the past two years that had been created back in 1966 by Ralph Baer, a supervising engineer at a company called Sanders Associates. Baer had come up with a game almost identical to "Pong" that played on a seventeen-inch RCA color set, but Sanders did nothing with the technology until Magnavox licensed it. Magnavox's Odyssey used Mylar overlays taped to the TV screen that depicted different game boards or playing fields. One hundred thousand Odysseys sold in 1972, its first year on the market.

Atari's home "Pong" had a sharper picture and more sensitive controllers than the Magnavox system, and it also cost less. Still, when Bushnell showed "Pong" off at toy shows, none of the major chains showed interest. Dejected, he returned to Atari with no idea where to turn next. Then the buyer for the sporting goods department of Sears Roebuck came to see him, and before he left, he had offered to buy every home "Pong" game Atari could make.

With the backing of Sears, Bushnell had the ability to boost Atari's production capacity. The retailer mounted a major television ad campaign, and Atari's 1975 sales shot up to almost $40 million. Bushnell spent as much of it as he could—on parties, expensive suits, and sports cars. "We were absolutely no more or less irresponsible or crazy than Ross Johnson and those guys at RJR Nabisco," Bushnell says. "The only difference is they were running corporate America."

Arcade games became more sophisticated when microprocessors dropped in price by the mid-1970s. Dan Van Elderen was part of the team that built Atari's first microprocessor-based game, "Sprint," a driving game with oncoming traffic that required realistic, quick reactions. Up until this point, Atari had basically been a hardware business. With microprocessors—which used stored information from programs as needed—software became integral to video games, and rooms full of programmers were hired.

With income from the arcade games, the deal with Sears, and more venture-capital money, Atari was poised to expand and take on the competition. By the end of 1976, twenty different companies, from RCA and National Semiconductor to Coleco, were making home video-game systems, each trying to outdo the next with marketing dollars and technology. When Fairchild Camera introduced the first full-color system with changeable cartridges (created by Alpex), Atari's entry had to be that much better. It was. Atari's engineers assigned their new products code names. Their programmable video-game system was Stella, named after a woman in the personnel department. Officially named the Atari 2600, it was a powerful and inexpensive machine, but the outlay required to manufacture and market it on a big-league scale was beyond Bushnell. He considered going public but decided to try to find a corporate investor first.

MCA and Disney declined. It would be a year or two before the likes of Sid Sheinberg fathomed the significance of the new video-game industry. Warner Communications, on the other hand, approached Bushnell.

Steve Ross, the company's silver-haired chairman, who later drove the Time Inc.–Warner Communications merger, heard from one of his executives that Atari was looking for investors. Ross knew about Atari. Once, at Disneyland, he had briefly lost track of his kids. When he found them they were gathered around an Atari video game called "Indy 8," an eight-player road-race game. "His family was hypnotized and he sat there and watched the machine suck up quarters," according to Manny Gerard, the executive who told Ross that Atari was looking for an investor in 1976.

Gerard saw the video-game business for what it was: the computer, entertainment, and consumer-electronics businesses rolled into one. "I saw the 2600 in an Atari lab and said, 'Holy shit! This is going to take over the world.' " He convinced Ross that Warner should not make an offer to invest in Atari but rather buy it outright. When Ross gave him the okay, Gerard negotiated with Bushnell and his highest-ranking cohorts.

At first, the Atari team said they weren't interested in an acquisition, but it was more a pose than anything. "We were exhausted," Bushnell says. "The offer from Warner was a relief." The size of the offer encouraged them too, and a deal was struck. Warner paid $28 million, a pretty good return on Bushnell's $250 investment. Bushnell and his friends made a fortune, and anyway, the deal had him stay on as chairman. Bushnell reportedly said, "I've always been telling people I was a millionaire. Now I am."

The relationship, however, was ill-fated. "I should have known it wouldn't last," Bushnell says. "It just wasn't fun anymore when it wasn't mine." He worked under Gerard for two years, but he had lost his focus. Gerard claims that Bushnell spent more time managing his personal investments than running Atari. "He wanted the business, wanted to run it, but didn't want to come to work," Gerard says.

When Bushnell did, he clashed with the Warner management on most issues. He says he wanted Atari's new computer, the 800, to blow the inferior Apple II out of the water (partly as revenge, after he passed up on the opportunity to be a founding partner in Apple). However, while Steve Jobs was encouraging developers to write programs for the Apple II, Atari threatened to sue anyone who tried to make software for the 800. If customers wanted a spreadsheet or word processor for the 800, they had to buy Atari's. Meanwhile, outside developers came out with software for the Apple II—VisiCalc spreadsheet, for one—that sold millions of the machines.

Bushnell also disagreed with Warner's handling of the video-game business. He felt their huge stock of 2600s should be dumped for cost because Atari would make its profits on software, but Warner management vetoed the idea.

Atari was poised for a big year in 1978, but so were National Semiconductor, Fairchild, General Instrument, Coleco, Magnavox (which had released Odyssey 2), and a dozen other companies. The Christmas season came and went, and few consumers, perhaps because they were confused by all the choices, brought video games home that year. Of all the entrants, only Atari and Coleco survived, and Atari was in shambles.

Manny Gerard, who had to answer to Steve Ross, put the screws to Bushnell, who was never known to respond well to anyone else's ideas, never mind anyone else's ideas about *his* company. "You can't rule by the divine right of kings," Gerard told him, whereupon Bushnell stopped returning Gerard's calls.

Gerard decided to bring in someone new to run Atari. He chose

Ray Kassar, a former marketing vice-president from Burlington Industries. Kassar was as buttoned-down as Bushnell was northern-California-casual, and a clash was inevitable.

Bushnell, his necktie flying over his shoulder, arrived late to the November 1978 annual budget meeting at Warner's headquarters at 75 Rockefeller Center in New York. Winded and pink-cheeked, he threw his jacket onto a coffee table and plopped into the only empty chair, at the far end of the marble conference table. He looked carefully around the table at the Warner brass, who began "the inquisition," as he remembers it. They wanted to know what Bushnell planned to do in the coming year with their subsidiary (which was generating $250 million in sales but no profits).

Bushnell let loose, attacking virtually everything Manny Gerard and Ray Kassar wanted to do. First, Atari should fold its sinking pinball-machine business. Second, Atari shouldn't even think about launching the 800 computer unless it changed its policy and encouraged software companies to create programs for the platform. Third, the 2600's price should be slashed. Warner, he said, should invest whatever it took in during the short term for the profitable long-term business. Greed would destroy the company.

Manny Gerard was outraged that Bushnell had aired his dissatisfaction in front of the Warner bosses. "Man, was he pissed off," Bushnell says. Gerard says he was countering Bushnell's "bullshit and lies." Gerard contradicted everything Bushnell said. He said that Bushnell was the one who was going to sink Atari. The two shouted each other down. In the end, as Scott Cohen writes, "Gerard yelled louder."

After the meeting, the two men met. Gerard said, "You don't believe in the program. Maybe you should leave."

Bushnell did: with $1 million cash, about $12 million in debentures (which Warner eventually bought back), a $100,000-a-year salary, and bonuses and options. The only condition was that he couldn't compete with Atari for the next seven years. "They saw me as an extremely creative gamer as well as a strategist," Bushnell says. "They knew that I at least *might* be right, and they didn't want me to be able to shove it in their faces. They were also afraid because I had tremendous relationships with all the engineers.

They thought the engineers would leave Atari and we'd go up against them and blow them away."

When he made his original deal with Warner, Bushnell retained some technology that Warner didn't want to pursue. One example was Chuck E. Cheese Pizza Time Theater. Relieved to be severed from any obligation to Warner, Bushnell charged headlong into building Chuck E. Cheese.

By 1981 there were 278 Chuck E. Cheeses, each bringing in more than nine times the profits of a typical pizza-chain restaurant (Nintendo's franchise opened in Vancouver in 1983). Bushnell was coining money, and his net worth escalated into the $100 million range. He began to collect jets and homes the way other people collect Depression-era glass. His first Learjet cost him $4 million. He named it *Danieli,* after the hotel in Venice where he and his second wife spent their honeymoon. Bushnell delighted in jetting around the globe, sometimes shuttling such famous passengers as George Bush and Francis Ford Coppola. When he visited a place he liked, he bought a home there.

But Bushnell became bored with jet-setting and only one business in the works. "I studied a lot of philosophy when I was in college," he says. "I learned a Kierkegaardian dialectic that postulates that the prime mover in the universe is boredom: God was bored, so he created the universe, and then man and woman. Woman was bored, so she ate the apple in the Garden of Eden. For me, boredom *is* a prime mover. I go from being bored to being completely hassled and harried, because I always bite off more than I can chew. And it gets worse as I get older, because now I know how to follow through on ideas. I'm in real trouble whenever I get obsessed with two or three new things."

This time boredom led to the founding of Catalyst Technologies in Sunnyvale later in 1981. Bushnell called it his "fifty-thousand-square-foot incubator," where over a dozen independent companies worked on projects he financed. The idea, he said, was to "fund a company with a key, not a check"—a start-up would immediately have an office, receptionists, copy machines, telephones, and everything else it needed. In return for all this, plus cash flow, Bushnell owned a large share of all Catalyst companies.

Cinemavision made a color monitor with four times the resolution of standard televisions. TimberTec was a computer camp for kids. ByVideo was a mail-order computer network company through which one could, say, order shoes via a video "catalogue."

Other Catalyst companies included ACTV, an interactive cable-television system in New York City which had a fancy remote control that allowed you to choose not only the camera angle from which you watched a baseball or football game but the fates of characters in TV movies. Another company, Compower, made computer products. Bushnell rescued a company called Axlon, and with it began manufacturing the Playskool Baby Monitor, which allowed parents to listen in on their sleeping infants. Axlon also came out with Petsters, robotic cats and dogs that ran around the house, and AG Bear, which echoed back spoken patterns in a gravelly grumble. IRO did skin care and color analysis electronically. Magnum Microwave made components for the government (for the Tomahawk missile) and other industrial customers. Then there was Sente, which created video games that incorporated holograms and video disks. Bushnell planned to pit Sente against Atari when the noncompetition agreement expired.

"I put the first dollar in all those companies," Bushnell says. It was a way for him not just to *have* dozens of ideas, but to give them life. His choice of the name Sente summed up part of what drove him. As Atari meant "check" in *go,* Sente meant "checkmate."

Some Catalyst companies were particularly brilliant. ETAK created the technology for TravelPilot, the first electronic car map to be sold commercially. A small TV-like screen displayed maps stored on CDs. When you typed in your present location and destination, street maps showing the best routes appeared. As you drove, the maps shifted to show the car's progress. It told where you were with an electronic compass, which tracked you, based on information from sensors on the wheels. Bushnell said he expected there to be a map system in every car in the country.

Of all the Catalyst companies, Bushnell's favorite was Androbot, which built "intelligent" robots for domestic use—robots that could learn from their experience. Information from nine or ten infrared and sonar sensors were processed through the robot's computer so that it would seek out human beings, learn the layout of a room, and perform an increasing number of useful tasks. Bushnell became obsessed with robotics. Androbot, he said, would be his "billion-dollar company that's going to last forever." But though his vision of a personal robot may someday come to be, it will happen without Bushnell. After he had spent $12 million of his own money, Androbot imploded.

So did other Catalyst companies. He sold IRO for a huge loss. "It turns out that women don't want scientific rigor in their beauty," Bushnell says. He ended up selling Sente for $3.5 million to Bally's Midway in order to raise cash for Pizza Time. Another failure was a liquid-crystal display business. ACTV was sold, although Bushnell retained some interest. Axlon, the toy company, went public the year that seventeen other toy companies went bankrupt—primarily because kids quit buying toys and started buying video-game systems made by a newcomer in the U.S. market, Nintendo.

Bushnell kept Axlon out of Chapter 11 but downsized it and made it into a licensing company. Eventually, an ETAK-like system *will* be in every car as standard equipment, or at least as an option, but Bushnell will not be the one to get rich from it. Rupert Murdoch bought ETAK (for $35 million) in 1985 and licensed the technology to Blaupunkt.

In the early eighties, at the height of Chuck E. Cheese's success, Bushnell, still fighting boredom, was spending less and less time at the helm of his company and more and more time at the helm of his yacht. "I was living rich," he admits. "A certain amount of hubris developed. All of a sudden you start thinking you can do no wrong. You've created one mega-enterprise. The next turns into another mega-enterprise and you have $200 million. Hell, no sweat. Why can't I buy whatever I want? You get sloppy. I took my eye off the ball."

That is, he went sailing. In 1983, he was racing in the Transpac and was out of touch with the staff that was running Chuck E. Cheese. When his boat sailed into Waikiki, taking first place, there was much fanfare, champagne, and cheering. Eventually Bushnell made it to a pay phone, but he was unprepared for what he heard. Chuck E. Cheese was going down the tubes—the company had to write off a $10 million quarterly loss. Bushnell didn't wait around to pick up his Transpac trophy and flew home, but it was too late. The company's board wanted to bring in a corporate doctor, and Bushnell resigned. Now he says it was the worst business decision he ever made. "I'm very tenacious," he says. "I believe I could have saved Chuck E. Cheese without going bankrupt." Six months after he left, Pizza Time was in Chapter 11 and his millions of shares were worthless. "It was like my luck was all used up," he says.

Chuck E. Cheese was merged with a hotel chain that eventually turned the company around (although the original Sunnyvale location became a dim sum restaurant). Bushnell, however, was left to face some harsh realities. He had to sell his jets, his yacht, and all but his Woodside and Aspen homes. "I was flying much too high to be brought down by a small arrow; it took a three-stage rocket, because I was in the stratosphere," he said. "I believed I could do no wrong." It was time for some soul-searching. He sat in his office in the dark, staring out the window. His cash flow, as he puts it, was tight. Chuck E. Cheese and Androbot were gone. He contemplated the future. "What's next, big boy?" he asked himself.

An industry insider who knew Bushnell believed that he was bitter about his sudden decline. "It's very sad," he says. "Steve Jobs *worked* for Nolan. Nolan should *be* Jobs, or Bill Gates. It eats away at him." Still, Bushnell was philosophical. "I've had a multiplicity of MBAs in the school of hard and soft knocks," he says. "I've had some tremendous lucky breaks and some unkind breaks, but it's never been dull."

As difficult as it was to imagine for many of those who worked with him, by 1990 the godfather of all the Silicon Valley boy-wonder entrepreneurs had to get a job—which is where Commodore came in. They wanted Bushnell on board to sell CDTV.

Meanwhile, the industry Bushnell had founded was thriving. By 1978, Americans were spending more than $200 million a year on home video games. By 1981, the amount had increased to $1 billion. Mattel had entered the fray with Intellivision. In 1982, video-game sales skyrocketed. Atari accounted for half of the revenue for Warner Communications and more than 60 percent of its operating net.

It seemed as if Atari could do no wrong. The 2600 was everywhere; 20 million units were sold, and there were 1,500 games available for it. Activision, Epyx, and many other independent companies began making millions manufacturing game cartridges for the 2600. Coleco came up with its strong competitor, ColecoVision, which promised, for the first time, arcade-quality video games on a home screen. With an expansion module, ColecoVision was able to play all the Atari 2600 cartridges. ColecoVision sold well, partly because it played a home version of one incredibly popular arcade game it licensed from Nintendo: "Donkey Kong."

Soon Mattel and Atari were making ColecoVision games for their own systems. Milton Bradley tried to keep pace with Voice Command video-game cartridges tied to the Texas Instruments 99/4A home computer, but the attempt fizzled. It was the exception in a rapidly expanding market: the home video-game business was bringing in over $3 billion a year.

Arcades, meanwhile, were bringing in even *more:* $5 or 6 *billion,* in spite of a backlash against them. Communities as far apart as Babylon, Long Island, Oakland, California, and Pembroke Pines, Florida, passed ordinances restricting play by teenagers of various ages. The United States' surgeon general, C. Everett Koop, issued a statement indicting video games for producing "aberrations in childhood behavior" and causing users to become addicted "body and soul."

But video games were sweeping the world. "Fascination with the games, often accompanied by cosmic brooding about their presumed bad effect on faith, morals and school attendance, seems to be universal," wrote John Skow in *Time* in 1982. The article reported that games such as "Asteroids," "Defender," "Missile Command," "Pac-Man," and "Donkey Kong" were consuming, in addition to all those quarters, 75,000 man-years in the United States alone.

Yet Nolan Bushnell's timing in leaving Atari was fortuitous. The company fell further and further behind Apple in computers, and then its bread and butter, the video-game business, crashed.

Bushnell winces when he remembers some of the mistakes Atari made under Warner. "The number of horrendous management decisions that went on in that place is amazing," Bushnell says. "A lot of people got involved with the company who really were underqualified, and there was a tremendous revolving door of vice-presidents. The company had been very successful, but nobody really knew why. All they were doing was pumping out cartridges and selling millions of units, but there was no strategic thinking going on."

By 1983, the $3 billion video-game industry had turned into a trickle—$100 million in sales for the entire industry—yet Atari and dozens of other companies were still churning out games by the millions. Bushnell says it was "an absolutely unconscionable screw-up" on the part of Atari that destroyed the video-game industry. "They expected the market to double when all rational thought said that it couldn't. The red ink poured forth." This devastated Warner Communications. Its stock went into a tailspin, plunging the company into a takeover battle with Rupert Murdoch. Steve Ross announced that Atari's troubles were responsible for Warner's announcement of a $283.4 million loss for the second quarter of 1983, "the worst in the company's history and triple even the most pessimistic Wall Street forecasts," according to *The Wall Street Journal.*

Inventory levels were mammoth. Atari built and then bulldozed almost 6 million "ET: The Extraterrestrial" games. Even more astounding, after licensing the game from Namco, Atari built more "Pac-Man" cartridges than there were players.

With the ludicrous number of games in inventory, prices were slashed. During the years of the decline, total unit sales actually increased, but the dollar sales went to a tenth of what they had been. "Atari hit two billion in sales, and that third year they were going for three," Bushnell says. "And that was why they hit the wall running as fast as they could." Bushnell made a substantial amount of money *shorting* Warner stock.

"Very seldom do you have an industry in which the dominant player not only abandons leadership but abandons the *industry,*" Bushnell says. "There was nothing left. Basically they retrenched, retrenched, retrenched, and didn't really try anything innovative. Nobody had enough cash to do anything for a long time. There was not a single innovation in product line at Atari after the day I left. Everything they did was just variation on the chip sets and the business I created. Atari abandoned the game market to Nintendo, pure and simple, and it abandoned the computer market to Apple and then IBM."

The dumping and discounting of cartridges eroded the market, and Atari and Mattel nearly went bankrupt. Coleco did better, but not because of video games. It released Cabbage Patch Kids and sold, in all, more than half a billion of them.

The year Atari recorded losses of $200–$300 million, Ray Kassar, who, as Atari's chairman, was running everything for Warner, needed to find a way to keep things going in a video-game market that had suddenly vanished, as well as in the rough-and-tumble computer business. He tried a number of things, several of them having to do with a certain company that had been doing phenomenal business in Japan.

Nintendo's dealings with Atari began when Atari licensed "Donkey Kong" for its home computer. Kassar and Skip Paul, Atari's senior vice-president, invited Minoru Arakawa and Howard Lincoln to their Silicon Valley offices for a meeting. Arakawa and Lincoln met with a dozen vice-presidents ("of everything imaginable," Lincoln says), twenty executives in all, in the corporate dining room. Kassar boasted that his chef was from one of the best restaurants in the world, and the meal was exquisite.

The meeting paid off; Atari licensed "Donkey Kong" for the 800. And because Hiroshi Yamauchi was pleased with the way the negotiations had gone, one day he called Arakawa and suggested that Nintendo approach Kassar to see if he was interested in buying the worldwide rights to the Famicom. Yamauchi's idea was that Nintendo would sell the machine in Japan, but Atari, which already had a worldwide distribution network, would sell it in the United States, Europe, and elsewhere. The benefit would be more than a per-unit royalty. Nintendo, which on its own might never be a contender outside Japan, would, as a partner of Atari, be able to sell software all over the world.

Atari's 2600 was outdated and the 5200 was going nowhere. There was a rumor that Atari was working on another, more powerful system, the Atari 7800, but Yamauchi was confident that his success in Japan carried far more weight than anything in Atari's R&D labs.

Arakawa made the pitch to Kassar, who decided that his proposition made sense. He either could release Nintendo's Famicom under Atari's name or sit on the Famicom and do away with a potential competitor.

A meeting was set up and Kassar told Arakawa he would send the Warner Communications corporate jet, a Gulf Stream, to collect him and Howard Lincoln. En route to the airport, Arakawa asked Lincoln if he expected lunch to be served on the plane. Lincoln said there would probably be no food on the short hop between Seattle and Sunnyvale. Arakawa was starved, so the two headed to a restaurant before meeting the jet at a private airport.

The jet, fitted with leather couches and gold-plated ashtrays, was empty except for Arakawa, Lincoln, and the crew. Once it was airborne, the pretty attendant set up dining tables with linen tablecloths and asked if the two were ready for lunch. Arakawa threw Lincoln a dirty look when she served pâté, fresh poached salmon, and Dom Perignon.

"Just eat the goddamn food," Lincoln muttered.

When the jet landed in San Jose, two chauffeurs escorted the Nintendo executives down a stairway into a waiting limousine. They drove to Atari's headquarters, where they were led to a conference room. The entire upper-management staff was assembled, from Warner's Manny Gerard and Kassar to Atari's Skip Paul and numerous lawyers and vice-presidents familiar from the "Donkey Kong" negotiations. In the middle of the meeting, Steve Ross poked his head in to say hello. He wanted to apologize for the fact that he had to use the company's jet to go back to New York; he had, he said, rented another jet to take Arakawa and Lincoln back to Seattle.

The meeting began with Arakawa's description of the Famicom. Questions came from Gerard and Kassar, at the far end of the conference table, but also from each of the dozen lawyers and executives. Scribbling notes and fielding most of the questions, Lincoln watched Arakawa. "I can always tell when he understands something or doesn't, when he's pissed off or when he's happy," Lincoln says. "This time Arakawa was just amazed by all those people, all that bureaucracy."

Arakawa and Lincoln left the meeting exhausted and uncertain of how much progress they had made. Back at the airport, where they boarded the smaller jet Ross had arranged for them, they were trying to relax and sort out the meeting when the copilot came back to tell them, "Mr. Ross left you some wine." An attendant served them bottles of a rare Bordeaux. "I don't know how we got home," Lincoln recalls. "We still had a buzz the next day."

The next step was for Atari to see the Famicom. Skip Paul, along with half a dozen Atari managers, joined Yamauchi, Arakawa, and Lincoln in a conference room at NCL in Kyoto. The system was demonstrated by Masayuki Uemura, who explained, through an interpreter, why it was better than any that preceded it.

Yamauchi came and went several times during the meeting. He used this disappearing act as a diversionary tactic: he wanted everyone to believe that he had far more important business going on elsewhere.

At the end of the first day of meetings, Yamauchi had so successfully confused the Atari delegation that Paul called Lincoln at his hotel to clarify some issues. Yamauchi had Atari manufacturing the machines at very low cost. They would have the worldwide rights outside Japan, and they would have software support from NCL. Nintendo would receive a relatively large royalty on each machine.

Throughout the week there was persistent haggling over percentage points of royalties, but Yamauchi was getting everything he wanted. Finally Lincoln used one of Yamauchi's dramatic disappearances to full effect: he said that Yamauchi was growing impatient. The deal had better be sewn up immediately or Yamauchi might decide to forget the whole thing. "You don't want Mr. Yamauchi to become annoyed," Lincoln said.

At the eleventh hour, the Atari negotiators retreated to a private office to telephone Ray Kassar in California, who was in touch with Manny Gerard in New York. Yamauchi came back into the room with Arakawa and Lincoln. Lincoln said, "Mr. Yamauchi, you shouldn't be in here. If the Atari people come back in and you're here, they'll take it as a sign that you are overanxious."

The chairman gave Lincoln a look that instantly humbled the cocky attorney. Yamauchi had his own negotiating tactics; he didn't need an arrogant young lawyer from America to tell him what to do. He remained in the room as the Atari team returned. Paul said that the deal was as good as done and asked Lincoln to write up the contracts. They would all meet again in a month at the June Consumer Electronics Show in Chicago to sign the papers. Yamauchi rose and left the meeting. Handshakes and backslapping signaled that the negotiations had been a success.

Yamauchi flew to Chicago for the CES, which he attended with Arakawa and Lincoln. In the convention center, they walked past Coleco's booth, where the company was showing off its new home computer, Adam, set in an artfully lit glass case. There, playing on the sharp color screen, was "Donkey Kong."

Coleco's stock shot up almost twenty points that day, and Atari was not amused when it saw the Nintendo game playing on its competitor's machine.

Ray Kassar's office sent a tersely worded letter to Arakawa threatening not only to cancel the deal Skip Paul had made, but also legal action. Atari, which owned the floppy-disk computer-game rights to "Donkey Kong," thought that Nintendo had double-crossed them and sold the game to Coleco.

Howard Lincoln arranged an emergency meeting with Coleco's president, Arnold Greenberg, that night in Nintendo's hotel suite. Present when the meeting began were Minoru and Yoko Arakawa, Ron Judy, Howard Lincoln, and, representing Coleco, Greenberg and several of his colleagues. There was also a translator.

Everyone sat around a table. Arakawa whispered to Lincoln, "Don't say anything. Mr. Yamauchi will do this."

Arnold Greenberg, a distinguished-looking man, gray at the temples, ready to celebrate because of the computer Adam's apparent success, asked where Yamauchi was. Yoko assured him that her father would be there in a moment.

Yamauchi entered the room abruptly and, without addressing anyone, stood at the end of the table. He became, as one of those present put it, "unglued."

He began with a breathy, high-pitched tirade in a Marlon Brando monotone and quickly became loud and abusive. With a piercing cry, he swung his arm in an arc in front of him, shooting his outstretched index finger toward Greenberg.

Yamauchi's diatribe, all in Japanese, completely stunned everyone in the room, with the possible exception of the Arakawas. Howard Lincoln says, "It scared the hell out of me."

The Coleco people weren't aware that they had messed up Nintendo's lucrative Atari deal—millions of dollars were in the balance—but they could see that they had somehow incurred Yamauchi's unfathomable wrath. When Greenberg turned to Arakawa for help, he was met with a cold stare. By the time Yamauchi wound down, no one in the room said a word.

The translator finally began to speak. "Mr. Yamauchi is very upset," the man said.

This understatement underscored the fact that the Coleco team could have no recourse but to roll over. The translator continued, calmly reciting the gist of the outburst, but Yamauchi had already won. Greenberg's excuses—he said that Coleco considered the Adam a computer with a video-game machine inside—were feeble. He then tried to turn it on Lincoln, blaming him for "the misunderstanding." This made Lincoln furious; he was about to jump up to respond when he felt Arakawa firmly grasp his forearm, holding him still.

Yamauchi spoke again, never wavering. He made it clear that there was nothing else to be said. No excuses would be listened to. Coleco had to refrain from selling "Donkey Kong" on Adam and announce the mistake, or there would be a lawsuit that would leave nothing of the company. There was no doubt that he meant it.

Greenberg and his colleagues retreated from the suite, shaken. Afterward, at dinner in the hotel's Japanese restaurant, Yamauchi, his tie loosened, turned to Howard Lincoln, who was still in a state of shock, and said, "Sometimes this is the way you have to handle people, Mr. Lincoln. What did you think about that performance?"

The Coleco imbroglio ended up being irrelevant to Nintendo's deal with Atari, which fizzled out of its own accord—mostly because Atari was itself unraveling. A month after the CES, in July 1983, Ray Kassar was axed from Atari.

In September, Manny Gerard hosted a meeting between Atari, Nintendo, and Coleco in his office at Warner in New York. Skip

Paul and his staff came in from California. Arnold Greenberg and the Coleco bigwigs were there, as were Minoru Arakawa and Howard Lincoln.

Gerard's office, with inlaid wood paneling, had a ticker-tape machine spewing out the latest market numbers and a bank of telephones with some sixty lines. Gerard explained that he was changing the office's decor—some new artwork was expected any day. As Arakawa took all this in, he knew that Warner was also laying off hundreds of employees and losing a fortune each quarter.

At the meeting a tentative compromise was agreed upon— "Donkey Kong" was divvied up so that Atari's deal with Nintendo could proceed—but the issues soon became academic. Coleco's Adam was a disaster and soon disappeared, and Nintendo learned (from an attorney who left Warner) that Atari never had the money to buy the Famicom; the negotiation was a charade orchestrated to tie Nintendo up and remove a potential competitor and perhaps to learn something new about video-game hardware and software. When the message reached Minoru Arakawa that the Atari deal was dead, he thought it was a disaster. He called Yamauchi with the bad news. Potential millions had slipped through their fingers, he felt.

The event, however, was seminal. Years later Arakawa said, "Can you believe that we almost sold the whole thing? If we had, no one outside of Japan would know about Nintendo."

Any remnants of the home video-game business in America all but disappeared. In 1984 Mattel sold off its electronics division. Arnold Greenberg folded Coleco. At Atari, to replace Ray Kassar, Manny Gerard brought in Jim Morgan from Philip Morris, where his background in marketing cigarettes hardly prepared him for the video-game business. Morgan bragged to employees that his seven-year, multimillion-dollar contract gave him the freedom to run the company any way he saw fit.

However, as a former Atari executive told *Business Week,* "Rome was burning and he was fiddling around." Atari reported a $536 million loss in the first nine months of 1983. Games that were meant to be priced at $40 were selling for $4. Morgan consolidated

Atari's forty offices around Silicon Valley to about twenty-eight and killed all but nine of Atari's new development projects. He let go a quarter of the company's employees. But the cuts were too little too late. Steve Ross, whom Morgan once called "the best man with numbers I have ever seen," had had enough. Atari was taken away from Morgan on July 6, 1984. He was never told that Steve Ross had decided to break up Atari and sell its pieces.

Atari's hardware divisions—the video-game systems and computers—were sold to Jack Tramiel, founder of Commodore Business Machines, for $240 million in notes (Warner retained 25 percent of the company). Tramiel believed that his new company, called Atari Corporation, could go up against Apple and Commodore. He had been virtually kicked out of Commodore, and he imagined a sweet revenge. Since Tramiel, who planned to run Atari with his three sons, had no interest in the coin-op business, Warner sold it to Masaya Nakamura, and Atari Games became a subsidiary of Namco. Under the agreement Atari Games could do anything except make hardware or software that competed with Tramiel's Atari Corporation under the Atari name.

"I look at it sadly," Nolan Bushnell says, surveying the devastation of the company he founded. "You can't help but have a certain feeling for a name that you chose out of the universe." He adds, "See, Atari could have been Nintendo and Apple under one roof."

The home video-game business was dead. The consensus was clear: no one in America wanted anything to do with video games.

But Minoru Arakawa, picking through the rubble, noticed that there was one group of people who were oblivious to all the death notices and eulogies. Video arcades were still packed, bringing in more money than first-run movies: billions of dollars. Perhaps, Arakawa wondered, it was not a lack of interest in home video games that had killed off the American industry. Perhaps it was the kind of sloppy business he had witnessed during his glimpses inside Atari and Warner that was to blame.

He decided to find out.

8

ENTER THE DRAGON

"The reason I have this terrific job," a buyer for a toy company began, "is that the guy before me was fired after he lost so much in video games. Do you think there is any way *I'm* going to make that mistake?"

Throughout 1984, Arakawa heard variations on that theme over and over when he met with toy- and department-store representatives to tell them he was considering entering the home video-game business. They thought he was nuts.

Arakawa marveled at the intensity of the hostility toward video games—even the phrase was taboo. In the horror stories about the industry, hyperbole was unnecessary. One of the legions of former Atari vice-presidents (who retreated into his father's pharmaceutical business after the crash) said he watched millions of unsold game cartridges being bulldozed into a landfill. Destroyed careers, divorces, and a suicide were blamed on the Atari crash. "It would be easier," one former toy-industry executive told Arakawa, "to sell Popsicles in the Arctic."

On the other hand, there was no letup in the sales of the Famicom in Japan. Were Tokyo and Darien that different?

Arakawa, Howard Lincoln, Ron Judy, and Bruce Lowry visited arcades, toy retailers, merchandisers, discounters, specialty stores, software developers, former Atari and Coleco managers and executives, and anyone else with experience or opinions about video games. "We kept trying to zero in on what we shouldn't do," Lincoln says. What they most often heard was that they shouldn't do anything at all. But there was a consensus that the "suck factor" was one of the biggest reasons for the industry's crash. The market had been glutted with terrible games. "Pac-Man" was a blast in arcades, but the home version "sucked." "ET," ridiculously hyped, "sucked." "Zombies from Pluto Stole My Girlfriend" *really* sucked."

Bad games such as these would never have survived in the arcades; kids would have tried them and deserted them. But there had been no easy way to test home games. Fancy boxes and expensive advertising campaigns made promises, and when the promises were unfulfilled, the customers stopped believing them. Systems and games went into the garbage.

Arakawa came to realize that it didn't matter how much money was spent on marketing, advertising, and promotion if the games weren't good enough. As a Nintendo slogan later acknowledged, "The name of the game is the games." Arakawa also knew that he had, if nothing else, great games. Sigeru Miyamoto's "Super Mario Bros." and "The Legend of Zelda" would blow these kids away. The question, then, was how he could get them to understand that Nintendo's new system was like nothing they had ever seen.

There was much to be done.

Arakawa felt it was vital that the Nintendo system be distinguishable from its predecessors. He decided it should be clear from the outset that the Nintendo system wasn't a toy. If it was marketed as a more sophisticated electronics product, the company could disassociate from the Atari, Coleco, and Mattel systems. There were other reasons to stay far away from the toy business. December 10 dating, which had helped devastate Nintendo when it was selling Game & Watches, was one. As a consumer-electronics company, Nintendo could take orders, deliver systems, and send bills that were due in thirty or sixty days. The marketing perspective would therefore be broadened: to include mass merchandisers, electronics stores, and discounters as well as the toy chains.

To interest an extended base of retailers, Arakawa wanted the system to be more than a game machine; it should have the capabilities of a small computer. NCL engineers were given the task of developing peripherals, including a keyboard, a music keyboard, and a tape-storage device. They came up with new, high-tech, infrared remote controllers and a cool Zapper gun to play shooting games. All these options indicated that the Nintendo machine was both a giant step forward from the old-wave systems and a new kind of system altogether. Parents would be more likely to buy it because it could do more for their kids (the keyboards, for example, promised educational and cultural value).

The R&D teams in Kyoto modified the system while Arakawa had some of his people in Seattle design the housing and packaging. A young designer named Lance Barr was assigned to make a system that looked high-tech sleek yet accessible. The main computer board and circuitry were nearly identical to the Famicom, but Barr fit them inside a slimmer and handsomer box. Gray and squarish, it looked more like a stereo component than the red-and-white plastic Famicom. The remote-control unit was understated; it could have been featured in a Sharper Image catalogue. The Zapper gun might have belonged to Luke Skywalker, the keyboards were slender and svelte, and the joystick looked like it belonged in a jet fighter. The system was given a name to reflect its maturity: the Advanced Video System, or AVS.

The major headache of counterfeiting also had to be addressed. The problem was the apparent impossibility of making an uncounterfeitable machine or uncopiable software. There was also a related problem. Ron Judy said that to avoid the "suck factor," Nintendo had to have a way of controlling the quality of software released for the AVS. Also, Judy pointed out, if the AVS could run the same games that ran on the Famicom, illegal Taiwanese games would flood the U.S. market. "We need a security system," he said, and Yamauchi and Hiroshi Imanishi set the NCL engineers on the task of creating one.

NCL's attempts to stop rip-offs in Japan, including the periodic system revisions, had been only partly effective. The licensing agreements also helped, but outfits that were going to counterfeit hardware or software didn't care about licensing agreements. Had NCL decided to put a security chip in the Famicom, it might not have lost some of the huge markets of the Pacific Rim. In addition, it might have been able to stop companies like Hacker International from releasing nonapproved games.

The security system the Japanese engineers devised was a complex implementation of a simple lock-and-key concept. The AVS wouldn't work unless a chip in the cartridges unlocked, or shook hands with, a chip in the AVS. The key was a kind of song the two chips sang to one another. If a cartridge was inserted into the machine that didn't know the song, the system would freeze.

Nintendo called the invention a "security chip," but it was referred to in the industry as a "lock-out" chip; it stopped more than counterfeiters because no one could manufacture their own games for the AVS without Nintendo's approval. Only Nintendo had access to the technology, including the specific computer code at its heart. Lincoln had copyright and patent applications filed for the security system.

While the security system was being developed, Arakawa asked Don James to recommend the best NCL games for the AVS. James and Howard Phillips played hundreds of games and gave Arakawa a list of their favorites. Arakawa chose forty and sent instructions to Japan to prepare English-language versions.

The AVS was to debut at the January 1984 Consumer Electronics Show. Don James designed a booth for the occasion—more substantial than the one in which Nintendo had shown Game & Watch. Nervously, James, Arakawa, Lincoln, and Phillips traveled to Las Vegas with AVS demos and boxes full of brochures. "The evolution of a species is now complete," the brochure announced. On the cover was a picture of three televisions. Playing on one was "Pong," a few dreary white lines on a black screen. Playing on another was a color tennis game, roughly animated blue-and-yellow stick figures on either side of a net. The third screen was veiled in a red cloth. Inside the brochure the system, with its numerous peripherals, was introduced. "Ninety percent of the Japanese market won't play anything else. Welcome to the future of American home video entertainment."

The show opened and Arakawa, James, and Lincoln excitedly manned the booth while Howard Phillips demonstrated games.

The AVS looked impressive, the Nintendo representatives were told. But almost all those who stopped by at the booth shook their heads when asked if they would consider placing an order. "The memories of Atari were too recent," Lincoln says.

Although Nintendo tried again at the industry's June show, it was clear that Arakawa had misjudged his ability to overcome skepticism. He hadn't been able to create the new category that combined computer power and entertainment. No one cared about the remote control, and they hated the keyboard—a turnoff to kids, industry executives believed (parents were irrelevant). The AVS had all the problems not only of the video-game business but of computers too. No one would touch it.

Back to the drawing board. Instead of attempting to improve on the video-game systems of the past, Arakawa decided that he should figure out a completely new way to sell it. He scrapped the computer peripherals. Kids wanted fun, not BASIC programming languages and cassette-tape storage drives. They tossed out the keyboard, the piano keyboard, and the remote-control unit as well as the name. R&D 1 was put in charge of a new peripheral that would make the system something other than a video-game machine.

In Japan, Gunpei Yokoi's team came up with ROB, or Robotic Operating Buddy. He was one foot high, gray, legless, and he really didn't do much. He was controlled by the video-game system. The flashing of the television screen activated a chip in ROB's head that caused him to move. Players controlled him at the same time they controlled action on the screen. In games designed for ROB, such as "Gyromite" and "Stack-Up," players would cause the robot to pick up chips from one stack and drop them onto a pad that triggered a door in an on-screen game. More than anything,

ROB looked cool. He would be used to sell the video-game machine.

James and Barr worked on a new design for the system—again high-tech gray, but boxier. Game cartridges slid into the front instead of the top, and revised controllers were attached by plastic cords. It wasn't toylike—it still looked like a consumer electronics product—but it was simpler than AVS. Nintendo de-emphasized the box in favor of ROB and the Zapper gun.

At the June 1985 Consumer Electronics Show, Nintendo debuted what Arakawa had renamed the Nintendo Entertainment System, or the NES. The operative word was entertainment. Everything Nintendo would do to sell the machine would emphasize this.

The reaction at the new show was somewhat better. Buyers liked ROB. Still, they were reluctant to place orders.

Arakawa stubbornly ignored the reaction. He said that the people in the industry were jaded. Kids would love it, he believed. To prove it, he commissioned focus-group studies in New Jersey. From behind a one-way mirror, he watched a random sampling of young boys play the NES and heard them say how much they hated it. Typical was the comment of an eight-year-old: "This is shit!"

Depressed, Arakawa wondered if he should give up, and in a conversation with Hiroshi Yamauchi, he said as much. Yamauchi denounced such fatalism. The market in America wasn't that different from the Japanese market, he said. "But the tests show . . ." Yamauchi interrupted him. "Ignore them," he said. "Try to sell the system in one American city. Then, if it fails, it fails. But we must get it into the hands of the customer. That is the only test that matters."

Arakawa, Ron Judy, Howard Lincoln, and Yamauchi considered the location of the test. Judy thought they should start a limited test in a small town, but Yamauchi shook his head. "What is the most difficult town to start in?" he asked.

The answer was obvious: New York City.

Yamauchi asked why.

Besides the obvious hurdles of New York's competitive market, it also had been hit the hardest by the crash of the industry in 1983. What's more, much of the excess inventories had been dumped there—not to mention that New York had the most savvy and cynical buyers in the country.

Yamauchi said that New York was where they should go. He gave Arakawa a budget of $50 million.

In late summer 1985, Arakawa leased a warehouse in Hackensack, New Jersey, bordered by a railroad and a cemetery. There were no windows. The only light in the cavernous room was from a few dangling naked bulbs, and it was spooky and depressing.

Arakawa brought about thirty NOA employees East. Ron Judy and Bruce Lowry were the first to arrive. Then a deputation of twelve more—they called themselves the SWAT team—flew out and landed in Newark in the middle of a hurricane.

At the terminal, twelve rented cars awaited the team. They traveled in a shaky caravan to the warehouse, which was flooded. As the group surveyed the dreary headquarters, Arakawa cheered them on. "If we can just get players to see it, it will be *really big,*" he said. "I know we can do it. It's a big job, but everything worthwhile is difficult. We just have to get it to the players. If we do, it will be *really, really big.*"

He got through to them. An ebullient chorus came back: "Yeah! It *will* be big."

Arakawa said, "It will be *really* big."

They echoed, "Really, *really* big."

Other employees arrived over the next month. From Japan there was Shigeru Ota, who did the books, and a technician, Masahiro Ishizuka. Cindy Wilson was an executive assistant. Others included Rob Thompson, who became the service manager. Howard Phillips, who had become shipping-warehouse manager of coin-operated games, was flown in. "We felt like the elite point team," Phillips says.

Judy, Ota, and Ishizuka lived in a New Jersey house Nintendo had rented. Their furniture consisted of wooden boxes and suitcases that they never entirely unpacked. The house doubled as a place to store spare parts, and it also became Nintendo's after-sale service center. Other employees lived in rented condos and apartments, two to five to a place, all furnished with junk-sale furniture and mattresses on the floor. Rob Thompson, Howard Phillips, and

Don James lived together in a townhouse in Fort Lee. Howard Phillips woke up every morning at six and made his way to the shower, where he sang opera. Thompson and James threatened to murder him.

The various groups carpooled to the warehouse in the morning and worked all day, breaking in the evening for dinner at a neighborhood coffee shop. Then they returned to work and stayed late into the night.

Howard Lincoln sent posters of Seattle for the SWAT team to put up in the warehouse so it wouldn't be so dreary. He and Arakawa, shuttling between Seattle and New York, made calls on retailers, as did Bruce Lowry. Judy met with advertising agencies and planned a promotion campaign. Other SWAT-team members manned the telephones, trying to convince buyers for large and small stores to see them. They pressed shopping-mall managers to allow them to demonstrate the NES. Arakawa used some of his budget to sign up professional athletes for the demonstrations. Mall managers were far more open to the idea of having famous ball players come by than businesspeople representing an unknown company with a strange Japanese name.

In October the push began in earnest. In pairs, the SWAT team hit the pavements, visiting department stores and large and small toy and electronics retailers. They worked to convince companies such as Toys "R" Us, Sears, Circuit City, and Macy's. Although Charles Lazarus, founding chairman of Toys "R" Us, and a very few others were receptive, most people could not pronounce *Nintendo* and were not interested in learning how.

Arakawa realized that the only way around the retailers' reluctance was to make it a risk-free proposition. Yamauchi, however, had trouble with this idea. He couldn't understand why he should offer a complete money back guarantee; Nintendo had never had to operate from a position of weakness. Arakawa argued that the ultimate sign of strength is to have so much confidence in your product that you would almost pay stores to carry it. Yamauchi didn't need to tell him how risky the tactic was.

Nintendo, Arakawa announced, would stock the stores and set up displays and windows. Nobody had to pay for anything for ninety days. After that period, stores would pay Nintendo for what they had sold and could return the rest. It was an offer store buyers couldn't refuse, although it was still greeted with skepticism. Then, one by one, companies agreed. "It's your funeral," one buyer said.

Many of the Nintendo team worked eighteen-hour days, seven days a week for the three months that preceded Christmas 1985. Ron Judy would load up a couple of rented vans with Don James's displays and drive to Long Island or Westchester County or some other New York suburb. In malls the team set up colorful booths with twelve or so monitors, each one attached to an NES. They arrived in the middle of the night and worked until four o'clock in the morning setting up the displays, then drove home to sleep for a few hours. The next day they headed back to the malls, where they stood next to Mets stars who were signing autographs and tried to get passersby to listen to their spiel. Mookie Wilson and Ron Darling even played NES baseball, projected on a large-screen TV.

"The trick was to get people to come over," Arakawa says. "If we could get it in the hands of the consumers, they would be convinced."

Howard Phillips turned out to be one of the team's best spokesmen; he had a knack for communicating his own enthusiasm, punctuating his sentences with words like *cool* and *neat*. He grabbed kids, old ladies—anyone—and before his victims realized what had hit them, he had them playing.

At one mall, the Nintendo team spent all night setting up its booth to be ready. Just as the crowds poured in, the mall's director came over and forbade them to turn on the games; "They attract the wrong sort of crowd," she said. She had only wanted to meet the baseball stars.

Days such as this made the Nintendo employees wonder what they were doing. "We're overworked, underpaid, we don't see our families, nobody wants what we're selling—what's the point?" they whined. Ron Judy would buy them all dinner and reassure them. "Just wait," he said. Arakawa would pat them on the back and cajole them: "It will be worthwhile. It will be . . ." They finished his sentence without enthusiasm—"really big. We know."

It was an uphill climb. Even with the guarantees, fancy in-store displays, and the promise of a $5 million advertising campaign, it took three sales calls to win over most stores. A buyer would finally

be convinced, but then a merchandising manager would say a flat
"No way." When he was convinced, a vice-president would say no.
But the Nintendo team was persistent, and more stores agreed.

The advertising campaign and press relations were run by Gail
Tilden, under the supervision of Ron Judy. Tilden had come on in
1983. She was a brunette with long hair, bangs down to her eyebrows, and gray-blue eyes. She was a combination of Annie Hall
looks and self-confidence; she looked down, stumbled over words,
and charmed almost everyone she met.

After a year at Britannia Sportswear, Tilden worked for a small
Seattle advertising agency. When a former boss of hers left Nin-
tendo to have a baby, she recommended Gail to replace her.
Within a year she was Nintendo's ad manager. With input from
Judy and Lowry, she hired an ad agency in early August 1985.

Although the Nintendo executives didn't know much about advertising a video-game system, they had learned from the company's botched effort to make commercials for Game & Watch to
trust professionals: no employees were in ads.

Tilden instructed the agency in the rules of NES advertising, all
designed to disassociate Nintendo from Atari. No-no's included
the use of the term *video game;* this was an *entertainment system.*
Software was never to be described as game *cartridges,* another
word associated with Atari. At Nintendo they were game *packs.*
The NES itself wasn't a *console* but a *control deck.*

The ads the agency came up with emphasized the variety of
games and featured ROB and the Zapper. It wasn't easy to advertise video games on television, because watching kids play the
games was about as exciting as watching someone read. The excitement was internal. They came up with ads that tried to convey the
feeling of video games: energy, color, danger, irreverence. In the
commercials, houses blasted off into space and kids explored a
spaceship where the control panels were game screens. The voice-
over asked: "Will it be you? Will your family be the first to witness
the birth of the incredible new Nintendo Entertainment System?"
A light of the sort in *Close Encounters of the Third Kind* showed.
"Now," the announcer exulted, "you're playing with power."

As promised, Arakawa began to blitz the New York area with
television advertising. Meanwhile, Tilden met with members of the

trade press, the reporters who covered the toy and consumer-electronics businesses. They were skeptical, even as they acknowledged that Nintendo's product had better graphics and games than the systems of the past; they simply didn't believe the company would be able to reignite consumer interest. Tilden tried to convince them that video games were an entertainment category, just like VCRs and stereos, but the reporters shook their heads; they had heard it all before. She explained the quality-control measures Nintendo had taken in order to prevent the market from becoming saturated with bad games that had plagued the industry in the past, but the journalists had stopped listening.

A second hurricane that swept through Hackensack one morning was a fitting metaphor. Through the warehouse door, the exhausted Nintendo staff watched the rain falling horizontally, from left to right. It became sunny for a while as the eye of the storm went over them. Then the rain poured down again, this time from right to left. Don James took photos.

Still, the extraordinary efforts began to pay off. A growing list of stores placed orders. But there was no time for the Nintendo staffers to congratulate themselves; success meant more work. Phillips, who managed the warehouse, received containers filled with systems from Japan and shipped them out to retailers. James had a team building window displays he designed for stores that had agreed to feature the NES for the holiday season. Everyone helped build them. One night, Arakawa and Lincoln raced Don James and Howard Phillips to see whose team could build more displays.

There was never a respite from the pressure, although there were some gratifying moments. When Howard Lincoln walked across Fifty-ninth Street in Manhattan and reached the FAO Schwartz store, he was stopped in his tracks when he saw the window display he and a few others had spent the previous night assembling. Midtown was crazy with holiday shoppers, and Lincoln got so excited he called Hackensack and insisted that the staff join him. Soon the members of the SWAT team were crowded together, standing in front of the toy store, staring at their window as if it were a barn they had just raised.

Howard Phillips, on the other hand, had a rather unexhilarating experience late one night while setting up a display at a Toys "R"

Us in New Jersey. A security guard on the graveyard shift came over and struck up a conversation. When he saw the video-game system, he asked, "Are you from Atari?"

Phillips explained that he was from Nintendo with a new and better system.

The guard said, "You're working for the Japs? I hope you fall flat on your ass."

They worked until the day before Christmas. Between 500 and 600 stores were selling the NES. The team members were spent. A group of them summoned the energy to drag themselves to Newark Airport to fly back to Seattle to spend Christmas with their families, but their flight was canceled because Seattle was fogged in. Most of the SWAT team spent a lonely Christmas in New Jersey.

The advertising and mall tours succeeded in building interest in the NES, and stores were racking up sales. The New York test wasn't quite as successful as Nintendo had hoped, but half of the 100,000 systems shipped from Japan were sold. Most important, retailers had decided that Nintendo had a viable product. It was enough to justify going forward.

Los Angeles was next. It was a tougher sell there because of the time of the year. They hit stores in L.A. in February, a bad month for retailers, particularly for toys. Still, enough systems were sold to encourage Arakawa. The L.A. sales were slow but steady, and retailers were enthusiastic for the most part. The team continued —to Chicago, San Francisco, and several Texas cities before going national. By the end of the first year, a million systems had been sold in America.

It was still slow going. Stores remained reluctant to commit much to video games, and most people in the industry assumed that Nintendo's limited success was a temporary aberration. In the second year, however, the company sold another 3 million systems.

When Arakawa had ventured into the food business with Chuck E. Cheese and the other restaurants, he got advice from Peter Main, his old friend from Vancouver, who had a wealth of experience in marketing and restaurant management. Nintendo's restaurants were making a profit, and Arakawa tried to convince Main to resign from his vice-presidency at General Foods and join Nintendo. He wanted Main to oversee the restaurants and help orchestrate what Main would later call Nintendo's "Invasion of Normandy," the NES launch.

Main, balding but with a trace of thinning gold hair on the sides of his head, wore big round glasses with thick root-beer-colored frames. Behind the glasses were dark eyes, jovial and frank. He worked in an office strewn with baseballs, a Hula-Hoop, and an electric train set.

A Canadian, Main had worked for years for Colgate-Palmolive. He headed their new-products group in Canada. He had spent years coming up with ways to convince people to buy Colgate toothpaste and various other products. In those businesses, a fraction of market share meant millions of dollars. "You had to be a street fighter," Main says. "You had to beat them on the curbs."

Main moved on to General Foods in Canada and managed its restaurants, including Kentucky Fried Chickens, steak houses, Burger Chefs, and White Spots. He also went into business for himself before returning to General Foods at about the time "this neat young Japanese couple" bought the house across the street. "It looked different from the other houses," Main says. "The shoes were outside the front door." Soon he got to know the Arakawas— "very, very warm people, even though their verbal skills were not considerable at the time."

In 1980, Arakawa told the Mains that he was quitting his job and going to work for his father-in-law in America to set up in New Jersey something called Nintendo of America. Main thought it was a pinball factory.

The Mains heard from the Arakawas through notes and Christmas cards and saw Minoru when he dropped by to check in on the property he still owned. Main later heard from Arakawa when he and Yoko had moved to Seattle after Nintendo returned West. In early 1982 Main helped Arakawa get into the restaurant business. Five years later, when Nintendo was completing its market tests for the NES, Arakawa asked Main to come work at Nintendo. Ron Judy had been in charge of marketing, but he was going off to Europe to promote the NES there.

Main knew nothing about the video-game business, so he studied all the documentation he could find about the Atari crash. "Nintendo," he concluded, "had a better mousetrap and a commitment to do this thing right." He came on as vice-president of marketing, with responsibilities that included advertising, promotion, distribution, and merchandising.

Main first sought to improve Nintendo's relationship with the retail community. He also began an assault on Wall Street, meeting key analysts who follow the toy and electronics businesses. With the success of the tests behind them, sales could have been expected to at least slog along, but there was no guarantee of bigger numbers or market longevity. Large discount chains and department stores were still not convinced. Space in their stores was at a premium; to minimize risk, they put Nintendo in only a few stores and carried just a handful of games. The toy stores were coming aboard, but the retailing base was still tenuous.

A forceful salesman, Main met with analysts and gave them "a hot tip." Analysts were always on the lookout for the next big thing. They were also looking for companies with strong balance sheets that might employ their organizations when it came time to offer stock. Nothing was known about Nintendo when Main sat down with these people, one at a time, to pitch them. He presented them with background on NCL, its history, and its financial status —a balance sheet with no debt. This caught their attention, and so did the numbers: Nintendo had a lock on 90 percent of the thriving industry in Japan.

The analysts checked with their counterparts in Tokyo, and after corroborating the information, talked to retailers. Having the top analysts asking about Nintendo gave the company a credibility it had never had in the United States. When analysts heard that other analysts were talking about Nintendo, it confirmed that they were on to something. It was a chain reaction. When buyers at Circuit City, Babbages, or the other electronics retailers told analysts that they weren't carrying Nintendo—that the electronic-game business was not in video games but in computer software— they heard back, "Are you *crazy*? These guys are already selling more of one title than you sell of all computer-game titles put together!" The buyers checked with other analysts and asked about the company. When Main went to Sears, a vice-president told him, "Funny you should mention Nintendo. I was just at an investment-analysis meeting and there were people asking me what we were doing about Nintendo." Sears was one of the toughest sells since it had been burned more than other companies by a huge investment in Atari.

The sleight-of-hand worked. Sears signed up, and Circuit City and Babbages (with its two hundred software stores) did too. Wall Street was inundated with stories about Nintendo. Kmart and Wal-Mart, conservative and cautious, expanded their commitment. To compete, the toy companies began buying more Nintendo products. "It became a self-fulfilling prophecy that something would happen," Main says.

In 1988, 7 million more NES units were sold, along with 33 million game cartridges. Two Nintendo games—"The Legend of Zelda" and "Mike Tyson's Punch-Out!"—sold 2 million apiece just as "Super Mario Bros. 2" was released (the original "Super Mario Bros." was included with the NES).

By 1989, there would be an NES unit in one out of every four American homes. By 1990, one third of American homes would have one—more than 30 million of them. By 1992, the video-game industry would be thriving again—it passed $5 billion in retail sales that year—and, for all practical purposes, it had all been Nintendo's doing.

If Nintendo had been an American company playing by the rules such companies follow, it would have given up long before there was any indication of success—that is, after Arakawa's original market surveys, when the AVS failed, or when there was resistance at the first trade shows. Many American companies are so wedded to market research that the devastating results of focus groups have signaled death knells. Had Nintendo been American, the company would probably have retreated when retailers in New York declined to place orders, or when it took more than a year for big sales numbers to appear. But commitment to an idea and pure tenacity are inherent in Japanese business philosophy—and certainly to Japanese business successes.

Arakawa's perseverance was vital—"I learned to set a goal and to do what is necessary to reach it"—but even more important was Yamauchi's commitment to back him. The money poured forth—more than the original $50 million Yamauchi had committed. He could afford to spend vast sums on the new product even if it meant fiscal quarter after fiscal quarter of weak profits or even losses. A CEO of a public company in America with stockholders to answer to four times a year would probably have withdrawn. Quarterly profit-and-loss statements do not tell the long-term story of a company, however. Heads of Japanese companies answer to investors, but they are not under pressure to deliver high dividends or dramatic short-term growth. The structure allows company heads to work toward long-term growth; they are not forced to abandon a strategy today because it didn't pay off yesterday.

Nintendo's success was proof of the superiority of a system that allows long-term commitment. This feature of Japanese business was one reason why Japanese companies ended up with almost 100 percent of the video-game hardware business, just as they had most of the television and VCR business and were on their way to having most of the business in products from flat-screen displays to certain high-capacity memory chips—all of them technologies pioneered in America.

As Arakawa succeeded in his conquest of the American market, he would be attacked by competitors and American politicians. Nintendo prevailed, they would charge, because of illegal practices, from price fixing to un-American monopolistic control to intimidation of retailers. But the companies (and the economy) that suffered because of Nintendo's success would have been better served if they had struck out against the American system which allowed Nintendo to stroll into a market that had been all but destroyed.

Still, Japan-bashers were correct in one sense. If the playing fields in Japan and America had been equal, Nintendo might never have destroyed the American competition. Even in its heyday, Atari never had much success in the Japanese consumer market. The Atari 2600 sold for about $120 in America, but by the time it reached Japanese consumers, after traveling through the trade barriers of middlemen and the many-tiered distribution system, it cost the equivalent of $380. At that price, few sold. Other American companies wrote Japan off, settling for small profits by licensing video-game systems to Japanese companies. Magnavox, for one, sold the rights to its Odyssey system to Nintendo. On the other hand, if the 2600 had been priced competitively in Japan, Atari might have become the standard there, as it had in America. Nintendo might not have been able to undersell Atari by so much, and so might never have tried to compete. The Famicom and NES might never have been developed.

After the Atari crash in America, almost nothing happened in the home video-game industry until Minoru Arakawa came along. Atari was in such bad shape (and had such a bad name) that its follow-ups to the 2600 sold a trivial number, and there remained no other American competition to speak of. The personal-computer companies could have come in at this point, but they weren't interested in that market. As a result of this miscalculation, Nintendo was soon making more money than Apple.

Arakawa also endured because he didn't care what anyone thought. Analysts rolled their eyes, but he refused to be dependent on the American industry's narrow view of the market. He kept slugging away because he believed, correctly, that kids in America were very much like kids in Japan. There were minor differences— gun games were popular in the United States, but not in Japan, while role-playing games did better in Japan—but the kids were similar enough to form a market that would buy more than 75 million Nintendo systems in the two countries by 1992.

The grumbling heard throughout the American industry could not diminish the fact that Nintendo did certain things better than companies in *any* industry. The company's products were good and the backing from its Japanese parent company was crucial, but Nintendo still would never have gained its enormous sales without phenomenal marketing—"the kind that America had never seen before," according to a competitor.

Peter Main and Arakawa led a multiphased assault that was meticulously planned and flawlessly carried out. Nothing was left to chance. Through the late 1980s, the company launched ad campaigns and the first organized merchandising program with interactive displays in stores throughout the United States. Anyone passing by a Nintendo display could stop and try it. TV commercials had piqued kids' curiosity, and soon anyone shopping with their children was dragged to Nintendo displays in stores. Once kids tried "Super Mario," Nintendo was put on Christmas lists. Peter Main wanted an even greater in-store presence. To get it he decided to bring in a professional merchandiser, John Sakaley, who knew the toy business inside out.

Sakaley had begun his career as a carpet buyer, then changed to become a toy buyer. He ended up working for Kenner, under Bernie Loomis, the company's well-known and respected president. Under Loomis, Sakaley formed Kenner's first merchandising department and introduced a series of innovations, including an approach begun by Mattel: stores within stores devoted to a single product (there was a *Star Wars* store in toy departments, with action figures, space vehicles, posters, and the like).

Eventually Sakaley left Kenner to become the group director of the retail sales force for the toy division at General Mills. Then Bruce Donaldson, the vice-president of sales for NOA, hired him.

When Sakaley was hired, he focused on developing a merchandising force that headed into the trenches and called on stores to make certain that the NES was prominently displayed. Eventually Toys "R" Us would feature full rows of Nintendo merchandise, and Macy's would incorporate NOA's ambitious store-within-a-store, The World of Nintendo.

To get stores to invest in huge Nintendo displays, Sakaley initiated the "merchandise-accrual fund." For each piece of Nintendo hardware or software purchased, the retailer was credited with a specific amount in a fund—a quarter for an NES system, a dime for a game—that was used to purchase displays Sakaley's staff created. Retailers' credits toward their merchandising-accrual funds doubled when they agreed to have a World of Nintendo. Of course this benefited Nintendo at least as much as the stores.

Eventually, 10,000 retail outlets had Worlds of Nintendo, where they showcased a growing cornucopia of products, all of which carried the Nintendo Seal of Quality, an idea Ron Judy had come up with.

Nintendo displays were elaborate. At some locations, laser-light beams shot through the air. Silver-metallic and fluorescent-yellow pipes and tubes snaked over and around girders. It was as if you were *inside* a Nintendo game. The displays won awards from the Point of Purchase Advertising Institute (POPAI) several years in a row.

This mammoth effort resulted in strong NES sales, but Arakawa saw that they were still held back by an inadequate distribution system. Some chains had signed on but still ordered cautiously. Others remained unconvinced. To get the holdouts, it seemed beneficial to hook up with a distribution network that already had a presence inside the stores.

Don Kingsborough was a legend in the toy business. He had been with Atari before founding Worlds of Wonder (WOW) to sell Teddy Ruxpin, a mechanical bear that told stories. Teddy's mouth moved when prerecorded tapes played on his built-in cassette player.

Teddy Ruxpin was the most popular toy around for a couple of Christmases; retailers wanted all they could get. To service them, Kingsborough developed a large, efficient distribution network.

Arakawa met with Kingsborough and made a deal that benefited both Worlds of Wonder and Nintendo. Joining forces with Kingsborough's group gave Nintendo immediate marketplace muscle. Teddy Ruxpin had brought in $93 million in revenues in 1985, its first year, and more than $300 million by the end of the next year. Overnight WOW was worth $550 million. By convincing Kingsborough to distribute Nintendo beginning in late 1986, Arakawa got presence and credibility. WOW had relationships with most toy, department, and discount stores. For its part, the Worlds of Wonder network gained a large new business servicing the Nintendo account. Revenues from the deal helped Kingsborough to expand.

Although the WOW operation did increase Nintendo's presence in the retailing world, Sakaley came to feel that Nintendo still wasn't getting enough "bang for the buck." He felt Nintendo could do a better job with reps committed *solely* to Nintendo, not to Teddy Ruxpin as well.

Sakaley discussed this with Arakawa, who gave him the go-ahead to start his own merchandising force. Sakaley had already started to organize it when, in the fall of 1987, Worlds of Wonder began to fall apart. The Teddy Ruxpin fad had run its course, and the company's costs were out of control. WOW had an assumed debt of $200 million and a tremendous inventory. "Worlds of Wonder has been in a world all its own," a toy-industry analyst said. Although a private investment group took over the company, Nintendo was told on a Friday in October that WOW would not be able to continue with its field-merchandising service.

Arakawa convinced Kingsborough not to lay off his field representatives for seventy-two hours. During that time he had Sakaley explore Nintendo's options. Sakaley and an assistant worked through the weekend, and on Sunday night he called Peter Main. In the morning, when the two men sat down with Arakawa, Sakaley told him that Nintendo should take over the WOW organization. After a call to Kingsborough, Arakawa had Sakaley hire the WOW reps as they were fired from WOW.

Sakaley had an instant force of 100 people and he hired fifty more. The new force carried cameras and photographed the Nintendo displays. Sakaley was able to send someone out almost immediately if a major store wasn't doing its part. The reps were also equipped with Panasonic hand-held computers with telephone modems that could relay sales information back to the main office. "Back in the Colgate days, it used to take two months to get a report after the fact to find out the mistakes you made—which you were compounding and which were leading you down the wrong road," Main says. "We were getting it daily."

Japanese companies in the automobile industry used efficient just-in-time inventory management systems. Essentially this meant that companies bought parts they needed only when they needed them. The merchandising feedback loop allowed Nintendo to instigate a kind of just-in-time inventory policy so that it ordered only what it needed as it needed it from NCL. Likewise, NCL could thus avoid over- or underproduction. Neither NCL nor NOA had to tie up money in inventory.

Main's and Arakawa's web sought not only to ensnare customers but to keep them. Nintendo encouraged customers to send in warranty cards with contests to win game cartridges. U.S. contest laws soon made it too complicated, so a new incentive was developed:

anyone who sent in a warranty card became a member of the Fun Club, whose members got a four-, eight- and eventually a thirty-two-page newsletter. Seven hundred copies of the first issue were sent out free of charge, but the number grew as the data bank of names got longer.

From the success of the magazines in Japan, Nintendo knew that game tips were an incredibly valuable asset. The bimonthly news-letter's crossword puzzles and jokes were fine, but game secrets were the most valued. The Fun Club drew kids in by offering tips for the more complicated games, especially "The Legend of Zelda," which had all kinds of hidden rooms, secret keys, and passageways. In the newsletter a secret code was revealed that led players of "Mike Tyson's Punch-Out!" to the last level—a bout with the champ. Without the code, it was extremely difficult to reach Tyson.

The Nintendo frenzy had now begun in earnest. Kids were com-peting with one another to finish games. When Nintendo's Red-mond switchboard received telephone calls from players who wanted tips, they too were enrolled in the Fun Club.

The mailing list grew. By early 1988, there were over 1 million Fun Club members, and this led to Arakawa's decision to start *Nintendo Power* magazine. In Japan, Nintendo had allowed other companies to make fortunes from magazines devoted to the Famicom. In America, with its long list of potential subscribers, Nintendo would keep the money and the control itself.

Unilaterally Arakawa decided that the magazine would take no advertising. His colleagues told him he was crazy—he was turning his back on a potential gold mine. With that subscription base, Nintendo licensees and companies with products geared for kids would have paid top dollar.

Arakawa, however, was emphatic, insisting that the magazine would be purely editorial. Companies would not be able to adver-tise bad games, as they had in the Fun Club. Of course the "edito-rial" content of *Nintendo Power* was really one long Nintendo advertisement—stories about game characters, lists of kids' high scores, and loads of maps and charts, as well as lots of game tips.

The company brought in a firm specializing in direct-response research to decide how to market the magazine, but after paying for a survey and a lengthy computer printout of advice, Arakawa threw the data away. "All the kids really have to do is feel the magazine, look at it, touch it, and understand it," he said, "so what I want to do is mail the magazine free to the people whose names we have. *Then* they will buy it." When he was told that this would cost $10 million, he was undeterred.

Gail Tilden had recently left the company to have a baby, but Arakawa wanted her back to run the magazine. Yoko said, "There was so much male power there that he needed Gail to diffuse it." Arakawa, who found Tilden talented, perceptive, and dedicated, called and convinced her to come back to work, although her baby was just a few weeks old. She prepared the first issue in January 1989, and it was sent out to all 5 million names on the data base.

There was something bordering on the insidious about *Nintendo Power*. Kids paid $15 for twelve monthly issues, which covered most of the costs of the magazine. The other costs, including mailing charges, were paid through the marketing budget. From the original mailing, 1.5 million people sent in $15 to subscribe. It was an audience that experts in the magazine business had almost written off: they didn't read and they had a million better things to do with $15. Nonetheless, *Nintendo Power* became the largest-circulation magazine for kids in America by the end of its first year.

Power had what Peter Main called an "ability to pre-sell product." It was as if Universal Studios owned *Premiere* magazine and other print media devoted to movies. Universal could then decide, well in advance, to trumpet a particular coming movie, building anticipation. As the movie neared completion, it could make grander and grander announcements. Just as the film was to hit the theaters, it could announce that it was the most incredible movie ever made, and that anyone who didn't see it immediately was missing the event of the season. The publications would then tell readers how much everyone loved the movie and push any holdouts to see it, all the while creating enthusiasm for the *next* Universal movie.

Power meant that Nintendo didn't have to waste money developing hundreds of games. It could develop a select few each year and be all but guaranteed that the games would sell at least a set minimum amount. Any advertising beyond the magazine was gravy, since *Power* guaranteed that Nintendo was in touch with millions of its most dedicated customers, enough people to create word-of-mouth demand for a game.

Editor Tilden assembled materials for each issue with the help of Howard Phillips, the most enthusiastic in-house player, a contributing editor, and a character in a regular comic strip. Planning a new issue, the two went through games that were coming out to determine how much and what kind of coverage they merited. There was no pretense of editorial independence; Arakawa, Main, and Lincoln approved the selections. The best games (or the ones Nintendo wanted most to sell) were covered in spread after glossy spread of maps, galleries of characters, and player tips.

"People sometimes just take kids for granted or act like they're really dumb," Tilden says. Nintendo did its best to speak to kids as peers. The voice they devised was perfect. The prose was a cross between the dialogue in *Wayne's World* and a Pee-wee Herman routine. *Nintendo Power,* in 1990, was scripture for up to 6 million readers a month. "Parents who complain that their kids don't read should pay attention—kids pore over every word of *Nintendo Power,*" according to Howard Phillips.

Tilden worked with a Japanese publishing company that designed and printed the magazine. Editors from Japan came to Seattle to discuss each issue. Thereafter the design teams put together game maps and layouts. Young writers hired by Tilden, many from within the company, wrote enthusiastic descriptions of the games and provided tips that often came directly from the game designers.

The magazine published Polaroid snapshots of the screens of kids' televisions to prove their outrageously high game scores. There were comics based on games, and a celebrity corner where kids learned that, for instance, Jay Leno loved to play "Contra."

Eventually twelve issues a year weren't enough to meet the insatiable need for insider knowledge, so additional quarterly *Nintendo Player's Guide* books—entire magazines devoted to a single game —were released. Some were given away as premiums to encourage kids to renew their subscriptions. "We have found that more and better information whets players' appetites for more games and accessories," Peter Main says. "Hungry consumers make happy retailers."

There was an unprecedented number of hungry consumers. Nintendomania was sweeping the United States. NOA's switchboard operators were deluged with calls from kids. They wanted more information than they could find in *Nintendo Power* magazine, things like how to set up their NES machines, so Peter Main came up with a way to take advantage of the calls. "The phone system is really the closing of the loop in a fashion that no other consumer company in this country has been able to do," he says. Phil Rogers, who oversees the consumer service department, says, "When we started, what did we know about consumer service? Not a damn thing. We knew about service to distributors because we'd been doing that for arcade games, but we'd never even talked to a consumer." When Main decided to set up the phone lines, Rogers figured that they needed four operators. They began with some six-button phones in January 1986.

Calls flooded in, so many of them that Rogers bought a $40,000 electronic call distributor in 1987. Within a year there were 550 people answering 150,000 calls a week on a new, $3 million phone system. Customers called an 800 number to reach consumer service representatives. If a store was out of a highly sought-after game, service representatives could advise a caller on its availability. Representatives took the callers' zip codes, and by accessing a data base, could tell the callers where games were available. Callers' names and addresses were added to the mailing list.

Many calls were from kids asking how to get past tricky villains in games. Consumer reps transferred the calls to Howard Phillips or other game players who worked for Don James. Some callers spoke Spanish and French, so bilingual representatives were hired.

The telephone company informed Rogers that their 800 number was backed up most of the time because half a million calls were coming in every week. Nintendo decided to initiate a 900 (pay-per-call) number for kids to reach the Captain Nintendo Hotline for tips and adventure stories about Nintendo games. More important,

Nintendo initiated game counseling. Kids by the hundreds of thousands called a separate 800 number with questions about games. Counselors manned the phones from 4:00 A.M. till 10:00 P.M.—to catch the early-morning calls from New York and the late evening calls from California—seven days a week. Hundreds of game counselors huddled in partitioned work spaces, each equipped with a Nintendo system and stacks of games, a computer terminal, notes, and "green bibles," bound volumes of game maps and secrets.

Many of the counselors were lured by a "PLAY GAMES FOR A LIVING" ad in the *Seattle Times*. When they were hired, they were given an NES and stacks of games to master. "It was better'n Christmas," says Phil Sandoff, one counselor. He and his colleagues fielded dozens of calls an hour, but the phone system still couldn't handle the load. Calls crashed into one another, and many were disconnected.

Some of the callers had detailed questions, but others just wanted to talk. "William, do you have a question about a game or what?" Sandoff finally asked a caller who wanted to talk about the problems he was having in school. A seven-minute rule was initiated; no call could exceed that time limit. Counselors developed ways to cut calls short when kids started asking about their favorite rock groups and movies. One caller sought marriage counseling. His wife, he said, was going to leave him if he didn't stop playing "The Legend of Zelda." Sandoff's advice to him was, "Shut off the game."

The phones were overloaded, and the 800-line service became so expensive that Nintendo discontinued it. Calls to game counselors became regular toll calls in 1990. Phil Rogers instructed the counselors not to speak with any child for more than three minutes without making certain that the parents knew that they were going to be billed for the call. After seven minutes, no matter what the child said, the counselors were told to gently get off the line. "Parents were not going to blame themselves for not controlling their kids," Rogers said. "They were going to blame us." Still the volume didn't slow down.

Blaine Phelps, a counselor who became a supervisor in the department, had answered the same newspaper ad before joining NOA (he had been living in his car after losing his last job). "We don't just give away secrets," he says. "We are trained in the Socratic method of game counseling."

For years, the most asked question was about "The Legend of Zelda." Callers couldn't figure out how to get past Grumble Grumble, a creature lurking on level 7.

"What does it mean when your stomach is going 'grumble grumble'?" Phelps would ask.

"I'm hungry," responds a puzzled boy.

"Right. And how do you make it stop?"

"I get something to . . . Hey! That's it! Feed Grumble Grumble!"

The game counselors did more than provide a customer service. First, they further bonded players to the company. The degree to which kids became obsessed with Nintendo amazed educators, psychologists, and parents. The magazine and counselors were part of the reason, encouraging kids to become immersed in it. Main says, "For our more youthful players, many of whom come home from school and find neither Mom nor Dad there, Nintendo came to mean more to them. It filled a larger role in their lives."

Second, Nintendo was gaining great insight into its customers: they were finding out which groups were excited by what games, and how games could be made better. Counselors gave pointers but also queried callers about their likes and dislikes. "We used those calls as market research," says Main.

The information about consumers—not from dated market research studies but from the daily input of diehard customers—gave Nintendo a living, breathing line to its customers every day, seven days a week, twelve hours a day. The feedback helped steer the company's product development and marketing strategies; the information went right back into the development process. Yamauchi had always boasted that he never let marketing people influence R&D, but this stuff was too precious to ignore. Best of all, since callers often asked Nintendo counselors what games were coming, demand was created for games months in advance through the phone network.

The counselors were some of the first people at Nintendo to realize that kids weren't the only Nintendo fanatics out there. Many of the early-morning callers were frustrated parents, some of whom had been up all night trying to beat a game. "It kills them that their children are better at something than they are," says Blaine Phelps. "They're obsessed with beating their kids." When Peter Main realized how many interested adults there were, he began directing marketing campaigns toward them. Similarly, calls from girls gave the company a better handle on what would make more girls buy systems and games.

Don James ran another operation that was part of the marketing loop. In addition to his work preparing Nintendo for trade shows (Nintendo's CES booths became the largest in the consumer industry, 60,000 square feet of light shows, lasers, rock music, and dancing girls) and overseeing design work done in-house, he headed product analysis, whose function was to monitor and maintain the quality of games. It was a way to be certain that the games the counselors and *Nintendo Power* recommended were good. More important, it could be used to direct customers to the better games.

When games were nearly completed in Japan, NCL sent them to Seattle, where James's crew reviewed them for the American market. They checked the text and characters in every game and fixed on-screen instructions and dialogue written in "Janglish," the Japanese designers' version of English. "GIVE YOUR BEST AND LAUGH A COAT," read the instructions to one game. "GET WEAPON. PECK CROSS TABLE."

In addition to improper English, the product-analysis people were on the lookout for (presumably) inadvertent racial slurs. The object of one game called "Gumshoe" was to brutally slaughter American Indians (they were transformed into generic bad guys). In "Casino," the only thief was black, so the man's skin tone was changed. In one game, the "bare-breasted snake women" were renamed Medusas and "hell hounds" were renamed Cerberuses, after the mythological dog that guarded the entrance to Hades. In another game, the enemy was called a Jew's Ear, which was the translation in a Japanese dictionary for a kind of starfish. The memo to Japan read: "Please remove Jew's ear."

Games also had to be extraordinary on their own merits, not dependent on characters that were well known in Japan but not in the United States. Eventually James initiated a formal evaluation process headed by himself, Howard Phillips, and Shigeru Ota, known collectively as the Big Three.

At first the evaluations were arbitrary and haphazard, but soon Ota (before he was tapped to go to Frankfurt to run Nintendo of Europe) adapted a system that had been used in Japan. He developed a forty-point scale on which each game was to be rated. The system had eight categories, each one worth up to five points. The Big Three played every new game until they got a feel for it. Then they evaluated it for attributes such as challenge, graphics, and fun. Some games were sent back for revision; some were killed. If there were doubts, a larger group of evaluators, mostly game counselors, gave their opinions. Phil Sandoff was part of the GC6 (which simply stood for "six game counselors"). "We're tough," Sandoff says. "First you think every game is the greatest. Then you get more critical."

After the evaluations, Arakawa had a good idea of how a game would do in the marketplace. However, there were occasions when doubt persisted; for example, if the Big Three and GC6 disagreed. If Arakawa wanted more feedback, the toughest critics of all were called in. Hidden in a room behind a one-way mirror, Arakawa and James watched kids play the game. "Sometimes you cannot get the honest answer by asking questions of children," Arakawa says, remembering the failed focus groups back in New Jersey. "But if you watch their faces while they are playing, you can tell very easily whether the game is good or not. We have more than 90 percent success in judging games."

The most important evaluator for the company in the early days was Howard Phillips, who played more often and better than anyone else on Nintendo's staff. Phillips now wore pastel bow ties and white oxford shirts. Since shaving off the Manson beard, he had begun to look even more like Howdy Doody. He never seemed to age. Arakawa called him one day to say that he had decided to upgrade his title in the organization: henceforth, Phillips would be NOA's Game Master, the Jedi of the video-game world.

The Game Master knew every game inside out. As part of the Big Three, he continued to play all new games and give feedback to the developers. His primary responsibility was, he said, as a player's advocate, influencing the designers. "These guys have been locked away for eight months or a year working on a project," he says. "It's their baby, but somebody has to tell them, 'There's a lot of great stuff in here, but this part just doesn't make it.'

"Within the company, all the people with straight ties—*sorry, guys*—who are not really active users don't necessarily know what makes a game work or not work." Phillips could critique a game as deftly as Pauline Kael dispatched movies, although sometimes his reason for liking or disliking a game could consist of no more than "just because."

Arakawa came to trust this response, for Phillips consistently judged games better than anyone. He mastered more than five hundred. ("Through perseverance and insider knowledge," he said.) Could he beat anybody? He answered modestly, "I don't know if I'm the best player in the world, but I know that on any given day I could beat just about anybody on any game." The key was that he wasn't trying to think like players; he *was* a player. His most common advice was a terse "Practice."

9

THE GRINCH WHO
STOLE CHRISTMAS

The Nintendo marketing blitz had the biggest companies in America aiding and abetting (and, of course, cashing in on) the Nintendo invasion. Promotions were directed by Peter Main and Bill White, whom Main had hired in 1987.

White, who wore gold-framed John Lennon specs, brushed his fair hair back around a youthful face. Although he was always dressed in standard corporate attire, he looked like he would be more at ease behind a drum kit in a garage rock band than inside the high-pressure upper echelons of Nintendo.

White's father and sister were in advertising. After college, Bill got a job at Carnation while studying for an MBA in the evenings. Later he worked in the packaged-goods industry. When his boss convinced him to follow him from the "stodgy" packaged-goods business to high technology, White moved to Seattle to work as director of marketing for a computer software company. Frustrated there, he picked up a copy of *Advertising Age* one day and saw that Nintendo was looking for a director of advertising and public relations. Two weeks later he was interviewed by Main, Arakawa, and Lincoln, and a week after that he had the job.

White saw that Nintendo was dominated by its high-powered marketing machine. Since the company had so few products, it made only a few commercials a year. This meant the quality could be—had to be—phenomenal. The budget could go up to $5 million for one commercial, easily four or five times more than most other companies spent. In spite of this, because the advertising was so selective and specialized, the total advertising budget represented only 2 percent of sales, compared to the 17 or 18 percent of many other companies.

In his dealings with the press, White quickly realized why working for a Japanese company was unique. The attitude of the press toward Nintendo seemed extra critical; it had to be on the defensive on many issues, but without appearing defensive.

Nintendo's sudden, pervasive infiltration of America was like nothing White had seen before. The game counselors, the magazine, the merchandising, and the TV shows helped make Nintendo far more than just a new hot product. It became a culture unto itself, and in spite of growing concerns about Nintendo in Washington, D.C., and a critical press, that culture kept growing.

White found advertising at Nintendo similar to what it must be like at a movie studio. Arakawa made the decision that software, not hardware, should be the focus of most advertising. "It is a software-driven business," White says. "The job is not so much to increase long-term brand equity as it is to build excitement around the next hit."

One of the first commercials made under White was the market introduction for "The Legend of Zelda," which received a great deal of attention in the ad industry. A wiry-haired, nerdy guy walks through the dark screaming for Zelda. The next commercial, in November 1987, was for "Mike Tyson's Punch-Out!" It had beefy Tyson walking into a room, sitting down, grabbing an NES with his mammoth hands, shoving in a cartridge, and facing a wall full of screens. Then he looks into the camera and breaks out laughing. In another, a Nintendo "Ice Hockey" commercial, a kid is playing the game in front of a TV set when a puck comes crashing through the screen into his living room.

In spite of these great commercials, in 1990 White and Main oversaw a shakeup in which they withdrew $20 to $35 million in Nintendo accounts from the firms of McCann-Erickson and Foote, Cone & Belding. Bill White said it was "philosophical differences" that caused the split. Leo Burnett, the large Chicago agency that advertised Miller beer and McDonald's, got the account. One industry analyst said the decision was made almost arbitrarily. "They shake things up so that no one becomes complacent," he says. "You earn your money working for Nintendo; they'll drive you crazy before it's over." Everything was calculated. Kids didn't buy the lion's share of video games; parents did. Nonetheless, Nintendo advertised almost exclusively to children and teens. Kids controlled not only their own spending dollars but their families'.

There were other ways to reach parents. In early 1988, there were discussions at Nintendo about the benefits of broadening the company's message to an audience beyond six-to-fourteen-year-old boys (if nothing else, to gain some respectability from parents who were skeptical about video games). Bill White and Peter Main determined that there was no need to ante up the many millions of dollars necessary to "buy another demographic target" with television advertising. Instead, they sought promotion partners who already targeted these broader audiences.

Pepsi had the right image and audience, so White went after it. Pepsi's promotion team was cautious. They studied Nintendo's market research and agreed to test an association with Nintendo with one of its smaller brands, Slice. The national TV promotion, which gave Nintendo systems and games away with Slice, worked so well that Pepsi executives said they wanted to plan a bigger tie-in for their enormous Christmas advertising campaign, this time with all the Pepsi products. Since Pepsi targeted the twelve-to-thirty-four-year-old audience, the key soft-drink consumers, Nintendo got vast amounts of exposure—and the credibility associated with Pepsi—for nothing. Commercials and in-store displays were only one way Nintendo benefited from the Pepsi tie-in. Pepsi bought nearly $10 million worth of Nintendo products at the same wholesale price Toys "R" Us and other retailers paid, and Nintendo was advertised on the outside of 2 billion cans of Pepsi. In return, Pepsi got the cachet of being associated with Nintendo.

The promotion was so successful that Nintendo looked for other partners to increase its exposure to parents and other adults. Procter & Gamble approached Bill White and suggested a "dealer loader" that retailers put up in stores throughout the country featuring Nintendo characters on Tide detergent displays. For Nintendo, the association with a product that was so well accepted in America brought more immediate credibility. Tens of millions of people were reached in the $20 million promotion.

The next partner Nintendo targeted was McDonald's. In 1989, White and Main sent a letter of introduction to the company's marketing division. The head of children's marketing at McDonald's was shown a preview of "Super Mario Bros. 3." After examining Nintendo's demographics and the results of the Pepsi and Tide promotions, she launched an entire campaign—including "Mario" Happy Meals—around "Super Mario Bros. 3."

Another promotion for "Super Mario Bros. 3" proved more valuable than any paid advertising ever could. Tom Pollack, the respected president of Universal Studios, met with Bill White and Peter Main and told them he wanted to make a movie around a video game. *The Jetsons,* a film that Universal had planned for a Christmas 1989 release, wasn't going to be ready and the studio needed a holiday film. When he heard about the various Nintendo competitions, Pollack came up with the idea of a *Tommy* for young kids that was called *The Wizard.*

Nintendo received a licensing fee from Universal, of course, and approval of both the script and the film's game footage. Bill White went to the film set in Reno to see how things were going, even though "all we were really worried about was just making sure that the game footage was spectacular."

It was. The "video Armageddon" scene was filmed at the theater at Cal Arts, the college in Southern California. Universal spent $100,000 on this set alone. As the movie's main character took the stage, the on-screen audience went wild when the feverish announcer told the contestants that the final, tie-breaking competition was to be on a game they had never seen before, the newest and best video game in the world. Dramatically, the curtains were drawn and a wall of monitors was fired up. "Here it is, ladies and gentleman," the announcer screamed, " '*SUPER MARIO BROS. 3!!*' "

Nintendo itself could not have dreamed up a better promotion. The movie was ready by November, four months before the game was launched. The excitement in the theaters was far greater for "SMB3" than for the movie itself. *The Wizard* made money even with its relatively small box-office gross, but the groundswell of anticipation it created for "Super Mario Bros. 3" was enormous.

When the game finally arrived in stores, the hype had been so intensive that the resulting rush for the game shocked even Nintendo. Bill White's team oversaw the creation of a television commercial that showed no game footage at all, but simply thousands of kids chanting passionately, "Mar-i-o, Mar-i-o, Mar-i-o . . ." The camera pulled back and the faces of impassioned boys and girls—decidedly white and clean-cut—became a sea of people calling out Mario's name. Then the camera zoomed back to a point out in space. Looking back toward earth, we see Mario's face in place of what should have been the North American continent.

Nintendo reaped the rewards when "Super Mario Bros. 3" went on to outsell any video game in history, and gross more than $500 million.

Event marketing, as an NOA report called it, was another tactic. Nintendo hosted a thirty-city, eight-month-long nationwide video-game competition pitting local champions and then regional champions against one another. The grand finale, the Nintendo PowerFest, was held at Universal Studios in December 1990. There was a Nintendo world championship in 1991 and, in the spring, the company held the Nintendo Campus Challenge, a two-day video-game competition on fifty college campuses. It was heralded as "a nationwide search for a college valedictorian of video-game play."

Al Kahn, heavyset, with sparse red hair, ran Leisure Concepts, a licensing company. In 1988, Arakawa, whom he knew from when he worked at Coleco, gave him the rights—for a standard royalty— to market Nintendo's characters. Kahn licensed game characters for everything from TV shows and records to lunch boxes and bed linens, and the Nintendo license became one of the most successful in history. There were "Zelda" board games, "Donkey Kong" watches, and "Mario" *everything*.

The first Nintendo television show, *The Super Mario Bros. Super Show,* first aired in fall 1988, and soon went into syndication. In fall 1990, *Captain N: The Game Master* became a network show, and eventually there was a third show, *Super Mario World.* For several years running, Nintendo shows were numbers one and two on NBC's Saturday-morning schedule.

The television shows brought Nintendo licensing fees, but this wasn't the primary benefit. "They're about trying to boost awareness of the characters," White says. "They're part of the endeavor to sell more Mario product by increasing the popularity and likability of the character, which, in turn, will help our character-licensing program—sales of T-shirts and all that—and of course the games themselves. There's really very little downside if the show is done in a quality manner." The fact is that no one could have confused the television shows with anything approaching quality, but they were no worse than other Saturday-morning fare.

In addition to all the other licensing deals, Bill White also contemplated the idea of a feature film for Mario. Arakawa told him to research it slowly and meticulously. White made some calls and heard back from the man in Hollywood who handles Nintendo's cartoon shows: Dustin Hoffman wanted to play Mario. Hoffman's agent, Mike Ovitz, had contacted them to arrange a meeting.

In New York for the newly devised Nintendo World Championships, White sat down with Hoffman in a hotel room on the Upper East Side, and the two spent an hour discussing the movie. Hoffman's kids were Nintendo maniacs, and he said he was dying to play Mario. White found himself in the awkward position of having to tell Hoffman that they were shopping the property. (Nintendo wanted Danny DeVito, as close to a dead ringer for Mario as Hollywood had to offer.)

There was no lack of interest in Hollywood when it came to cashing in on the Nintendo craze. Arakawa rejected an offer from Fox because "they didn't understand the character." A feature deal was finally ironed out with Jake Eberts and Roland Joffe's company, called Lightmotive. They hired Barry Morrow, who cowrote *Rain Man,* to write a script, and a series of rewrite men were brought in. The directing team of Rocky Morton and Annabel Jankel, creators of the original *Max Headroom* film for British television, was signed up. The production designer was David Snyder, whose impressive credits included *Blade Runner* and *Pee-wee's Big Adventure.*

The Nintendo movie went into production in May 1992 for a May 1993 release—without Danny DeVito, who had ultimately turned down the part of Mario. After all the talk about Hoffman and DeVito (Hoffman's agency, Creative Artists, says he never considered the part), Nintendo had to start from scratch. Although Tom Hanks had agreed to accept the role for $5 million, Nintendo went with Bob Hoskins, popular after his role in *Who Framed Roger Rabbit,* but much less expensive. The bad guy, King Koopa, was played by Dennis Hopper.

There were a host of other promotions. In 1991 MCA released a record of "Mario" songs that included a "Mario" comic book. Among the musical numbers was the last song Roy Orbison recorded before he died. There was another feature movie in the works, too, an animated movie similar to (but hopefully better than) the TV cartoons.

Long before any talk about movies was heard in the halls and offices of Nintendo, Peter Main had told Arakawa that he felt it wise to market video games like movies—released cautiously, rationed so that demand outpaced availability, and then withdrawn from circulation as soon as interest began to wane. This rationing tactic, treating games like priceless objects, worked. After all the hype about a new game took hold, kids dragged their parents to stores, but outlets couldn't keep the games in stock. The rush to get games such as "Super Mario Bros. 3" or "Link," the sequel to "The Legend of Zelda," caused near riots of excited game-buying.

The competition to acquire games rivaled that for tickets to Michael Jackson's last concert tour. Ultimately more product was sold. A kid who was absolutely dying to get "Link" would arrive at a store, only to find it sold out. Maybe he would try a few other stores without success, but then he would buy another Nintendo game, so that his parents would end up paying $30, $40, or $50 for a second or third choice. Then, a week or month later, a new supply of "Link" would come in. The kid wanted "Link" more than ever then, and unless his were the most iron-willed of parents, they would succumb. Even the kids whose parents held out still managed to get games; in 1989, in a survey of what kids in Sandwich, Illinois, bought with their allowances and other money they earned, the near unanimous choice was Nintendo games.

The editor of one toy-industry journal noted that "Nintendo has become a name like Disney or McDonald's. They've done it by doling out games like Godiva chocolates." In 1988, *Fortune* observed that "so far the strategy looks like a winner."

"Inventory management" is what Peter Main called it. The Atari wave had floundered in large part because of a flooded market. Main made certain that scarcity whetted the public's appetite and sustained demand as Coleco had done in 1984, when there was a shortage of Cabbage Patch Kids. By design, Nintendo did not fill all of the retailers' orders, and it kept half or more of its library of games inactive. In 1988, for instance, it sold 33 million cartridges, but market surveys showed it could have sold 45 million. That year retailers had requested 110 million cartridges, or nearly 2.5 times the indicated demand. Main said that retailers exaggerated demand and Nintendo would rather have them pleading for more than have to worry about excess inventory.

In contrast to the prerecorded-video business, in which an average tape had the longevity of 90 to 120 days, some Nintendo games were popular for a year or more. If a game came through the evaluation process with a thirty-six or thirty-seven on the forty-point scale, Nintendo viewed it as "a potential grand slam," according to Main. A hugely successful game only had to sell 500,000 copies, and grand slams could sell millions.

When Nintendo released "Dr. Mario" in 1991, Charles Lazarus, head of Toys "R" Us, called Peter Main. "It looks like we're going to be short on 'Dr. Mario,' " he said. "What are you going to do about it?"

Main made certain his largest customers got a healthy share of the games, but the company refused to cave in to their demands;

no company had all of its orders filled all the time. By then, the toy and electronics as well as department stores were dependent on Nintendo, not the other way around. NOA accounted for an inordinate amount of the revenue of some companies. For Toys "R" Us, it meant 17 percent of its sales and 22 percent of its profits. Nintendo called the shots even when it came to companies used to throwing their muscle around.

The fact that the parent company and the president of its American subsidiary were Japanese exacerbated NOA's problems when it played rough. Tactics that would have been called aggressive if used by an American company were viewed in some quarters as unscrupulous when the company was Japanese. In only thirty-six months Nintendo had gone from near anonymity to the point where it accounted for 20 percent of the toy industry and large percentages of other retailers' gross sales. It was felt that "by definition we must be doing something illegal," Peter Main says. No industry wanted to be so dependent on a single company, especially a Japanese company. In Main's view, "Most of the criticism is from those who have decided not to join the club." Some, however, hadn't been invited.

NOA's reputation began to be suspect around the time of the most severe game shortages. Although Nintendo orchestrated some of the shortages, they were worse than it had planned, and consumer demand was higher than the most ambitious forecasts. In 1990, retailers were furious when Nintendo couldn't deliver as many systems as they could have sold that year.

Nintendo had long since terminated its money-back-guarantee policy and replaced it with tough terms: order, receive shipment, and pay. The sales reps in the front lines made certain that retailers kept small inventories. By keeping customers on a short leash, Nintendo made enemies, but cash kept rolling in. This policy proved its worth in 1990 when the second largest toy-store chain in the United States, Child World, had serious financial difficulties and was on its way to bankruptcy. A listing of Child World's creditors was announced in December. Companies such as Hasbro and Mattel were owed up to $25 million, but though it had accounted for almost 20 percent of Child World's business, Nintendo wasn't on the list. "These other guys were pumping all this product into a bottomless pit," Peter Main said. "They are still owed that money today." Because Nintendo was working so closely with Child World and knew, from constant inspection of financial statements, that the company was in trouble, the retailer was told that it had to pay for its Nintendo product as much as a year in advance. "We are not loved for that," Main says.

Nintendo also made retailers furious with its return policy. American retailers proudly boast, "Customer satisfaction guaranteed or your money cheerfully refunded." However, when consumers return TVs and toasters to Macy's or Sears, the retailers return them to the manufacturers.

In contrast, Nintendo said it would guarantee low defect rates. It claimed its defect rate was 0.9 percent for hardware and 0.25 percent for software, numbers similar to the VCR industry's high standards. Still, retailers sent systems back to Nintendo that had been returned by customers for reasons other than defects; they claimed to be dissatisfied or, typically, that their dog had chewed up a controller.

Nintendo's reaction was to initiate a service network around the country. Service would be free for ninety days; after that, customers would have to pay. Because of their extensive service network and the promised low defect rates, Nintendo announced a new policy: no returns. Once a game cartridge was opened, a refund was out of the question.

Pandemonium followed. One of the largest retailers in the country threatened to stop carrying Nintendo Systems and products. Nintendo refused to change the policy and the retailer refused the products. The retailer held out for three months; after that it crawled back and agreed to Nintendo's terms.

Piles of cash poured forth from America to Japan. NCL's net sales figures shot up, mostly because of its U.S. subsidiary. Arakawa became responsible for up to 60 percent of Yamauchi's business, according to Hiroshi Imanishi. NCL's net sales in 1987 were $1 billion. In 1988, they went to $1.5 billion. In 1989 and 1990 they topped $2 billion, and in 1991 they shot up to more than $3.3 billion. In 1992, Nintendo foresaw sales topping $4.5 billion. Pretax profits had risen from $186 million in 1987 to more than a billion dollars in 1991 and 1992.

Nintendo's stock soared. In 1991 the company reached number eighty-six on *Business Week*'s Global 1,000. In U.S. dollars, the company's market value was $14.56 billion, and it was ranked twenty-ninth among all Japanese companies.

The relationship between Japanese parents and their American subsidiaries is, notoriously, one of master and slave. Hiroshi Yamauchi, however, after his initial reluctance, gave Arakawa far more autonomy in running NOA than most chairmen of Japanese companies gave the heads of their subsidiaries. He still made the most significant decisions affecting the future of NCL, but he did so in consultation with Arakawa. Minoru Arakawa, for his part, let down his guard somewhat; he had come to trust his father-in-law's instincts. He was in regular communication with Yamauchi; they often spoke several times a day. As they came to appreciate and respect one another, the relationship between the two men changed and they learned how to take advantage of each other's strengths.

Arakawa still became angry when Yamauchi crossed the line and his advice seemed more like orders, and Yamauchi remained frustrated that Arakawa wasn't better at communicating his plans and motives. But Yamauchi's doubts about Arakawa dissipated in proportion to the enormous amount of money he was receiving from NOA. Yoko Arakawa breathed easier; she had fewer fears about finding herself in the middle of clashes between her father and husband. If anything, she was sometimes concerned because the two men were getting along so well.

As Yamauchi had determined the style of Nintendo in Japan, Arakawa imprinted his personality on NOA. His philosophy evolved into a unique management style. There were bumpy moments as the number of employees grew with the sales figures, but he rarely lost his temper and never lost sight of his goal of the moment. He worked obsessively but he managed quietly, almost in a whisper. His mind may have been racing, but he said so few words that it sometimes made his associates uncomfortable. His silence was a distancing device; people squirmed while he deliberated. He never pontificated or lambasted, and his ego never seemed to lead him; in fact, he did what he could to keep out of the limelight. Peter Main and Howard Lincoln made most of the required public appearances. Part of the reason was that Arakawa was embarrassed by his slow and cautious English; also, he felt he had better things to do with his time. (This attitude occasionally backfired. When *Frontline,* public television's investigative program, did a piece on U.S.-Japanese trade tensions, Arakawa declined to be interviewed, sending Howard Lincoln in his place. It looked as if Arakawa had something to hide.)

In striking contrast with Nintendo's tough image, Arakawa was almost always quiet and even self-deprecating. He ruled by force of will rather than decibel level. NOA reflected his good humor. There were video games to play in Café Mario, the company's dining room, and there was lots of joking around. Arakawa himself instigated some of the better practical jokes. Once he circulated an old photograph of Howard Lincoln throughout the company as if it were an urgent memo. In the picture, taken in the seventies, Lincoln wore a plaid suit and thick glasses. On it Arakawa had written, "Would you buy a used car from this man?"

On another occasion, Nintendo employees decided to "initiate" a new employee (formerly with Atari, coincidentally) who had developed the habit of parking his Porsche diagonally across several parking spaces so that no one would park too close to his prized auto. One day, everyone, Arakawa included, parked haphazardly around the lot. It was the new employee's worst nightmare: his car was in the middle of hundreds of cars, some of them perilously close.

Professionalism was laced with northwestern ease. AT&T's executives showed up on a Friday for a meeting with Arakawa. No one had told them about the no-suits-on-Friday dress code; they sat at a conference table in their standard corporate attire opposite Arakawa, who was wearing jeans, a T-shirt, and green felt sneakers. (After Howard Lincoln told him he looked like an elf, Arakawa never wore the shoes again.)

In the lobby of NOA's headquarters is a smoky glass coffee table and a crystal horse's head in a glass case. Three receptionists answer telephones and greet visitors. After visitors are cleared, their names are typed into a computer and name tags are printed out before they are allowed inside.

The office looks like most other offices in high-tech or communications companies. There are nondescript lithographs of nature scenes on the walls and pastel-pink wall panels, gray carpets, and partitions. Still, Arakawa's invisible imprint pervades. Enthusiastic employees—many of them, including some department heads, conspicuously young—buzz in and out of open-doored offices (a memo was once distributed castigating managers who worked behind closed doors). The managers' offices at the periphery of the building all have large windows with open blinds. It is, Arakawa says, "for the light from outside and for clear thinking." Everything and everybody is connected—the repair center is near where new games are tested, which is near where the marketing team huddles to preview new commercials, which is near where the rows of game counselors field telephone calls from across the country. Arakawa admits there is nothing greatly innovative about his management style. "All I have done is taken the walls out between our managers and workers," he says.

The corporate offices are also imbued with the sense that Nintendo deals not in circuits or microchips but in fun. A dash through the marketing department means running past the game counselors, glued to TV screens and their "green bibles." In other corners are the game testers and evaluators. On the second floor is a sample World of Nintendo. Nearby is a mini-arcade of Nintendo games. In individual offices are reminders that the people of Nintendo are consumed by some wackier pursuits. Offices are decorated with baseballs, wind-up toys, stuffed animals, basketball hoops, and at least one blow-up doll.

At the same time, there is a sense that something important is happening. It is customary for employees to work late, often past midnight. There are rumors that the pressure, albeit subtle, is behind more than a few martinis consumed in the evenings behind closed doors that are never supposed to be closed. Employees seem to work as hard for Arakawa as they work to sell Nintendo products. In a way, Arakawa has succeeded in creating, in Redmond, Washington, the mentality of a Japanese company.

Salaries are about average in comparison to similar positions in similar industries for the Seattle area. Nintendo refuses to release salary figures. Since it is not a public company in America, no one knows the executives' salaries. This is significant because Nintendo is thereby not forced to compete with the inflated salaries in the industry. Although his company makes more than American companies from Chrysler to IBM, Arakawa and his vice-presidents claim they take home relatively modest salaries. Nintendo follows the Japanese model of controlling executive compensation in relation to salaries for starting workers, and all employees do well when the company does well because of a significant bonus program; employees can earn up to an additional 50 percent of their salary each year, depending on the company's and their individual performance.

Arakawa and Howard Lincoln spend a weekend twice a year going through all the employees' salaries and awarding bonuses based on evaluations made by managers. But besides a decent retirement plan and good health benefits, that is it. "Nobody around here is going to make zillions of dollars," Lincoln says. "That's one of the things that you give up, even though we sure as hell have made a lot of people outside Nintendo millionaires."

If not rich, at least they are suntanned. NOA bought four parcels of land on Hawaii, where Arakawa built two 9,000-square-foot oceanfront homes at a cost of $20 million. They were ready in the winter of 1991, when the Arakawas and Lincolns christened them on a vacation. The huge houses each have four bedrooms and open common areas with 180-degree epic views, a private swimming pool, and access to the Mauna Lani Golf Course. Employees can rent the houses for a subsidized fee on a first-come, first-served basis. After three months at the company, any Nintendo employee can rent either a room or a whole house.

Still, more than bonuses and perks push the employees. With rare exceptions, Arakawa fills high-level positions from within the company. Nintendo employees cite this as the greatest benefit of working for him. The company's personnel director began as a receptionist. Howard Lincoln's former secretary went on to run the licensing operation. Counselors have become testers, and evaluators and executive assistants have become finance managers. As in Japan, most employees enlist for the long haul. Attrition is low, and with a few notable exceptions, defections are rare; no one leaves Nintendo.

Arakawa's primary concern is that Nintendo does not become the victim of out-of-control growth. This is reflected in small but meaningful touches (meaningful particularly when compared with companies that run so much less efficiently than Nintendo). There are no reserved parking spaces. There is no executive dining room. Café Mario serves burgers, a special of the day (chili-mac), pizza, salads, and frozen yogurt sundaes. There are no corporate jets ("The day that happens I'm out of here," Lincoln says), and there is no executive suite of managers. Every office—including Arakawa's and Lincoln's—is a ten-by-ten-foot square. Department managers' offices are adjacent to the areas where their people work.

Ever mindful of Atari's huge number of employees—layers of bureaucracy, vice-presidents of everything imaginable—Arakawa has initiated a system to control hiring at Nintendo. Either he or Lincoln have to sign an approval before a salaried employee can be hired. "We've made it so difficult for anyone to hire a new employee that they better darn well have a good reason to do it," Lincoln says. The company remains lean and mean, as Arakawa likes to say. For him, Japanese tradition applies here: he believes that when an employee is hired, he or she is hired for good.

To keep communication open, Arakawa has initiated weekly staff meetings. One week's meeting will be dedicated to the administrative staff, the next to operations, and the next to marketing. Arakawa, Lincoln, Rogers, and Main always attend the staff meetings. Afterward, the four of them meet in a conference room to discuss whatever issues have been raised.

As Nintendo grew and he feared the company would become unwieldy, Arakawa adopted a system of "action memos" and an authority chart, used in Japan for organizing companies. The levels of responsibility and levels of communication that resulted are as good as carved in stone. Department heads each know what the others are doing, and Arakawa knows what they all are doing.

Expenditures are also carefully controlled. Most managers can spend up to $5,000 without approval, but an officer has to approve expenditures from $5,000 to $50,000, and Arakawa has to approve anything above that. Action memos about smaller transactions and other business decisions still cross Arakawa's desk. It is a way for him to know what is going on and to be able to track who makes which decisions. Anyone who signs off on an action memo is accountable. Nothing—from major purchases to press releases—falls between the cracks. It is important to Arakawa that there be no surprises.

Arakawa feels he has to be a better organizer than his father-in-law. He does not have Yamauchi's sixth sense about products, and occasionally he is wrong. He believes that no one can learn that sort of judgment. "You are born with it," he says. "No matter how I study and how I train myself, I will never have those talents." But, he says, "my talent is maybe to find somebody who *does* have those talents." He does have a knack for choosing exceptional people. "Arakawa is held up by the top people around him," an employee says.

Nintendo took over the toy business, a volatile and cut-throat multibillion-dollar industry, in 1988.

So much money was at stake in the industry that careers routinely were made or destroyed based on one product. It cost millions to develop and manufacture a new toy and pre-test it with kids, and then tens or hundreds of millions to go into full production and to market. Risks were huge, and except for staple toys like Barbie and Matchbox cars, the biggest hits came and went with the whims of millions of kids.

The toy industry was dominated by a few giants. In the race for product and market share, several companies gobbled up many of the smaller players. Hasbro, which had already acquired Playskool and Milton Bradley, bought Tonka, which also owned Parker Brothers and Kenner Toys, in 1991. ("G.I. Joe and the Real Ghostbusters join forces," the press said.) Hasbro then went off to battle against Mattel, taking over the number-one spot. To acquire new companies, the majors took on tremendous debt and interest commitments. This situation—plus the added effect of the recession, which reduced discretionary spending—meant that in the late 1980s and early 1990s, toy companies had to cut back on R&D.

The result was that nothing dramatically new came from the toy business. In turn, no innovation meant bored customers who went where the excitement was: Nintendo.

Nintendo ate up a larger and larger share of the toy business, becoming far bigger than Hasbro and Mattel combined. Of the estimated $11.4 billion spent on toys in 1989, 23 percent was spent on Nintendo products. Of the thirty top-selling toys in America, twenty-five were Nintendo or Nintendo-related. NOA held every spot in the top ten—and this was during the greatest shortage of Nintendo products.

Arakawa couldn't fill orders. Before Christmas 1989, Peter Main was threatened with bodily harm and lawsuits by company managers and presidents who blamed him for single-handedly destroying their business. The only stores that had Nintendo products at Christmas that year were retailers that had sufficient cash reserves to allow them to begin hording systems and games that summer. Toys "R" Us and a few other chains had invested heavily in Nintendo products for that Christmas and had a reasonable supply. Most companies, however, blamed Nintendo when their businesses plummeted. In one pre-Christmas month in 1989, toy-company stocks tumbled 42 percent, *Fortune* reported. *Financial World* said, "Nintendo . . . is once again sticking out its tongue at the rest of the U.S. toy industry."

The toy industry *hated* the fact that Nintendo didn't play by its rules: it didn't use December 10 billing; it didn't belong to any toy manufacturers' associations; it didn't bother showing its products at most toy-industry trade shows. Charges began to surface that it was cutting into the toy industry, *taking over* the toy industry, manipulating it by illegally fixing prices and intimidating retailers, threatening to cut off their product supply if they carried competitors' products or dared to discount Nintendo machines and games.

Arakawa and Peter Main became the target of complaints—first from Nintendo's competitors, then from retailers, and eventually from members of Congress. Although Peter Main was named 1989 Marketer of the Year by *Adweek* magazine, a profile in *The New York Times* had this to say of him: "To his admirers . . . he is a master seller of children's entertainment. . . . To his critics he is an aspiring monopolist, squeezing supply and jacking up profits."

Since the toy companies couldn't beat NOA, they tried to cash in on the Nintendo craze. Five major companies got licenses from Nintendo to release character-related toys such as Mario and Zelda board games and action figures. Mostly, however, other companies hoped that Nintendo's invasion of their territory was temporary. They prayed for (and in many cases, expected) Nintendo to go the way of Atari as they bided their time with old standbys: Slinky, Mr. Potato Head, Lincoln Logs, Barbie. But by 1991, Nintendo had become so well entrenched that the toy industry realized it wasn't about to disappear. Hasbro's chief acknowledged the troubles the country's economic recession had brought to his industry, but complained that "the toy industry has been much more impacted by Nintendo."

By 1991, Toys "R" Us, which had a 22 percent market share of the toy business with its 450 stores (plus 100 more abroad), had sales of $5 billion, and Nintendo products represented almost one fifth of its total sales. Discounters became Toys "R" Us's biggest competitors (Child World and Lionel Kiddie City shared 7 percent of the toy market). Kmart and Wal-Mart controlled up to 10 percent each. The largest, Wal-Mart, didn't even carry Nintendo's competitors. Although Wal-Mart had no World of Nintendo in its stores, Kmart, which had held out for years, agreed in 1991 to open five hundred of the stores-within-a-store.

The toy industry had been hurt by the recession as well as deep discounting of toys in pre-Christmas sales. Child World was on the brink of collapse (it finally went into Chapter 11 in 1992). The industry did better in 1992, but Nintendo still accounted for seven of the top ten toys of the year. It was, once more, the Grinch who stole Christmas.

Whether or not Nintendo actually absconded with the toy business, many people see the company as an evil force because it deals in video games, hypnotizers of youth. Since video games first appeared, they, like pool tables and pinball machines, have been viewed as contributors to juvenile delinquency. Nintendo is considered guilty by association.

The more Nintendo became a force in the lives of millions of kids, the more concerned parents and educators became. The worries were exacerbated by the degree of devotion kids seemed to have for the games and for Nintendo's culture. It was as if America's kids had joined a cult and were under the spell of a cartoon character who did the bidding of a hidden Japanese lord. "Notice the way 'Super Mario' is drawn," a worried parent wrote in a letter to a magazine. "He has the eyes of someone who has been brainwashed."

Fads had come and gone before, but this was different. Kids played video games, conspired with one another about game strategy, drew pictures of video-game characters, and for their homework wrote video-game adventures. While kids also loved *The Simpsons, In Living Color,* and MTV, the intensity with which they played video games was noticeably different from that of the attention they paid to television. Hospital staff saw how dramatically different Nintendo playing was from TV watching. Some seriously ill children in a hospital who played Nintendo required half as much pain medication as those who didn't. Television had no effect on the amount of medication required.

According to doctors, playing Nintendo games has the power to alleviate pain for two reasons. First, the intensive interaction with video games requires a degree of concentration that acts as a diversion and distraction from pain—and everything else. Secondly, the highly excited state achieved during this interaction generates a steady flow of endorphins into the bloodstream. Endorphins are the naturally occurring proteins that mask pain and cause a sense of euphoria. Nintendo playing can cause a sort of high, but so can jogging.

Parents, educators, and psychologists are concerned that Nintendo may be worse than television because of the relentless and hypnotic attention that is required. Some kids who play for long intervals complain of headaches and blurry vision. Some are diagnosed with "Nintendinitis," severe muscle cramps in their game-playing thumbs. Experts worry that Nintendo may train kids to be aggressive and inure them to high degrees of violence. Others simply feel that any excuse to be in front of the television encourages antisocial behavior.

Parents and teachers have noticed that kids who play a lot of Nintendo have increased levels of both determination, on the one hand, and of frustration, on the other. Some kids are out-of-control "Nintendo zombies," as Oprah Winfrey dubbed them on her television talk show. There were worries that video-game playing could be damaging to kids' cognitive and social development. Parents on *Oprah* complained that their kids played Nintendo games every free hour of the day. One guest on the show was an eleven-year-old boy who admitted that he was an addict who played twelve hours a day, before school and afterward until bedtime. Why his parents allowed him to play so much was another question, but the example wasn't unique; left to their own devices, kids play compulsively.

Nintendo weakly answered this litany of concern with hollow-sounding arguments that claimed playing video games can actually be beneficial. It increases hand-eye coordination and response time. Unsolicited independent confirmation proved how much: the U.S. military found that recruits who had played a lot of Nintendo did disproportionately well in flight-training programs.

The problem with Nintendo's retort was that there were many other ways for young people to improve hand-eye coordination—by building with blocks or playing catch, for instance. And how much hand-eye coordination does anyone (besides future fighter pilots) really need, anyway? The company's arguments did little to stem the growing concern about a generation of "vidiots," hypnotized kids who blindly follow a pied piper in the form of Super Mario.

JUNIOR AN ADDICT? a *USA Today* headline read. The answer, according to the article, was: not to worry. An expert from Pomona College, Brian Stonehoil, was quoted: "These games provide children with a strong sense of dramatic victory without the slightest bit of physical danger. People who play will simply keep at it until it loses its appeal." A mother said, "I'm scared to death of computers and [my kids] won't be. Sure, part of me wishes they'd be like my generation, happy simply swinging or jumping rope. But they're not . . . [and] at least they're not watching lots of TV."

Many researchers believe that Nintendo is superior to television because it is interactive. Television is a one-way medium, with viewers passively devouring the programming and relatively little brain activity occurring. Video games, on the other hand, stimulate keen responsiveness. Many professionals feel that as a result Nintendo is far healthier.

In addition to the active involvement demanded by Nintendo games, some experts feel that the games also require useful kinds of thinking. Video games, they claim, help develop problem-solving abilities, pattern recognition, resource management, logistics, mapping, memory, quick thinking, and reasoned judgments. "Knowing when to fight and when to run applies to other life situations," says a game maker. The lessons learned during video-game sessions can, he argues, be used in everyday life.

Video-game violence is far less harmful than television violence, some professionals maintain, because the blood and gore in video games is less realistic. Even though game designers seem to spend inordinate amounts of time and energy dreaming up bizarre ways to kill and maim, for the most part the violence is the stuff of science fiction and fantasy. The concern that the games anesthetize kids to real violence and increase their aggression is debated by some psychologists who feel the opposite is true—that the games actually release aggression, and that all the beating up of video-game villains is cathartic. According to one San Francisco child psychologist, video games build self-esteem: "You accomplish something, you are rewarded. You may even save the princess."

Howard Phillips notes that video games help build kids' confidence by rewarding them for success and resilience. "The number of times in school a teacher asks any one child for an answer is pretty limited," he says. "Most of the time kids raise their hands and respond and get back a quick 'Right' or 'Wrong.' If they're wrong, they've lost their chance and someone else is called on. But with video games, if a child makes a mistake—bang, he's immediately back into it, trying again. They have constant, immediate negative and positive feedback."

In her book *The Second Self,* Sherry Turkle writes that video games have hidden benefits. "You have to follow rules and be logical and patient," she says. "Working out your game strategy involves a process of deciphering the logic of the game, of understanding the intent of the game's designer, of achieving a meeting of the minds with the program." But the program came from the designer, not from the individual child. The creative choices in the current generation of Nintendo games are few.

Peggy Charren, the founder of Action for Children's Television (ACT), a watchdog organization that looks at the effects of television on children, examined Nintendo and came away with mixed conclusions. Yes, video games helped make kids comfortable with computers and machines, but since most games were geared specifically to boys, girls were excluded from that education. She was also concerned about the sexism and violence of many games. She concluded that although video games had some benefits that television couldn't offer, they were a passion that shouldn't be indulged to the exclusion of other activities. "Too much video-game playing is a problem not because of what you are doing but because of what you are not doing," she said. Even Hiroshi Yamauchi seems to agree with her: "I mean, too much eating is bad, too. This is the age of the computer. No one can stop history—kids are interested in playing computers." Howard Phillips is more defensive in the face of parents who charge that Nintendo is addictive. "Nintendo is not responsible for raising your children," he says. "Parents who blame Nintendo are passing the buck."

When Sigeru Miyamoto, the creator of Mario, Luigi, and Zelda, heard the complaints about video games, he simply shrugged his shoulders. "Video games are bad for you? That's what they said about rock 'n' roll."

Arguments that justify video-game playing are unconvincing to many people. Nintendo could have done better at offering games with socially redeeming subject matter and games that educated. If kids are going to spend all that time playing Nintendo, their parents would feel better if they were learning something concrete and valuable.

Nolan Bushnell believes it is incumbent on the game companies to offer more. Video games, he says, could be one answer to America's education crisis. "Let's assume that 1 percent of the teachers in this country are absolutely fantastic," he says. "That means that only 1 percent of the students are able to take advantage of those excellent teachers. Teaching is the only environment in which that is true. Everybody gets to benefit from the top 1 percent of the people in sports and in entertainment, but most kids go through school with no contact with the great teachers. And of course, great teachers aren't often able to do much teaching, since they are put in situations that are not conducive to education. Technology is the only thing that can change that. Put those great teachers on a cartridge, pop it into a game system, and kids everywhere will have access to them. Programs can react to the way a child learns. If a child shows himself to be an object-oriented, visual learner, the system can detect that and continue to teach him in that way."

There is another bonus to video-game machines: the message comes via a medium that kids enjoy. Computer-game companies such as The Learning Company, Broderbund, and LucasArts have made fun games that are also educational. Inherent in the games are geography lessons, math, science, and reading. Kids who have access to computers can play them, but there are almost no educational Nintendo games. The reason, as a software-company executive noted, is that "kids like them like spinach."

Nevertheless, Nintendo has seen the public-relations value of creating educational uses for the NES. If nothing else, it would appease parents, who control the pursestrings. "The view was that Nintendo *must* be bad; only cancer grows that fast," observes Peter Main. "We were less well prepared than we should have been in that period because we hadn't really devoted the time to providing the offsetting evidence. Beyond arguments about hand-eye coordination, we stumbled through that period and looked a little more naive than we would have liked."

Since then Nintendo has tried several things. First, it has encouraged software developers to make games of socially redeeming value. "*Sesame Street* Learning Games" and "Donkey Kong Math" are examples. (They are also examples of why companies don't make educational games; they were dismal failures.) One more exciting product was released by a software company, Software Toolworks, in 1990. Dubbed the Miracle, this piano-teaching system is an electronic keyboard that plugs into the NES. The cartridge that comes with it is filled with piano lessons. A child, following on-screen lessons, will practice on the Miracle. The program can monitor the lesson and point out mistakes.

The Miracle is one answer to critics who say Nintendo has no higher purposes than blasting enemies and wasting kids' time. NOA's evaluators awarded the Miracle the highest rating ever given to a product developed outside the company, although it has sold only marginally well in spite of a $6 million ad campaign for Christmas 1991. By then the Software Toolworks designers had completed versions that work with IBM, Amiga, and Macintosh PCs, and sales picked up through the next year.

More "educational," or at least creative, games were released beginning in 1992. Spectrum Holobyte's addictive "Wordtris" and a terrific animation program, Nintendo's own "Mario Paint," indicated that things might get better.

Besides trying to push select educational software and hardware, Nintendo has tried to win over its critics with good PR. Nintendo in Japan had been criticized for contributing little to the community, but Arakawa understands the value of community and social involvement. In 1987, NOA was contacted by a young girl who had been an enthusiastic Nintendo player until she was in an automobile accident that left her paralyzed from the neck down. The company responded by developing a system that could be used by handicapped players and sold some two thousand of them at cost ($175).

Andrea Miano, a therapist who works with handicapped children at Shriner's Hospital in San Francisco, says the Nintendo Hands Free system "enables kids to play games other kids are playing; it builds a great deal of self-esteem. It's not simply diversionary recreation but therapeutic recreation. For a child who has very little ability to move at all—to do anything—to be able to do well in a game like this—it's fairly extraordinary what it does for them."

Nintendo has also tried to alleviate some criticism by supporting research into ways that video games can be better teachers. In 1991, Hiroshi Yamauchi awarded $3 million to MIT's world-renowned Media Lab, specifically for research on how children learn while they play. The Lab was working on high-tech learning tools that, according to Dr. Seymour Papert, "look and feel more like Nintendo games than schoolbooks."

It was considered unlikely by some that the Lab would come up with any negative conclusions about video-game playing as long as Nintendo was its benefactor. It was a misguided criticism, however; MIT was not researching the effects of video games, but of new fields such as "constructivism," based on the Piagetian thesis that knowledge is built by the learner, not supplied by the teacher. It may well be that Nintendo games of the future will apply this principle—build knowledge by creating—and still be as much fun as the current generation of games. Marshall McLuhan once said that anyone who tries to make a distinction between entertainment and education doesn't know the first thing about either. There is no reason why Nintendo's popularity with kids cannot be exploited for higher ends.

In a *New York Times* piece on education, Morgan Newman, cofounder of a multimedia publishing company, AND Communications, was quoted as saying, "We think that quite probably what's going on with the crisis in education out there is that kids are often just bored. They're going home in the afternoon and watching MTV and then they go to school and the teacher says, 'Open the book to page 225,' and their eyes just glaze over. . . . We're embracing instead of denouncing the language that human beings in the 1990s want to hear to keep them engaged." Nintendo was one of the "languages" he spoke of.

In a search for NES software that had the insidious secret agenda of education, groups have been formed around the country. Dave Hammond, who spearheaded one effort, says, "If we don't produce more young people who know how to think and problem-solve we aren't going to be able to compete in the global economy. We need a way of attracting kids in in the after hours. More than 40 million kids play Nintendo. A system is in place. The games we're working on will be fun and educational. We plan to stage a Super Bowl of Nintendo on these games, an Olympics. There will be rewards—money and TV commercials. Kids at home will watch and say, 'I can do that.' We want it to be as attractive to them as football and wrestling."

Nintendo has supported some outside projects and has initiated others. It underwrote a first-ever symposium on "Video Games in Popular Culture," a forum for researchers and academics to discuss the impact video games have on their players and on the culture as a whole. The conference's organizer, Bowling Green University popular-culture professor Dr. Christopher Geist, said, "I don't believe anyone ever expected video games to have such a fundamental effect on our society in so many areas. [They] have become an integral part of the fabric of American life, changing the way we think, the way we learn, and the way we see the future."

Nintendo has also teamed up with *Junior Scholastic* magazine, which held a contest that had students designing the "ultimate" video game. Schoolkids in classrooms around the nation wrote their ideas up in essays. Nintendo awarded students with college grants and scholarships for the best ones. Teachers were impressed that kids who previously had been reluctant to write at all were now turning in fifteen-to-thirty-page entries. The students discussed their work with the kind of enthusiasm that teachers had not seen on any other subject.

People who complain that Nintendo-obsessed children are missing out on social skills don't understand the Nintendo cult. The exclusive club is a social network for millions of kids. To get in, you don't need to be a star athlete or the coolest or most popular kid in class. All you needed is a Nintendo system, or access to one (at a friend's, a clubhouse, or at school). Grown-ups may not understand it (or may feel excluded), but a Nintendo generation is emerging. "In the future we will be living, in part, in virtual reality," Nolan Bushnell says. "To survive and make it in that dimension, we are going to have to be mentally awake. We are going to have to learn how to live and be comfortable and maneuver in a computer environment. These kids are in training."

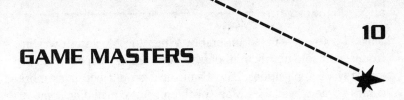

10

GAME MASTERS

For all its lofty high-tech pretensions, the video-game industry is a razor-and-razor-blade business. Just as Gillette wants consumers to buy their Atra shaver so they will then buy lots of Gillette Atra blades, video-game makers want consumers to buy their hardware so they will buy lots of their software. A consumer in America who invested in a Nintendo system ended up buying, on average, seven game cartridges (compared to an average of twelve in Japan). The average price of a cartridge was about $40. Software accounted for more than half of Nintendo's profits after 1989.

The first software available for the Nintendo system in America was made by Nintendo itself, but Arakawa knew from the beginning that a wider variety of games than Nintendo could produce would be needed. It was no different from the computer business, where it was almost always understood that the more companies developing software for a hardware platform, the better. A Texas Instruments machine with closed architecture—that is, which ran only TI software—failed miserably. When the Macintosh was introduced, it sold poorly until outside software developers released an array of applications. The Famicom also did better as more games became available. Many of these games, including some of the bestsellers, were made by third-party licensees—by Namco, Capcom, Data East, and others. Those games, in turn, helped sell more hardware, which created the demand for more software. Also, American companies could be expected to bring in new kinds of games, ones particular to the sensibilities of American players.

From the outset Arakawa planned on initiating a licensing agreement based on the one developed by NCL. In the United States, as in Japan, Nintendo would make money on all the games sold no matter who developed and marketed them, but Yamauchi had encountered problems that Arakawa wanted to avoid. In Japan, in spite of NCL's controls, the overall Nintendo business was hurt by a glut of games, many of them of inferior quality. Arakawa wanted a licensing agreement that would prevent this from happening in America. The built-in lock-out chip, Yamauchi admitted, was to censor games as well as to stop counterfeiters. He says, "It was our way of assuring consistent product quality and to keep the taste level high—no dirty games, no games with bugs or bad design." It also protected Nintendo's profit from every game sold that played on the NES—an estimated 700 yen, or $5 per game developed by other companies. Had Gillette been able to come up with a patented, copyrighted system to lock out non-Gillette-made razor blades that fit the Atra, it too could have made money on blades made by competitors.

In addition to the security system, Arakawa wanted a contract similar to NCL's to control the licensees. He and Howard Lincoln finally concocted one with restrictions so severe that even Lincoln anticipated problems. To assuage his fears—and to cover Nintendo —he had NOA's outside attorney, John Kirby of Mudge, Rose, research precedents to determine if they could get away with it. Kirby gave Nintendo the go-ahead.

For the "privilege" of being allowed to make games for Nintendo's machine, developers had to grant Nintendo the approval of the games, packaging, artwork, and commercials. These terms— which, in spite of Mudge, Rose's verdict, would eventually be chal-

lenged in court and questioned by the Federal Trade Commission
—gave Nintendo the right to reject games, or portions of them.
For instance, NOA could censor such games as "Custer's Revenge," which included nude Indian maidens tied to stakes, or a
game that had, as targets, dancing babies that melted into quiver-
ing, bloody blobs when hit by a bullet.

The plan had Nintendo evaluating every game and giving it a
rating on the forty-point scale. Licensees then would place an order for at least 10,000 cartridges. The finished cartridges, manufactured by NCL in Kyoto, would be sold back to the developers.
Depending on the memory capacity required, Nintendo charged
$9 to $14 per cartridge. The agreement stated that the price "includes both the cost of manufacturing, printing, and packaging the
[games] and royalty for the use of [Nintendo's] licensed intellectual properties." Licensees ended up taking on all inventory, distribution, and sales responsibilities—and all the risks.

This was all fairly standard, based on the earlier NCL agreement, but Arakawa and Lincoln added more. An "exclusivity
clause" was designed, whose purpose, they insisted, was to further
encourage developers to make good games. Licensees could only
make up to five Nintendo games a year and they could not release
them to play on any other video-game system for two years from
the time they were introduced. The games couldn't be sold outside
the United States and Canada. "If they could only make it for the
NES and only make a limited number of games, then it might dawn
on them that they had better make a good game," Howard Lincoln
says. "They couldn't afford to make many mistakes because they
only had five shots a year."

Lincoln, a master at licensing agreements since his Coleco–
"Donkey Kong" coup, ironed out the wrinkles. It was as airtight
and restrictive as anything ever seen in the industry, but Arakawa
insisted that there was still room for huge profits for everyone.
Licensees could sell the games to dealers for more than twice the
price they paid for them. (Dealers would double the price again
and sell them for between $30 and $55.)

Arakawa offered licensees access to Nintendo's marketing, development, and customer services: promotion in the Fun Club
newsletter (and, later, in *Nintendo Power*), development advice

from the game evaluators, and consumer service through the Nintendo game counselors. Approved games would have the Nintendo quality seal and could therefore be sold as part of Nintendo displays in retail outlets. Licensees could also show their wares in the Nintendo booth at the most important industry trade shows. At the January 1991 show, Nintendo and its licensees occupied the most exciting CES exhibit anyone in the industry had ever seen; it filled a tent big enough for a three-ring circus.

In 1985, however, no companies jumped at NOA's offer. Software companies that survived the Atari debacle felt as burned by home-video games as retailers did. They believed that the video-game business had shifted—for good—to computers. Computer games were safer to gamble on, too, since they were so much cheaper to manufacture. Publishing games on floppy disks cost nothing compared to the astronomical costs of cartridges, which required expensive computer chips. Marketing computer games also required less since the target audience was more defined.

It took half a year for the first licensees to sign up. They were American subsidiaries of Japanese companies that, for the most part, imported video-arcade games. In many cases, they had been instructed by their parent companies to work with Nintendo; many of them had already dealt with NCL in Japan.

These companies cautiously signed the agreements and set about developing home versions of their arcade games. They placed conservative orders of ten or twenty thousand cartridges and received their first shipments at the end of 1986. Everything sold.

One of these companies, Data East, was run by Bob Lloyd, a former professional basketball player. Data East was a successful arcade game company when Minoru Arakawa called Lloyd to ask if he was interested in becoming a Nintendo licensee. Lloyd considered it, since he had seen Nintendo's sales taking off through 1985.

Before meeting to discuss details, NOA sent Lloyd a draft of the third-party contract. In a subsequent meeting, he said he had a problem with the contract; he could commit to buy a minimum of 10,000 units of a game, but he wanted to be able to buy smaller reorders.

Howard Lincoln shook his head: there would be no exceptions. Arakawa later teased Lloyd when Data East was placing orders for hundreds of thousands of games at a time. "It was a license to steal," Lloyd says. Data East and all the early licensees sold an average of 75,000 copies of every game they put into the market. Kids scooped them up as soon as they hit the stores. Soon Data East was bringing in $100 million a year, about ten times its total sales at the time Lloyd joined the company.

Capcom, another early licensee, became a $160 million business, with 240 designers and programmers, because of its Nintendo games. Capcom's "Mega Man" series was a huge hit in America. It also made a lucrative deal with Disney. Games such as "Mickey Mousecapades," "Chip 'N Dale Rescue Rangers," and "Duck Tales" sold millions of copies. It wasn't always easy working with Disney, however: the Disney people had to approve every aspect of games based on their characters. Importantly, Mickey could never die; players lost "tries," not "lives."

Capcom was one of the few companies kids knew by name (most kids knew only Nintendo games, not Data East or Konami names). Capcom also had its own game-counselor system, based on Nintendo's model. It encouraged kids to buy Capcom games by offering rebates for multiple-game purchases. Fully costumed actors made promotional appearances as "Mega Man" in stores. Capcom also publicized its donations of NES and software to children's hospitals throughout the country.

Konami Industry Company, Ltd., of Kobe, Japan, established a U.S. subsidiary in 1982. It had been successful in the coin-operated video-game business with such classics as "Frogger," "Super Cobra," and "Scramble." In 1986, the company got its license with Nintendo of America and released its first game, "Gradius," an arcade hit, in February 1987. It was a big seller. The first Nintendo multiplayer game, "Rush 'N Attack" (guess what that was about) was introduced next. Other Konami releases included "Top Gun," based on the movie of the same name, which sold 2 million copies and won design awards. "Double Dragon" was another hit.

In 1987, Konami convinced Arakawa to break the rules and allow it to form a new company, Ultra, in order to get a second license; it could then release five more NES games a year. One

Ultra game would become the second-biggest moneymaker of all time.

A comic book called *Teenage Mutant Ninja Turtles*—"heroes on a half shell"—was created by freelance artists Kevin Eastman and Peter Laird in May 1984. It featured four turtles named Donatello, Raphael, Michelangelo, and Leonardo who brandished "numchucks" and other ninja weapons. According to the story, they had been real teenage turtles until a radioactive "mutigant" transformed them. As Teenage Mutant Ninja Turtles they said "dude" and "cowabunga" a lot and ate pizza whenever they could get their hands on it. They lived in the sewers under New York City.

Eastman and Laird borrowed the thousand dollars they needed to publish their black-and-white comic book. They founded a company called Mirage Studios (because the studio existed only in their minds) and traveled to a comic-book convention to hawk *TMNT* number 1. They sold 175 copies.

A UPI reporter heard about *Teenage Mutant Ninja Turtles* and wrote a syndicated story that helped sell another 3,000 comics. Amazed that they had made a profit (of $100), Mirage Studios released *TMNT 2,* which included an ad for a T-shirt and buttons. The reaction stunned the pair. They signed a licensing deal for more Turtles products, and by the end of 1990 had sold $1 *billion* worth of TMNT merchandise in thirty countries.

The Turtles was one of the most successful cross-licensing campaigns in history. There were Turtle movies, comic books, live rock 'n' roll stage shows, toys, a breakfast cereal, and individually wrapped Hostess pies "straight from the sewers to you," that were filled with slimy green pudding. The *TMNT* cartoon show was the highest-rated Saturday morning show in CBS's history. The first *TMNT* movie made $250 million.

Ultra snared a license with Mirage to make Turtle video games. In 1989 and 1990, the first two years on the market, Ultra's parent company, Konami, took in $125 million from the first "Teenage Mutant Ninja Turtles" game. More than 4 million sold.

There were various sequels. Konami sent a memo to retailers in early 1991: "Cowabunga! Check out what's ahead with your favorite reptilian moneymakers in March . . ." The list included the release of the movie *Teenage Mutant Ninja Turtles 2: The Secret of the Ooze,* in 2,500 theaters, a Ninja Turtle interview with Barbara Walters on Oscar night, a soundtrack album featuring Vanilla Ice, ads for the movie on television, radio, and in magazines (over $60 million was spent on advertising and promotion), and a tie-in for a free pizza with the purchase of the "TMNT 2" NES game.

With the help of the Ninja Turtles, Konami, which had earned almost $10 million in 1987, took in $300 million in 1991. It became the largest of the NES licensees, the eighth-largest software publisher (Microsoft was number one), and the ninth-largest toy company in the United States—almost all based on the Turtles, who also starred in hugely grossing arcade games. (One coin-operated TMNT game in a good location could bring in $1,000 a week at the peak of the craze.) Laird and Eastman's Mirage Studios became very real, earning $10 million a year.

In the spring of 1987, Howard Lincoln was contacted by the first American companies that were interested in Nintendo's licensing program. The first American licensee was formed specifically for the purpose by Greg Fischbach and Jim Scoroposki, veterans of the Atari video-game boom. The two men, who knew each other from when they worked at a games company called Activision, had laughed at Nintendo a few years earlier when they saw Arakawa and Lincoln in a minuscule booth at a trade show hawking the NES with ROB. In 1987, however, Scoroposki went to the CES convention and observed how well Nintendo was doing. It had had a healthy Christmas, and Scoroposki told Fischbach that the company just might make it. They decided to enter the Nintendo business. Scoroposki ran a sales organization that sold to toy stores on the East Coast, and with the growing number of Nintendo systems in homes and an existing shortage of good software, they felt they couldn't lose—certainly not very much. They decided to give themselves until July of that year to see if they could make some money. They teamed up with other Activision alums, Robert Holmes (former vice-president of marketing), and, to represent them in Tokyo, Hiro Fukami. To choose a name, they picked up a thesaurus. Fischbach said it could be anything as long as it started with an *A* or a *Z*, so Scoroposki opened the book and read some words aloud. They settled on Acclaim.

The new company released its first game, "Star Voyager," in August 1987, and one of the first 3-D games, called "Tiger-Heli," soon after. Orders poured in, more than their most optimistic projections. "The market was absorbing anything," Fischbach says. In the first quarter of 1988, the brand-new company made more than $1 million in profits. Acclaim grew, merging with Gamma Capital Corporation, and Fischbach and Scoroposki took it public. Analysts pushed its stock, touting it as one of the best-run video-game companies.

In 1990 Acclaim embarked on one of the most extensive campaigns of advertising, corporate sponsorship, and promotional tie-ins of all the licensees. On the back of Jell-O Pudding Pops were game tips. Free games could be won with coupons from Chips Ahoy cookies, and there were in-theater displays to promote the video-game version of "Total Recall," the Arnold Schwarzenegger film. Acclaim began its own smaller version of *Nintendo Power* (mailed to only 250,000 players), and the company copublished (with Scholastic) a book based on the game "Wizards & Warriors."

As Acclaim grew, it made a smart deal with WMS, the company that oversaw both Williams Electronics Games and Bally's Midway. Acclaim bought the right of first refusal to make Nintendo games out of all WMS arcade titles (games such as "Narc," which included digitized action of live actors, and "Arch Rivals," a "basketbrawl" game). Acclaim then became the second licensee to be able to produce more than five games a year when it bought another Nintendo licensee, LJN Toys, which MCA had unloaded. LJN had strong games in its catalog, including "Roger Rabbit," several "Spiderman" games, and "NFL Football." When the deal was being negotiated, Fischbach asked MCA's president, Sid Sheinberg, "Why don't you guys buy us instead of us buying you?"

Sheinberg pushed up his horn-rimmed glasses and responded, "We need another company like a mouse needs a hat rack."

It was a time of enormous growth for all the licensees. When LJN was sold, its chairman, Jack Friedman, founded another toy company called THQ. In late 1991 THQ's stock almost doubled in value when it received its license to sell Nintendo games. "Based on conservative estimates, we see sales of $40 million this year and $80 million in 1991," said an independent money manager in *Business Week*. He said he expected the stock to double again the following year, after which, when it hit $100 million in sales, "the likes of Mattel will want to acquire it."

Acclaim and the other companies that had signed on early were there when the number of NES users jumped from 2 to 3 and then to 10 million. By the time other companies woke up to the video-game boom, hundreds of millions of dollars were being made by licensees. Twenty-five had signed on by the end of 1987, and there were forty by 1988. Companies that waited lost the opportunity to make fortunes. Trip Hawkins, founder of the computer-game company Electronic Arts, realized that his biggest mistake was in waiting so long to enter the Nintendo business. His company missed out on sales of hundreds of millions of dollars.

Through those early years there was a sense pervading the industry that any games would sell well and that excellent games would sell enormously well. The potential was so high that the pursuit of hot games was as fervent as that of blockbusters in Hollywood. But the potential profits were higher than for movie companies because the development costs were much lower. A smash movie like *Raiders of the Lost Ark* could bring in $200 million, but it could cost a quarter of that to make and millions more to market, whereas a Nintendo game cost about $1 million to develop and, for the biggest games, several million to market. The only significant expense was NOA's central requirement: cash, payable in thirty days. After the money-back offer expired, Nintendo originally required 50 percent up front and the balance on delivery. Then most companies had to offer a secured letter of credit with their orders so that Nintendo was guaranteed to be paid on delivery. Nintendo therefore had no accounts receivable to speak of.

For the companies that could meet Nintendo's terms, the market remained voracious. In the record industry, releases with sales of a half million CDs or records were awarded a gold record and were handsomely profitable at a sale price of $12 or $15. Nintendo games were going gold every day of the week and sold for three and four times as much. "We were all fat, happy, and stupid," says Bob Lloyd.

Minoru Arakawa took the brunt of the fury when, starting in May 1988, licensees couldn't get the volumes of the games they ordered because, Nintendo claimed, of a worldwide microchip shortage. Nintendo was accused of fabricating the shortage, or at least exaggerating it. Arakawa *et al.* were keeping the licensees on a tight leash, doling out games like a child with a bag of M&Ms. Critics went so far as to accuse Nintendo of filling the orders of companies that were in favor and holding out on companies on the outs.

At this time, there was a chip shortage that affected the entire electronics industry. Nintendo was manufacturing almost 2 million hardware units a month and 6 million cartridges, and all of each contained various numbers of chips that were in short supply. When the chip shortage was at its most acute, in the middle of 1988, Nintendo responded by scrapping a dozen games in the catalog. "Donkey Kong" was one victim. It also postponed producing new games, including Sigeru Miyamoto's "Link," and, later, his "Super Mario Bros. 3." Nintendo also began the controversial practice of parceling out cartridges among the licensees. Sean McGowan, a toy-industry analyst for the investment firm of Gerard Klauer Mattison Co., told *The Wall Street Journal,* "The scarcer something is, the more status gets attached to having it. You've got to keep [games] scarce." Still, he conceded, Nintendo had not manufactured the shortage to that degree. "Demand is far higher than they thought it would be, and they would like to see a lot more shipped. It's their strategy to undership demand, but this is ridiculous."

Valid questions remained about how severely the shortage affected Nintendo and whether the company could have found alternative chip sources. In addition, there was vehement criticism about the way Nintendo allocated cartridges. Hiroshi Yamauchi and Minoru Arakawa held the fates of licensees in their hands. Hiroshi Imanishi explained the allocation system when reporters questioned him in Japan. "Licensees tell us how many cartridges they want, and then we evaluate the games. Based on that review, we decide how many cartridges should become available." Representatives of American companies that were heavily (or completely) invested in the Nintendo business were outraged. The shortage came when customers were clamoring for all the games they could get.

Arakawa, Lincoln, and Juana Tingdale, who was in charge of the licensees, fielded the callers, who ranged from pleading and frustrated to threatening. There was nothing the Nintendo brass could say to appease company presidents who had placed orders for, say, a million cartridges and had received only one or two hundred thousand. Some charged Nintendo with sabotaging their business. Some of the attacks were anti-Japanese: "they" were doing it to "us." The argument was that Nintendo could have gotten plenty of chips but refused to deal with non-Japanese semiconductor companies. Arakawa fanned the flames when he all but said that non-Japanese chips were inferior. "If Americans can make good-quality chips at cheap prices, we are prepared to buy them at any time," he said. "But we haven't been able to find the good quality at a good price here." Howard Lincoln said, "The licensing program has always been run with the approach that we would treat every licensee the same. We have never deviated from that. We want to help our friends, but we have never made exceptions."

Bob Lloyd, who was one of Arakawa's and Lincoln's closest friends, affirmed that he received no special treatment, no matter how hard he tried. Greg Fischbach of Acclaim called and pleaded with Lincoln. There was nothing to be done, he was told. "The shortage was the great equalizer," Fischbach found. "Every company sold out every game no matter how good it was, no matter how well the company was managed. Anyone with product was able to sell it."

Nintendo could have purchased more chips as the shortage eased. However, the prices rose; certain chips quadrupled in price. In some cases, Nintendo had committed to sell cartridges for a fixed price, and rather than accept a smaller profit margin, the company chose to wait. By the time chip prices fell in the latter part of 1989, a crucial year and a half had come and gone, and companies had lost a chance to make millions of dollars.

This was only one of the controversies that turned some licensees against Nintendo. Gail Tilden at *Nintendo Power* personified NOA's ability to make or break companies. Each month she and her staff looked at a list of new games and checked their evaluations. She, Howard Phillips, and several others tried many of the games themselves and determined, after consulting with their bosses, how much coverage they would get in the magazine. Tilden looked at how "deep" they were; as she put it, "how much coverage it took to demonstrate the games through maps and text." Coverage in *Power* was a significant factor in whether or not a game would hit. Licensees that had a game coming out trumpeted *Nintendo Power* coverage in connection with any other advertising and promotion strategies.

Charges surfaced that Nintendo used the magazine to manipulate and control the industry. Good games from companies that were not in Nintendo's good graces were being ignored while some terrible games received pages of coverage. The head of one Nintendo licensee said, "If I pissed Nintendo off, I would get less product. My games would get hit in *Nintendo Power,* and they'd get low ratings." He would "piss" Nintendo off, he said, by releasing games for Sega or other competing systems, or by criticizing the security chip or licensing agreement.

Tilden insisted that the licensees' own games had the same chances of extended coverage in *Power* as Nintendo's. However, as a spokesperson for one licensee said, "Their games always get covers, always get page after page in the magazine, while we are made to feel privileged to be mentioned in passing." Tilden's counterargument was that the games Nintendo released had superior evaluations and earned the coverage. Nintendo did have a sound philosophy when it came to the games it released. They followed the model Yamauchi had pioneered at NCL; only several highly rated games were released a year. NOA culled them from NCL's list and chose only a few of those released in Japan. The ones that survived the cut were more likely to be great games. That, coupled with NOA's marketing barrage, resulted in major successes one out of three times, compared to one out of twenty for the licensees.

To prove that Nintendo had the licensees' best interest at heart, it offered to assist them in game development. Howard Phillips met with licensees' designers and gave them critiques and suggestions. The Game Master's insights were deemed invaluable by some companies. "I responded as if I were a player out in the marketplace getting this one new game, comparing it with all the games that I've seen," he says. However, sometimes they were less than appreciative. "They thought we were telling them their business," Phillips says.

Tony Harman, a manager in the game-evaluation group, also worked with licensees. Besides giving advice, the group made certain that the games in development met Nintendo's standards. They didn't catch everything, however. Harman worked with a licensee called Jaleco on an NES version of "Maniac Mansion," which Jaleco had licensed from the game's creators at LucasArts. He gave the go-ahead and thousands of games were sold before someone at Nintendo noticed a quirky touch: a player could place a hamster in a microwave oven and the hamster would explode.

Nintendo informed Jaleco that the exploding hamster had to be deleted in future cartridges. In a press release, Jaleco defended the original version of the game: "Although Jaleco USA does not condone the placing of rodents into a cooking apparatus, the feature added a degree of fun to the already wacky and wild game."

With Capcom USA, Phillips's team edited some of the grislier games that came in from its Japanese parent company, although Capcom's own censors weeded out the most offensive touches. The American version of the brutal "Final Fight" was released without some of the original's flourishes: blood oozing from wounds, and villains who were exclusively black and Hispanic. When a Capcom USA representative suggested that it was tasteless to have the game's hero beat up a woman, a Japanese designer responded that there were no women in the game. "What about the blonde named Roxy?" the American asked. The designer responded, "Oh, you mean the transvestite!" Roxy was given a haircut and new clothes.

Less dramatic modifications were made in "Mega Man." In the Japanese version, the hero got strength when he ate sushi. The Americans had it changed so that he devoured hot dogs. They also changed his eyes to make him appear less Asian.

Some of the licensees tried to cash in with products other than games. They developed peripherals, or add-ons, to the Nintendo system. Besides Software Toolworks' Miracle piano, Bandai Corporation, a Japanese toy company and American licensee, made the Power Pad. It was an answer to parents who worried that kids spent too much time in front of the television, sluglike, playing Nintendo. The Power Pad, a three-foot-square flat plastic sheet, could be attached to the NES in the place of one of the controllers. Sensors on the pad "read" the footsteps of stocking feet. In combination with games such as "Track Meet," players controlled on-screen characters (a sprinter, a long jumper) with *their* running in place and jumping on the pad. Nintendo made a deal with Bandai to sell the Power Pad with the NES in America, and half a million units were sold.

Mattel released the Power Glove, a space-age piece of armor that kids strapped on to their hand and forearm. It was developed by JPL, a company that dealt in futuristic virtual-reality technologies. Using the glove in place of a controller, players could use their arms and hands to play games. They fought Mike Tyson with full punches and drove a car with their arm outstretched, fist clenched, pointed at sensors that attached to the TV. Its first Christmas, the Power Glove sold out immediately, although interest in the product evaporated once kids realized how difficult it was to make it work well.

Peripherals and games continued to be released by the hundreds a year as the number of licensees grew. There were more than sixty in 1990, when one of the most diehard holdouts, Electronic Arts, finally signed up.

In a new industrial park along the freeway that connected Silicon Valley with San Francisco stood a three-tiered building that looked like a flattened Guggenheim Museum wrapped in blue ribbon. Inside, above the receptionist's desk on the second floor, were three monitors showing off video games such as "Skate or Die," "James Pond," and "John Madden Football." To reach the office of the company's founder, one had to wade through a sea of Technicolor Nerf balls. There, behind a small desk, sat Trip Hawkins, boyishly dressed in a polo shirt, jeans, and Converse All Stars.

His straw-colored hair was stuffed unsuccessfully into a San Francisco Giants cap.

There was a philosophy behind the Nerf balls, Hawkins explained. "Whenever things get too serious around here, every employee is obligated to load up with Nerfs and lead an attack, screaming, 'NERF ALERT!' " On many evenings, EA employees would be working in front of their computer screens when someone would flip a switch and the place would be cast into darkness except for the muffled beams from emergency lights and the glow of computer screens. It was a signal to employees to take position and be on guard. Their colleagues were crawling around partitions, under desks, or around Xerox machines. Each armed with five Nerf balls, they set out on the attack. Anyone who was nailed was out. "The idea," Trip Hawkins says, "was to keep people loose and remind them why we're here. They know what's going on here when they see that you can bounce a Nerf ball off my nose in a meeting. It sets a tone for the company. . . . It's from playfulness that you get your best creativity."

Hawkins's office had Nerf balls in a fruit bowl among the oranges and apples. On the wall was a mock "Dewar's profile" and schedules of all the major-league baseball teams. Because of—or in spite of—all this, Hawkins's company had grown to be one of the largest and most respected entertainment *computer* software companies in the world.

Hawkins had grown up in Pasadena, California, where he spent a childhood devoted to playing games. In high school, he designed board games. During his first year at Harvard, he made a football simulation that used actual sports statistics.

Hawkins created his own interdisciplinary major: strategy and applied game theory, combining social science and computer courses. Back then, in 1975, he decided that one day he would start an entertainment software company. He even knew when: 1982. "That's how long it would take for the technology to get into enough homes so that there would be enough people who would want to buy it," he says.

Hawkins enrolled in the MBA program at Stanford, and after graduating joined Apple as its manager of market research. The

Apple II computer had been out for a year; Hawkins was the company's sixty-eighth employee. He helped put together a service program and instituted the industry's first field training for dealers. He also worked on the first accounting and mailing-list programs, and on a word-processor, Apple Writer. It was extremely exciting to be part of Apple during those early years. "We didn't know what we were doing," Hawkins says. "But we believed in it with everything we were."

Regardless of how well Apple did during Hawkins's tenure, he considered the company a brief stopping-off point. In May 1982, according to his plan, he left Apple at the age of twenty-eight, and arranged a meeting with a hand-picked team, renegades from Apple and other Silicon Valley companies, and Bing Gordon, a college buddy. Gordon was working in San Francisco at Ogilvy and Mather, the advertising agency. After graduating from Stanford, he had worked for a string of agencies, and then was product marketing manager for an industrial electronics firm. Tanned, with thick hair brushed to the side, he was a dandy, dressing in outfits that included expensive white shirts under a vermilion double-breasted wool vest, flowered rainbow tie, gray flannel pants, and black loafers.

Hawkins held a powwow at his house with Gordon and his group of potential founders. "Hypothetically, if there *was* a company, what should it be like?" he asked. As they brainstormed, Hawkins told the group that Don Valentine, the venture capitalist who had helped finance fledgling companies such as Apple and Atari, was ready with a $2 million investment. Before the meeting was over, the group decided to go forward. They chose the name Amazing Software. Later, inspired by the movie studio United Artists, they changed it to Electronic Arts.

At Apple Hawkins had seen that the most creative developers didn't want to work on staff. "When you put them on staff, they lost something." He believed that EA needed to find out a way to motivate people to do their work independently. He believed that the programmers were artists who ought to be treated, motivated, and marketed as such.

Hawkins used the old Hollywood studio system as a model and put software designers under contract. He supported them but gave them the freedom to work in whatever eccentric ways they chose. He was the first to credit them as authors on their games. The packaging was designed by graphic artists, like at record companies. Bing Gordon came up with an advertising campaign that summed up the company's driving principle. A handful of EA's independent software designers were photographed in a soft black-and-white shot; their faces were youthful, intense, and individual. A headline asked: CAN A COMPUTER MAKE YOU CRY?

Hawkins built the company with his rare combination of techie nerdiness and business acumen. His management style was quirky and creative. Staff meetings were something between a church service and an NFL huddle. There were impassioned speeches about "the mission," as well as lots of clowning. Before the meetings, Hawkins asked his department heads if they had any "praisings" for him. If he was told, for example, that an employee had put in a seventy-hour week in order to close the books for the month and, in the process, had discovered a glitch in the accounts-payable process, Hawkins singled the person out with lots of thanks and congratulations. He also gave out performance awards. At year-end meetings, he awarded most-valuable-player and rookie-of-the-year prizes.

EA's first year brought creative success—the Studio, as it became known, produced its first games—but it was selling less than half the number of games it could have if it had better distribution. Larry Probst, another Stanford graduate, was hired as vice-president of sales. Probst had been national sales manager at Activision and national accounts manager at Clorox, and had held various positions at Johnson & Johnson before coming to EA. He worked with Hawkins to create what they named the Electronic Arts Affiliated Labels, modeled after the distributing companies in the record industry. The idea was to hire more sales reps and build a bigger organization to help EA distribute more software. Other software companies, such as Mediagenic and LucasArts, became EA Affiliated Labels—that is, EA distributed their games—and the consortium soon became the WEA (Warner-Elektra-Asylum) of the computer-game industry. This distribution business grew to represent a third of EA's earnings.

Hawkins continued to oversee the company when it grew to

three hundred employees and three divisions. Besides the Studio, where the games were created (eventually more than a hundred game designers were under contract, managed by in-house producers), and the Affiliated Labels, there was a growing international division. Electronic Arts had many hits, but Hawkins was careful not to rely on them. No one game accounted for more than 6 percent of revenues at any time. The company had a sound reputation with good games, innovative marketing, and the most effective public-relations department in the business. The Electronic Arts spokespeople—Hawkins, Gordon, and Probst—were probably the most quoted experts on entertainment software in the industry.

EA's games were as diverse as "Skate or Die," "Populous," and a complex strategic historical war game, "Patton vs. Rommel." Many of the most successful were sports games. Hawkins recruited sports stars (Larry Bird, Michael Jordan, John Madden) and other celebrities (Chuck Yeager) to endorse games. The big names carried marketing weight. Retailers might not have wanted to hear about yet another football game, but they were interested in "John Madden Football." The endorsements came in the beginning for relatively small fees; Dr. J. signed on for only $20,000. When John McEnroe's agent wanted $350,000, EA passed.

A decision Hawkins made from the beginning that proved a successful policy was to design software for many computers, from the Apple II to Amiga to Commodore 64 to IBM. The one platform for which Electronic Arts did not make games was video-game systems. Hawkins felt that the post–Atari-boom video-game business would all land back on computers, where it belonged, and when Nintendo made its appearance in the United States, he expected that it would quickly disappear. Other companies that had been in the video-game market, such as Activision, had switched over to make games for IBM and Apple PCs. Many people in the industry felt the future had gone to PCs forever.

Hawkins felt that computers were superior in every way. The one was relatively boundless; the other was rinky-dink. He incorporated his prejudice in the original Electronic Arts business plan, a commitment to "stay with floppy-disk-based computers only." He believed there was a growing number of computer users who were hobbyists, primarily interested in entertainment, and that they bought their computers to play games. In addition, there were the people who bought their home computers for business but who would also buy a game or so every year. Hawkins believed that computers would soon become the all-purpose machines for everything from spreadsheets to "Pac-Man."

He was wrong. Most people bought computers for business, not entertainment. For their leisure time they wanted to get as far away from computers as possible—at least as far away as the family living room and the television set, to which a video-game machine was attached. In focus groups, kids affirmed this. The message EA's researchers heard was that computers were boring. Mom and Dad wanted kids to use one, their teacher wanted them to use one —computers were in the category of things that you *had* to do. When the researchers asked the kids what they *wanted* to do, the answer was nearly unanimous: Nintendo.

"The best companies and the best programmers were making computer games," one of Hawkins's game designers says. "But the Nintendo player didn't care about the sophisticated leaps we were making on computers—the frame rate of the images or incredible sound. They just wanted fun. It was like we were making gas guzzlers and the Japanese were making subcompacts. Our competitors saw the writing on the wall and started making subcompacts."

The NES sounded the death knell for the dream of the personal-computer revolution, that of one computer per family. Bing Gordon, Hawkins's buddy, compared it to the dream of James Watt, inventor of the steam engine. Watt had believed that one day there would be a single engine in every household that would connect to all kinds of pulleys and gears to run everything from the washing machine to a food mixer. Instead, technology progressed rapidly and motors were soon so cheap that every household could have many—everything from the washing machine to the Cuisinart. Microprocessors got cheaper too. Instead of one central computer running everything in a household, there were many microprocessor-based tools.

Electronic Arts had backed the wrong horse. Nintendo became gargantuan as EA stood by and watched, and Hawkins was in trouble: he almost lost his company. At one high-level meeting in 1989, when financial advisers complained about him, Hawkins pulled what one of his partners calls "his Nikita Khrushchev"; in the middle of a discussion of the fate of EA, he removed his shoe and pounded it on the table.

EA had had its worst year ever, and in the face of this downturn Hawkins had decided to expand the company's overseas operations. In the span of a year, he founded operations in England and Japan and acquired companies in Australia and France. It was too much too fast. The companies in France and Japan had to be shut down, and the British and Australian operations had to be trimmed back. When EA wrote off a lot of mistakes and finished with its first quarterly loss in six years, the board finally stepped in. Board members threatened to remove Hawkins as head of the company. One told him plainly, "You're not qualified to be the president of a company this size."

Hawkins knew they were wrong. He believed firmly that he could build EA into a profitable $100 million company; he just needed time. After he got the meeting's attention with his shoe, he says, "I ate a lot of humble pie. I wasn't a prima donna. I said, 'Fine. What is it you think I should do?' "

Lots of ideas were voiced that day, but Hawkins already knew the answer. "We had to go into the video-game business," he says, "and that meant the world of mass market; there were millions of customers we were going to be trying to reach." He spoke eloquently, and the more he spoke, the more animated he became. He addressed the group, admitting that it galled him to realize that he had been wrong. Now it was time to catch up.

Hawkins spread the news to his troops. "Basically," one engineer says, "he read us the riot act."

There were some serious concerns. The inventory risk—the need to place such large orders with Nintendo—was dangerous. EA needed capital.

In August 1990, a headline on the business pages asked: WILL ELECTRONIC ARTS SIZZLE OR SLUMBER? The article revealed that the computer-game company was going public at $8 a share. With money from the public offering, the company charged into the video-game business. Its first "buy" of cartridges from Nintendo cost $4 million, equal to the entire finished-goods inventory of all of the company's floppy-disk products—five hundred of them—on a single game. The risk was enormous.

Moreover, creating video games was different. Previously Electronic Arts had been targeting sophisticated computer gamers, and their designers weren't much interested in making games for twelve-year-olds. The company had made almost no action games.

Hawkins turned his missionary zeal toward designers, setting them the task of creating Nintendo games. Many felt the effort was beneath them, and that the 8-bit system was a giant step backward; they were used to 16-bit systems with sixteen-color high-resolution EGA and VGA displays and 640K of RAM. Such games had two, four, six, or more megabytes of instructions. To write for Nintendo meant working with its slower processor, 128K of RAM, fewer colors, and a lot less storage.

Although most Electronic Arts developers sneered at video games, some young designers were ecstatic. "At last," said Michael Kosaka, the author of "Skate or Die."

Kosaka sat at a desk so cluttered with computers and video-game systems and monitors that his office looked like a control panel at a hipper version of NASA's Mission Control. In addition, there were toys, a Darth Vader poster, a stereo, a Raleigh twenty-one-speed bike, and books about karate in English and Japanese. Kosaka was deep into his first Nintendo game, "Skate or Die 2."

Another designer contracted to EA's Studio was Will Harvey, who had founded his own company when he was only sixteen. Harvey had been an honors student from Foster City, California, an Eagle Scout, and a football player when he thought up the idea for a computer game that transformed his Apple II into a music studio. The program, "Music Construction Set," was remarkable for its time. A joystick controlled a movable hand on the video screen that picked up notes, sharps, clef signs, and other symbols, then set them down on a staff. The computer played them back when the tiny hand pointed at an icon of a piano. The program required no computerese and no previous knowledge of musical notation. Trip Hawkins saw it and decided "in about three seconds" that he wanted it. Reviewing it in 1983, *Time* said it was "one of those rare pieces of software that open up the computer market to a new class of consumer."

After further versions of Music Construction Set, Harvey came up with a game called "Zany Golf." Then he ambitiously set out to create an adventure game like none he had ever seen. The game, "Immortal," was unique because of the perspective from which the player experienced the game—as if he were looking down on the world from heaven. The main character wasn't a typically youthful, virile warrior but an ancient wizard. The best part of the game, Harvey believed, was that "when you got to the end, you realized that what you probably thought was wrong."

When Harvey's game was completed, EA sent "Immortal" to the Nintendo evaluators. This was the first time EA worked with Nintendo, and the computer-game purists were skeptical of the feedback they would receive because they assumed they knew a lot more about computer games than anyone at NOA.

Weeks later, when Nintendo came back with suggestions, Harvey was surprised at how sensible they were. The evaluators wanted him to add a more substantial musical score. They said that the wizard should have more than one life per game. The wizard's battles should, they said, take place on the screen; they should not be conceptual battles; the hacking and pummeling was an excuse for kids to press the buttons on the controller a lot.

Nintendo wanted Harvey to add a scoring system, which he resisted. "This was a quest," he said. "The only score is survival." Harvey also resisted the addition of more than one life. "Just like in life, you should have to figure it out the first time—slowly, cautiously—or die," he said. Still, he agreed to all of Nintendo's suggestions except for the scoring.

Bing Gordon said that in spite of his reservations about Nintendo, "over the course of working with them I've been highly impressed with the integrity of their people. The rating system is fair. On a scale from zero to a hundred, where zero meant the system was totally manipulated for Nintendo's self-interest and a hundred meant that it was absolutely democratic, they'd probably get a ninety. I've seen a little bit of self-interest, but this is America, the land of self-interest."

PC games, formerly EA's mainstay, became less important. EA's PC business shrank from 93 percent to 66 percent of total software sales. The PC software business remained solid (overall industry sales were up 13 percent in 1990 after declining in 1989), but it was dwarfed by video-game sales. Within a year of going public, EA stock more than quadrupled in price, reaching over $35. In late 1991 the company was trading at up to thirty-four times earnings.

In December 1990, Hawkins turned the day-to-day management of Electronic Arts over to Larry Probst, and Bing Gordon took over a larger and more visible role. Hawkins, who retained the title of chairman, had other business to attend to. In an item appearing in *The New York Times* in June 1991 it was revealed that Hawkins had stepped down at EA because he had a new project in the works. "Industry sources say that [Electronic Arts] . . . has engineers hidden in the California woods." They were, the article said, working on a new kind of video-game machine, and Hawkins was personally in charge.

Electronic Arts was one of an increasing number of Nintendo licensees that thrived. By 1991, a hundred companies had Nintendo's quality seal. Many did well, but in exchange they were turning over large amounts of money, as well as tacit control of their businesses, to Nintendo. One company, however, refused. To fight back, its executives set a plan in motion that would, they believed, permanently break Nintendo's choke hold on the American video-game industry.

11

THE BIG SLEEP

"You have no idea what you have taken on: *a tiger who will skin you piece by piece.*"

—Howard Lincoln

Minoru Arakawa had an odd habit: he fell soundly asleep at the most inauspicious moments.

On the way back from Japan one time, Arakawa and Howard Lincoln stopped off in Honolulu, where they checked into the Kahala Hilton. As it happened, the two Nintendo bosses, both passionate duffers, arrived while the Hawaiian Open was in progress at their hotel.

Arakawa and Lincoln put on their swimming trunks and headed for the pool. En route, Arakawa suggested a detour; they should go check out the tournament, he said.

Lincoln told him he was crazy; one does not simply go watch the Hawaiian Open on a whim. Tickets sell out months in advance.

Arakawa shrugged and said, "Come on. Let's see."

Along one fairway, the two men found a place to watch from behind a roped-off area and stood there for a while as a succession of powerful drives arced past them. Lincoln remarked on one of them. "Nice shot!" he exclaimed.

When he heard nothing back, he turned and discovered that Arakawa was gone. Looking around, he saw that Arakawa, who had snuck under the rope, was sitting in the middle of the fairway under a palm tree.

The golfers, including Jack Nicklaus, Tom Watson, and Lee Trevino, continued to tee off, their golf balls flying over Arakawa's head. Then, as Lincoln watched, Arakawa stretched out, his arms under his head, and fell soundly asleep.

Looking around to make sure that no one was watching, Lincoln ducked under the rope and headed over toward Arakawa, who was comfortably snoring. "Mino, for Christ's sake, wake up," Lincoln called out, shaking his friend.

The Nintendo president was unwakable, so Lincoln lifted him as best he could and dragged him off the course. Walking by at that moment, Tom Watson shook his head in disgust.

They were safely off the course when Arakawa awoke, smiling up at Lincoln. He stretched and sat up. "What's wrong?" he asked.

Most of Arakawa's catnaps were harmless, but there was one notable exception. In August 1988, Mino and Yoko threw a small but lavish dinner party in their beautiful house in Medina, an exclusive suburb of Seattle. Yoko, a studied cook, had prepared an elegant dinner: an appetizer of scallops, a mixed salad, and barbecued salmon. Their guests were Howard Lincoln and Hideyuki Nakajima and Randy Broweleit, executives of Atari Games, which had just become a licensee.

After the meal, they took their drinks outside. They sat on the expansive deck, from which they had a view of the lake and Seattle's skyline. Arakawa fell asleep.

Hide (pronounced *He*-day) Nakajima, a compact man whose conversation was punctuated with a euphonious giggle, became visibly impatient. Yoko said that it was not unusual for her husband to conk out in the middle of a dinner party, but Nakajima looked at Arakawa with disgust. He would not forget this moment.

Hide Nakajima had a bulldog stance accompanied by a deceptively gregarious grin. In a business where players never showed their cards, Nakajima appeared to be unusually open. He was a spark plug in a world of staid and sober executives.

When the original Atari was formed in 1972, Nolan Bushnell had hired a Japanese-American businessman to form a subsidiary in Tokyo. The businessman asked an attorney to help him find a general manager, and the man had recommended his brother, Hide Nakajima.

Prior to that, Nakajima had been with Japan Art Paper Company for seventeen years, working his way up the corporate ladder. (Coincidentally, for generations Japan Art Paper had sold paper to Nintendo for *hanafuda* playing cards.) In spite of a string of promotions, Nakajima was disillusioned with the company. "I saw that I was just a gear, and that no matter how much I accomplished, I was replaceable," he says. Leaving the security of the large firm and the three-hundred-year-old paper industry, he jumped at the chance to work at the small, entrepreneurial Atari, in an industry that was just being invented.

Atari Japan imported games such as "Pong" and "Gran Trak" from its U.S. parent company. Competing with dozens of new entries in the fledgling coin-operated video-game business, including Nintendo, Nakajima met the same kind of resistance to video games in Japan that Bushnell found in America, but sales slowly grew. No matter how much came in, however, the money disappeared. According to Nakajima, employees were stealing large amounts of cash. The result was that Atari's suppliers were not being paid and the company was nearly bankrupt. Nakajima's boss left the sinking, debt-ridden company and Nakajima was, by default, left in charge. Nolan Bushnell would later single out the Japan debacle as one of his first huge mistakes.

In meetings with Bushnell and other Atari executives, Nakajima argued that the company was salvageable, but the Japan subsidiary had already lost several hundred thousand dollars, which Atari could ill afford since it was struggling to establish itself in the United States. Nakajima used some of his own savings to pay off suppliers and keep Atari Japan afloat.

His friends from the paper industry suggested kindly that Nakajima return to his former job. He declined. He planned to do what he could to stop the company from sinking; failing that, he would go down with it. Bushnell felt he had no choice but to sell the subsidiary, debt and all, so Nakajima made inquiries.

Atari's name was valuable, and several coin-op companies made offers. Sega, which then made jukeboxes and pinball machines, bid $50,000. The head of Namco (then still called Nakamura Manufacturing Company) wanted Atari too. Masaya Nakamura, the founder, saw the acquisition as a way to instantly expand his small kiddie-ride company. Nakamura shocked Bushnell and all other potential bidders by offering to buy Atari Japan for $800,000, sixteen times more than Sega had offered. After negotiations, the bid was adjusted to $500,050, still an astronomical amount. Bushnell was delighted to take the money, and in 1972 Nakamura got Atari Japan and a debt that took two years to pay off.

Hide Nakajima planned to quit. He had no intention of working for Namco; it would be worse than working for the paper company because at least Japan Art Paper had prestige. Nakamura, however, convinced Nakajima to stay on for six months.

Assigned to build up Namco's international business, Nakajima succeeded, increasing sales from $5,000 to $500,000 within that brief time, and an astonished Nakamura convinced him to stay on longer. Three years later, Nakajima decided he was in the video-game industry to stay. He found working for Nakamura to be instructive. "He could be difficult, but he could foresee the future," Nakajima says. "That was his destiny. Everyone thought he was mad when he paid so much for Atari, but it turned out to be a very wise investment."

The deal with Bushnell allowed Nakamura to be the exclusive representative for Atari products in Japan for ten years, and as a result Namco became one of the largest video-game companies in Japan. Nakamura bought rights to other games, had others developed in-house, and sold them by the thousands. He also opened arcades that featured Atari games, and his earnings quintupled.

Nakamura promoted Nakajima to executive vice-president of Atari Japan in 1978. He also asked Nakajima to join Namco's board of directors. Later that year, Nakajima convinced Nakamura to open a subsidiary of Namco in the United States, and he was put in charge of the project.

Nakajima flew to California and, with a small movie camera,

filmed sites around the Bay Area. With Nakamura's approval he chose an office across the street from Atari's old headquarters in Sunnyvale. Namco America opened its doors in 1978. To help him run the new company, Nakajima hired a young attorney, Dennis Wood, away from Hewlett Packard. From Sherman, Texas, Wood had graduated from the University of Montana's law school in 1974. After passing the bar, he became a justice of the peace in Missoula, Montana, when he was only twenty-three. Later he worked in the legal department of Hewlett Packard until he accepted Hide Nakajima's offer. Namco America had started with two employees, Nakajima and a secretary, and it was ominous that the third employee should be an attorney.

Nakajima did the coin-op deals and Wood did merchandising and licensing. Wood licensed Namco's Japanese arcade games (including "Pac-Man") to companies such as Atari (in America) or Bally's Midway. He also opened a merchandising department that licensed "Pac-Man" pillowcases, pajamas, and the like. He brought with him his homespun wit and amicable demeanor. The elfish attorney became Namco America's vice-president, second to Nakajima. As the company grew, Wood was also put in charge of personnel, administration, and legal affairs. The American subsidiary brought in a significant income for Namco. Virtually all its income was pure profit, for there was no overhead. As Wood says, "What overhead? The lights. Our salaries."

Namco America grew through the early 1980s, when Nakamura visited Nakajima in the United States. Apropos of nothing, Nakamura one evening said that he wanted to buy Atari. Nakajima eyed his boss warily.

"Are you serious, Mr. Nakamura?" Nakajima asked. "I don't think it's possible. They are about a hundred times bigger than Namco."

Nakamura said, "Hide-san, you will see. Soon the sun will revolve around Namco."

In 1985, after the Atari crash, Warner sold off the scraps of the company that had once grossed more than any of the other businesses in its multitentacled organization. Steve Ross, Warner's opprobrious chairman, couldn't get rid of it fast enough. Nonetheless, Ross knew that video games and related technologies would be back, and he retained interests in both Atari Corporation (25 percent) and Atari Games (40 percent). He sold Atari Corp. to Jack Tramiel, the former chief of Commodore. Tramiel wanted a computer company but wasn't interested in Atari's game division, which he could have had for almost nothing.

The result was two Ataris, "both of which don't like each other," says Dan Van Elderen of Atari Games. "Our claim always was that we're the *real* Atari—the original Atari." Van Elderen says Tramiel resented Atari Games because it was living, breathing, thriving proof that he had been dead wrong in his decision to stay out of coin-operated games and software.

When Atari Games was on the block, Nakamura told Nakajima it was time. Nakajima, who knew the Warner executives, negotiated and got Atari Games for Nakamura for a little over $10 million. The company had assets—talented engineers, a plant, and some successful coin-op games (such as "Marble Madness" and "Gauntlet")—but it was losing money. "We were buying the potential," Dennis Wood says. Price Waterhouse accountants advising Namco warned against the purchase, but Nakamura went ahead. The size of Namco America increased dramatically; some 230 people came with the new company. So did a distribution network and a factory in Ireland with another seventy employees. Most important, though Namco was acquiring a company that had been badly managed, it made some of the best games in the history of the business.

The new owners of Atari Games wrestled the company into financial shape, helped along when "Gauntlet" became a huge seller. Throughout, Nakamura and Nakajima sparred over the best way to run it. Nakajima felt that his boss was holding him back. Rather than viewing Atari as an investment, Nakamura still saw the American company as a competitor that he didn't want to see become too powerful. He also didn't like sharing ownership with Time Warner.

Nakajima found negotiating with Nakamura increasingly frustrating. He also was angry because of Namco's shoddy distribution of Atari's games in Japan: Nakamura wouldn't sell Atari games to competing arcades. On the other hand, Nakamura was fed up with an American subsidiary outside his immediate control, so he agreed to sell Atari Games to Nakajima and Time Warner in 1987. Time Warner, Nakajima, and other employees absorbed the Namco shares so that Time Warner had roughly 80 percent and Nakajima and the employees' group 20 percent. Nakajima then resigned from Namco's board and from his position as president of Namco America.

Wood and Nakajima, running Atari Games without Nakamura's meddling, had been watching the Nintendo market, and in 1987, Wood proposed that they make the leap. The Nintendo business was growing, and it would be easy to take advantage of it with conversions of Atari's arcade games. Atari Games was prevented from entering the home video-game business under its own name because of the agreement Warner had made with Jack Tramiel. They came to a simple solution: they would create a new company.

Nakajima and Dennis Wood met in the chairman's office at the Atari Games headquarters in Milpitas, California. They were joined by Dan Van Elderen and another executive, Randy Broweleit, and together ironed out plans to start a subsidiary that would make games under license for the NES. It was named by Nakajima. The Japanese describe the *go* board as the universe. The central point of the universe, the point of the creation of all things, is the *tengen*.

Broweleit was to run the day-to-day operations of Tengen, which he did for over a year before leaving to start a company that licensed games that played on the NES. When he left, Van Elderen took over as Tengen's chief.

Dan Van Elderen had been with Atari longer than anyone, beginning, when he was only twenty-three, as a technician—before Atari even had an engineering department—and working on Nolan Bushnell's assembly line building coin-operated "Pong" games. He stuck with Atari through the sale to Warner, Warner's sale to Namco, and Namco's buy-out. He had been the senior vice-president of research and development before being put in charge of Tengen.

Tall and powerfully built, Van Elderen appeared intimidating when in fact he was a soft-spoken, kind man who had an easier time blending in with his engineers than with his business team. He talked engineer's talk more easily than profit margins and market share. An outdoorsman, he had reminders of his favorite pastime throughout his office. There was a wire sculpture of a fisherman and a sign that read: THE LORD DOES NOT SUBTRACT FROM THE ALLOT-TED TIME OF MAN THE HOURS SPENT FISHING. But in his office there was also a glimpse of another side of his nature. On the wall was a sticker with the word NINTENDO on it. The word was slashed through in red.

It was not difficult for Tengen to convince Nintendo to allow it to become a licensee. Atari Games made some of the best arcade games, and it was the name most people associated with video games. Arakawa believed it was significant: even Atari had suc-cumbed to Nintendo's dominance.

In mid 1987, Nakajima and Dennis Wood met with the top Nintendo executives at a coin-op industry show and said they were interested in a license. Nakajima and Wood said they wanted ex-ceptions to the licensing agreement—to make more than five games a year, for instance.

As composed as always, Arakawa said that changes to the basic agreement were impossible. He shrugged as if to indicate it was not personal. "All licensees have to be treated the same," he pointed out. He did agree to have attorneys negotiate some of the minor points in the agreement, but that was as far as he would bend. Still, the Atari Games executives agreed to play by Nin-tendo's rules.

Atari Games' attorneys worked with Nintendo's on the agree-ment—"tweaking it," as Lincoln puts it. Although it was signed in January 1988, it was a charade. It would turn out that by then Hide Nakajima had already begun his efforts to "put Nintendo in its place," in the words of an Atari executive.

In the spring, Tengen representatives, including Nakajima, came to Redmond for a meeting. Trying to coddle Nakajima—to smooth feathers that had been ruffled by Nintendo's refusal to grant him special terms—Arakawa shared report after report, divulging "the jewels of our business," Lincoln says. He supplied Nakajima with details of how the business ran day to day, of how individual retail-ers should be handled, and much more. Arakawa says he went out of his way to befriend Nakajima and advise him.

The first Tengen games for the NES, announced at the June

1988 CES, were a conversion of the ever popular "Pac-Man," a terrific baseball game called "RBI Baseball," and "Gauntlet," converted from the hit arcade game. The games had received high evaluations from the Big Three and GC6, and Nakajima was poised to sell many of them. His timing, though, was unfortunate, because his entry into the Nintendo business occurred during the worldwide microchip shortage. Prior to the CES, Nintendo had made an announcement to all its licensees that the chip shortage meant that NCL would not be able to fill their orders. Nintendo would apportion game cartridges, treating all companies equally, but they all would get fewer games than they wanted.

Licensees were asked to estimate the number of cartridges they would be ordering. From this list Nintendo calculated the number of cartridges they should reasonably expect. The allocations depended on several factors, including a game's ratings and the size of a licensee's distribution network. An elaborate system had been devised, NOA claimed, with rules that applied to its own games too; orders of the company's own games would be cut back or would be postponed in order to supply chips for a licensee with a better game.

The first time the calculations were made, Lincoln and Arakawa decided to personally call the licensees with the results. They thought they would run through the list in a single morning, but it took days; the people at each company begged, pleaded, argued, flattered, and argued some more. It was such a difficult process that the decisions about subsequent allocations were sent in the mail. Lincoln jokes, "Arakawa and I would run away and hide so that we wouldn't have to take the heat."

The allocation system, Arakawa insisted, was as fair as it could have been. Some of the licensees understood, but others felt Nintendo was using the shortage to manipulate—and, in some cases, strangle—them.

Hide Nakajima's orders for games were halved, then halved again by Nintendo. Tengen received less than 25 percent of its initial order and, finally, a tenth of what it claimed it could have sold. "They are keeping supply low to keep prices high," Randy Broweleit charged.

The head of the independent Software Publishers Association, Ken Wasch, charged in December 1988, "The SPA believes that Nintendo has, through its complete control and single sourcing of cartridge manufacturing, engineered a shortage of Nintendo-compatible cartridges. Retailers, consumers and independent software vendors have become frustrated by the unavailability of many titles during the holiday season, and believe that these shortages could be prevented by permitting software vendors to produce their own cartridges."

Dan Van Elderen asked if Nintendo would allow his company to receive larger orders if he found sources for the scarce chips. Nintendo agreed, on the condition that Tengen pay the difference if the chips were more expensive. Other companies went the same route; Acclaim searched for chips too. Van Elderen found a source and informed Nintendo, whose representative said the company would evaluate the chips in Japan and, if they were acceptable, inform Tengen how much it would pay for them. "*They* would tell *us*," Dennis Wood says. But the pricing issue never came up, because Nintendo decided the chips were unacceptable.

Wood claims that Nintendo rejected the chips because they were not made in Japan, and that NOA said that American or Korean chips were not of a high enough quality. "We're talking about chips for games, not for a Cray computer," Wood steamed. "You don't have to wear a conical hat with the sign of the zodiac on it in order to make these chips nowadays. It's not quite as simple as going to a hardware store and picking up a bag of nails of different sizes, but we are not far from that." He further charges that Atari Games contacted Sharp, the Japanese electronics company, which said that chips were available—until Sharp learned they were to be used for Nintendo-compatible games. Then Sharp recanted; there were no chips after all.

Acclaim, however, did find chips that NCL approved, although not until 1989, according to Greg Fischbach. Nintendo insisted that the chips most other companies found were of inferior quality. Yamauchi and Arakawa refused to accept them, insisting that they were more concerned about the long-term integrity of products with the Nintendo quality seal than the short-term profits of licensees. "From our point of view, we gave everything we could," Lincoln says. "People were going to have to make millions, not zillions."

"It really set us off," Wood says. "We knew then that we were being jerked around."

Van Elderen was incensed. "Frankly, the historical roots of this company are Atari. *Atari!*" he says. "We created this industry eighteen years ago, and we weren't going to be told by anyone that we couldn't play at our own game."

"We didn't know what was going to happen," Wood admits, "only that Nintendo could literally strangle us with a silk scarf." Van Elderen says his company was backed into a corner. "It became obvious that we had to make some different arrangements, and that's when we decided to work around Nintendo."

In fact, the Atari Games/Tengen chiefs had decided long before to work around Nintendo. The efforts, begun even before the chip shortage, were kept secret for almost a year, Van Elderen admits, even though Atari Games continued to work with NOA as if everything were "all sweetness and light," as Howard Lincoln puts it.

In August 1988, Hide Nakajima and Randy Broweleit met Arakawa and Lincoln in the Donkey Kong conference room to discuss the strategy for selling Tengen's three initial games. Then the representatives of the two companies played golf at Arakawa's club, and the Arakawas gave the fateful dinner party for Nakajima and Broweleit at their new home in Medina.

The dinner progressed, but something odd was in the air. When Howard Lincoln accompanied Yoko Arakawa to the kitchen to help with drinks, she whispered to him, "What's going on with these guys?"

"We couldn't put our finger on it," Lincoln says.

Apparently oblivious, Arakawa, who had had a couple of glasses of wine, fell asleep when the gathering moved outside after dinner. Even though he awoke in time to say good night to his guests, Nakajima viewed the Nintendo president differently. Everyone pretended that nothing had happened, but something was deeply wrong.

In October Nakajima called to invite the Arakawas to play golf at Pebble Beach. There were only subtle hints that anything was amiss. Nakajima was chatty and amiable, but amidst the social-izing, he asked Arakawa more questions about Nintendo's business. Arakawa was wary, but he answered them because he was trying to make up for the dinner party.

Back in 1986 Dennis Wood had had attorneys go over Nintendo's licensing agreement to determine if there was a legal way Tengen could produce and sell games that would play on the NES without going through Nintendo. The lawyers deduced that Tengen could do so if there was no infringement on Nintendo's patents or copyrights. This meant that Tengen would have to come up with its own chip that defeated the security system—that is, Tengen would have to unlock the lock-out chip.

On the assumption that customers wouldn't care whether or not Tengen was a Nintendo licensee, that stores just wanted product, and that customers would buy good games regardless of the Nintendo quality seal, Nakajima had a group of engineers analyze the Nintendo security system. They found that the security chips in the hardware and software were identical chips that, basically, communicated with each other. As long as they were communicating, the system operated; if they weren't, the system froze up. The engineers then tried but failed in their efforts to replicate the technology. Atari engineer Pat McCarthey concluded, "Unless there is a specific profit motivation . . . I recommend that the investigation end here." There *was*, the court would later find, a profit motivation, and Atari did not discontinue the project.

Atari had outside engineers try to reverse-engineer the chip—that is, "deprocess" it. Engineer Donald Paauw was assigned the task of analyzing "peeled" or dissected chips in an effort to understand the program embedded within, but he did not succeed. Finally, after the engineers failed to figure out the chip by reverse engineering, they were given some help.

Nintendo's security system was the subject of two protected intellectual properties. NOA had filed for a patent on the lock-and-key system itself in 1985—number 4,799,635, entitled "System for Determining Authenticity of an External Memory Used in an Information Processing Apparatus."

The other piece of intellectual property that Nintendo protected was the song the security chips "sang," the computer code, which was known as the 10NES and was registered at the U.S. Copyright Office. Copyright protects original works—songs, literary works, computer programs—as opposed to inventions or formulas, which are covered by the Patent Office. Initially Nintendo had no plans to register the copyright because the code itself would have to be placed in the Copyright Office's unpublished-works storage facility, in Landover, Maryland, but an attorney specializing in copyright law advised Howard Lincoln that the code would be secure. Although individuals could examine files in the Copyright Office, no one was allowed to remove anything; even note-taking was prohibited. Reassured, Lincoln decided to submit the program, so the entire computer-generated code, indecipherable incantations to anyone who didn't speak computer language, was locked away in a file inside a storage vault in Landover, Maryland.

There was, in fact, one legitimate way for someone to get access to a copyrighted code. An affidavit could be filed indicating that the work was the subject of litigation. With the affidavit, a copy of a work could legally be removed from the office. The odd logic went like this: a company being sued for violating a copyright couldn't defend itself unless it could review the copyrighted material. It was explicitly stated that no one was allowed to use the material for any purpose other than this.

Atari Games hired a local Virginia law firm to get the code. The firm may have believed that the request was legitimate, for it had been told that Nintendo had sued Atari Games. Ten days after Atari Games signed its agreement with Nintendo, on January 28, an employee of the law firm filed an affidavit in the U.S. Copyright Office. The affidavit, filed on behalf of the Virginia firm's client, Atari Games, indicated that a copy of the code was required for pending proceedings under way in U.S. District Court, in the Northern District of California. The form indicated that the program was "to be used only in connection with the specified litigation."

The code was sent over from the Landover storage facility to the Copyright Office. An employee of the law firm headed up to room 402 and waltzed out with a copy of the 10NES copyright in his briefcase.

No suit had been filed against Atari, and none would be filed until November 1989, almost two years later. Atari would later defend the action by claiming that litigation was imminent at the time, but the judge who would preside over the lawsuit refused this defense. The court said, "Atari's purpose in obtaining the program in early 1988 was commercial rather than legal." Tengen's Van Elderen later told reporters that his team had succeeded in reproducing the lock-out chip by the process of reverse engineering—unraveling the design—when, in fact, they had taken it from the copyright documents.

Tengen's Silicon Valley R&D division was in a series of back offices of the Atari Games building, which was set atop what had once been an orchard of prune trees. There a team of engineers pored over the illegally obtained code and worked to create their clone chip. By comparing the information obtained from the Copyright Office with copies of the binary code read through microscopic examination of "peeled" chips, the engineers were able to correct and verify their version of the program.

With the documents in hand, the Atari engineers had only a slightly more difficult time making an exact copy of the Nintendo security system than they would have had, say, assembling a bicycle that came with clear instructions. Atari engineers re-created the Nintendo chip with the embedded incantation that allowed it to sing to the chip inside the NES, and dubbed their version the Rabbit. Installing a prototype Rabbit into a game, they plugged it into an NES system, and pushed the power button to fire it up. Immediately the TV monitor lit up with the Tengen logo, and by August 1988, when Hide Nakajima was at dinner at the Arakawa home in Medina, Atari Games was producing cartridges on its own that incorporated the Rabbit.

At the end of the year, on December 12, Atari Games filed suit against Nintendo in U.S. District Court in San Francisco. Essentially the suit claimed that Nintendo was succeeding at the expense of all potential competitors by means of monopolistic and exclusionary business practices. "Through the use of a technologically sophisticated 'lock-out system' . . . Nintendo has, for the past several years, prevented all would-be competitors, including Atari, from competing with it in the manufacture of video-game car-

tridges compatible with the Nintendo home video-game machine,"
the document charged. "The *sole* purpose of the lock-out system is
to lock out competition." The suit further claimed that the lock-
out chip and Nintendo's monopoly of the industry interfered with
competitive pricing, allowing Nintendo to control the supply and
prices of cartridges available to consumers. "The impact of Nintendo's conduct has been to block any competition in the manufacturing market for video-game cartridges compatible with the
Nintendo machine," the suit stated.

In the complaint, Atari Games announced that it had developed
the functional equivalent of a key to "unlock the lock-out system,"
and said it was beginning to compete with Nintendo.

In early December 1988, Nintendo held its grandest Christmas
party ever. There were tables of freshly carved roast beef and
turkey, and bowls of spiked punch and endless champagne. Employees and their spouses danced to the music of a big band. The
following morning, when the staff was suffering a collective hangover, one of the company's public relations people called Howard
Lincoln. "You're not going to believe this," he told Lincoln, "but
there's something coming over the wire that you ought to know
about. Tengen is having a press conference and is announcing that
they reverse-engineered your security chips and are going to start
making games for the NES without a license. They've filed a lawsuit for $100 million against you for violation of the antitrust laws."

In the press release, Atari Games' Dennis Wood took the offensive. "Who gave Arakawa, Lincoln, and Main the power to decide
what software the American public can buy?" he asked.

Arakawa, whose hangover put Lincoln's to shame, heard from
his partner. "Guess what those sons of bitches did," Lincoln said.
The two of them remembered the golf game at Pebble Beach, the
dinners—all the times Nakajima had pumped them for details
about the Nintendo business—and all of the information they had
stupidly provided. "We'd been scammed," Lincoln says.

Faxes flew back and forth across the Pacific Ocean. When he was
told, Yamauchi said he wanted Atari stopped, whatever it took.

Arakawa, Lincoln, and Peter Main met to decide how to respond. They agreed that an immediate response was critical; if they

could stop retailers from selling Tengen's games and simultaneously go after it for violating the agreement, Nintendo would suffer little damage. They also concurred that they had to learn how Atari had broken the security system. Later, when they did, Lincoln said, "They entered into a contracting relationship with us knowing they were going to screw us. They were trying to reverse-engineer from the start and they were getting nowhere. They abandoned the reverse-engineering project and all of a sudden our source code pops up. Bang. There is a lot more to this."

The morning of the Atari Games press conference, Nakajima called Arakawa. He said, "I guess you heard what's going on." He said he wanted to talk in person. "We expected to be sued," Nakajima said afterward, "but I thought I might be able to settle things before they escalated." When the two men met at the Seattle-Tacoma airport the following day, Arakawa held back his anger in order to hear what Nakajima had to say.

Nakajima seemed tense and awkward. He told Arakawa he was against the tactic his company had used. He blamed others at Atari Games and said that it would not have gone this far if Nintendo had been more flexible. "Let us do our own manufacturing," Nakajima said. He indicated that he was still open to being a licensee if Nintendo would give in, and that Atari Games would withdraw its lawsuit. Arakawa walked away.

After the meeting, Arakawa reported to Lincoln what Nakajima had said. Lincoln shook his head. "Now I understand the look on Nakajima's face when you fell asleep [at the dinner party]. The *contempt*," he told Arakawa. He said Nakajima thought he was the stronger man, and that he saw Arakawa as nothing more than the vain and disrespectful son-in-law of Hiroshi Yamauchi. Lincoln now says, "I thought to myself, you have no idea what you have taken on: *a tiger who will skin you piece by piece.*"

A battle for hundreds of millions of dollars ensued. Tengen tried to sell its games—notably lacking the Nintendo quality seal—and NOA pulled out all stops in its attempt to stop it. Posing as a David against the Goliath of Nintendo, Tengen fought viciously.

Nintendo filed a twofold countersuit almost immediately, charg-

————————————————————————

ing Atari Games with "the fraudulent inducement" of Nintendo to enter into the licensing agreement and with the sale of unauthorized and unsupported games. It claimed that Atari Games violated the Racketeer Influenced and Corrupt Organization Act (RICO) by creating Tengen as a front company in order to defraud Nintendo. In good faith NOA had given substantial marketing and technical support for the Tengen NES cartridges. Then, the complaint continued, "having achieved its goal of strong public identification with the games and packaging with Nintendo, beginning sometime in late December [1988] or early January, Tengen commenced selling unauthorized versions of these same games."

Part of Nintendo's lawsuit claimed that Atari Games, needing more than the security system in order to go after Nintendo's market, had also had to learn how the Nintendo business worked. Because Atari Games had a Nintendo license, it was given all the information the other licensees had, but because of Nakajima's personal relationship with Arakawa, Atari Games was also given detailed information about retailers. As a licensee, it had no trouble acquiring distribution. As a result, Tengen's Nintendo business grew to about $40 million a year for Atari Games—although, says Dennis Wood, unimpeded by Nintendo it would have been *hundreds* of millions of dollars a year.

The press saw larger ramifications in this litigation. "A verdict against Nintendo might prohibit the company's use of the lock-out chip in its control decks, creating opportunities for any independent software developer who wished to design games for use on the Nintendo machine," said an article in *The New York Times* in March 1989. "A verdict in favor of Nintendo would probably have a spillover effect into the personal-computer industry, where it could have a chilling effect on the free flow of ideas and innovations that have characterized that market since its inception."

The lawsuit was only the start of Nintendo's effort to fight Tengen. Nakajima was informed that his company would not be allowed into the massive Nintendo booth at the January 1989 Consumer Electronics Show. Nintendo had its largest presence yet at this show: its "booth" was a 40,000-square-foot, black, high-tech structure that passersby called the Death Star. Inside were lavish displays by Nintendo and its licensees. It was as if the Death Star housed the legions of Nintendo loyalists while, outside the gates, the infidels—Tengen and other Nintendo competitors—were poised in ramshackle encampments.

Nintendo executives also decided to tighten the screws on Tengen by going directly to retailers, sending them a letter in which NOA threatened to sue any company carrying Tengen games. On behalf of Nintendo, John Kirby wrote Charles Lazarus, chairman and CEO of Toys "R" Us, on January 24, 1989, "If your company handles products which infringe Nintendo's patent or other intellectual property rights, Nintendo intends to avail itself of the full range of its legal remedies." When he did not hear from Lazarus immediately, Kirby wrote again six days later. Toys "R" Us had to "cease and desist" its sale of Tengen cartridges and reply by February 1. An attorney for Lazarus wrote back that the company would comply. An article in *The American Lawyer* in April 1990 reported that Kirby faxed back still another letter. "I understand from your letter that Toys is immediately removing the product from the shelves in all its stores. . . ." it read. "It is vitally important that this be confirmed to me immediately. I will call you at 4 P.M. today . . . to confirm this fact. Needless to say, we have had investigators purchasing the product at various Toys 'R' Us stores and they are being instructed to return to those stores at 5 P.M. today."

Atari Games sought and won a preliminary injunction to stop Nintendo from threatening dealers, but the decision was reversed on appeal. Either way, Nintendo had made its point, and it continued its campaign of intimidation. Some of it was carried on in court, some of it by less overt means. "Companies would not carry our games because there was pressure from Nintendo which could jeopardize their business," Nakajima says. "Even the big companies like Toys 'R' Us couldn't stand up to them." Dan Van Elderen claims the intimidation was subtle but effective. A Nintendo rep would say quietly, "You know, we really like to support those who support Nintendo, and we're not real happy that you're carrying a Tengen product." Then, after a pause: "By the way, why don't we sit down and talk about product allocations for next quarter? How many 'Super Marios' did you say you wanted?" Van Elderen says he had distribution in the top fifteen retailers in the country before these threats, but that every one of them subsequently dropped Tengen.

It was certainly true that few retailers would risk Nintendo's threat of a lawsuit against them. Beyond that, retailers were beholden to Nintendo for the steady supply of product, which in some cases represented their entire profit margin. If NOA held back supply of a hot game, customers would flock to competitors across town, or even in other towns. No company had ever been in this position, which meant that none had ever had such clout. Despite repeated denials, Nintendo exercised it without much delicacy, even when the court enjoined it from suing retailers while the suit with Atari Games was pending. "If a retailer carried Tengen games, their Nintendo allocations would suddenly disappear," a representative of one retailer charged. "Since it was illegal, there were always excuses: the truck got lost, or the ship from Japan never arrived."

The upshot of real or imagined intimidation was that Toys "R" Us, Bradlees, Target, Wal-Mart, and most other large retailers refused to carry Tengen's or any other unauthorized games. Since retailers wouldn't carry them, it was pointless for companies to make them. Nintendo had made its point. The head of one software firm told *The American Lawyer* that he had been "at numerous meetings of conspirators" who wanted to get around Nintendo's system and that they all "chickened out." The reason: "You don't fuck with a nine-hundred-pound gorilla."

Al Chaikin, CEO of Circus World Toys, with 328 stores across the country, who admitted that Nintendo was "the most sought-after product in the business today," also revealed how the company worked. Tengen's games were popular, he told the court in a deposition in 1990, and he felt that they were good products. He nonetheless succumbed to pressure when Nintendo threatened to sue companies that carried the games. After receiving a threatening letter, Chaikin responded to John Kirby in a letter he sent in June 1990: "Your threats to take action against us before Nintendo's dispute with Atari Games can be decided by the Court leaves us no choice but to discontinue carrying Tengen cartridges. . . . To that end, and to insure uninterrupted supply of Nintendo products to us in the ordinary course of business, we are discontinuing our purchase of Tengen cartridges. Inventory already on store shelves will not be replenished."

Kirby was not placated, writing back to Chaikin's attorney: "I am very pleased that Mr. Chaikin has consulted with counsel with respect to my letter. . . . However, both my client and I find the tone and substance of his response thoroughly inappropriate. . . . I do not accept his statements which indicate that he intends to respect these rights in the future only to please Nintendo rather than as an acknowledgement of the legal rights of Nintendo as to which you and he have apparently made a decision."

Chaikin felt the squeeze in a way that threatened his entire operation. In the deposition, he charged that Nintendo had changed its credit terms with Circus World because he carried Tengen games. Nintendo had previously allowed Circus World Toys credit, but then changed to cash in advance. An attorney for Atari Games asked him, "[Then] was there a time when you ceased dealing with Tengen products, and shortly thereafter Nintendo afforded Circus World better credit terms?"

Chaikin responded, "Yes, there was."

Stuart Kessler, a vice-president for the Ames department stores, was also deposed by Atari Games. Ames (and its Zayre division), with 461 retail outlets, had been doing about $10 million a year in Nintendo business. In a letter, after NOA threatened Ames for carrying Tengen games, Kessler wrote to Howard Lincoln: "We value our relationship with Nintendo and would do nothing to jeopardize our future together."

Lincoln had written to another Ames vice-president, Earl M. Spector: "Ames has known for more than one year that Nintendo considers the sale of unlicensed Tengen video game cartridges for play on the Nintendo Entertainment System to be a violation of its patent rights. Ames has continued to sell Tengen cartridges in blatant disregard of Nintendo's rights. In light of Ames' decision to sell infringing cartridges, Nintendo has decided that it will cease doing any business with Ames. Henceforth, any purchase orders submitted by Ames will be rejected." The letter was signed by Lincoln, "Very truly yours . . ."

Ames was cut off in August 1989.

Meanwhile, in preliminary hearings of the lawsuits, Atari maintained that the code it had received from the Copyright Office had nothing to do with its ability to reverse-engineer the security system. The judge—Fern Smith, of the U.S. District Court for the Northern District of California in San Francisco—refused to believe this. In a strong opinion she wrote in March 1991, when granting Nintendo's request for a preliminary injunction against Atari, she lambasted Atari for thievery. Her conclusion was made after she compared the source code that Atari's engineers supposedly arrived at with Nintendo's code as received from the Copyright Office. They were almost identical. Atari's Rabbit code included far more information than was necessary to make the chip work. Had Atari actually created it independently, there would almost certainly have been more variation.

Atari Games had hurt itself, perhaps irreparably, by securing the code from the Copyright Office under false pretenses. "Atari lied to the Copyright Office in order to obtain the copyrighted 10NES program," Judge Smith wrote. "The court disapproves of Atari's argument that Nintendo's subsequent claims of infringement retroactively justify dishonesty. . . ."

In the conclusion of her opinion, Judge Smith prohibited Atari from copying, selling, or using Nintendo's copyrighted computer program in any way. She ordered Atari to halt marketing, distributing, or selling its NES-compatible cartridges, and to recall all its product in stores. Although it was far from the end of the litigation, the court also indicated that Nintendo had the right to "exclude others and reserve to itself, if it chooses," the right to sell game cartridges. The entire industry watched the decision spellbound, viewing it as a body blow to competitors who had hoped to challenge Nintendo.

In early 1989 Atari Games filed a countersuit to Nintendo's countersuit, this one accusing NOA of infringing on one of *its* patents. "Nintendo built its business on borrowed technology," Dan Van Elderen says. The original Atari had patents on all kinds of devices in *all* video-game systems—motion technologies and circuitry, for example. Although the company had never pursued the myriad companies that had built systems based on those technologies, it now tried everything it could against Nintendo. "The first time we hauled the patents out as a weapon was with Nintendo," Van Elderen admits.

Atari Games was not alone in suing Nintendo for patent infringements. Magnavox, which had patents from its early research in video games, charged Nintendo with infringing a patent that had to do with on-screen game play (the technology that caused the caroming of balls or bullets off walls or enemies). Another company, Alpex Computer, sued over infringement of a patent related to the relationship between video-game microprocessors, memory chips, and the television screen; and an inventor named Jan Coyle sued for infringement of *his* patent on technology behind the NES's color encoding. Nintendo settled with Magnavox and Coyle. Alpex claimed that Nintendo was using its patented technology in its system and in more than 150 games. This suit, unresolved through 1992, could prove expensive for Nintendo. The Atari Games suit, however, could cost Nintendo even more and set a precedent because its ability to control software for the NES was at stake.

Howard Lincoln's adversary at Atari Games, Dennis Wood, planned obsessively the many fronts of his attack. He sat in his office sipping the *kukicha* poured by his assistant, June Yamamoto, lifting the fragile porcelain cup to his lips and blowing across the hot liquid. "We will not give up," he said, even though, as he admitted, "they keep hitting us every time we stand up."

Another significant case was brought against Nintendo by Jack Tramiel and his son, Sam, respectively the chairman and president of Atari Corp. The Tramiels sued Nintendo for $160 million for violation of antitrust laws. As a hardware manufacturer (it released follow-up systems to the 2600 that never caught on, as well as a hand-held video-game system), Atari Corp. claimed that the Nintendo licensing agreement preventing licensees from releasing Nintendo games on competing systems for two years amounted to unfair restraint of trade. Because of that clause in this agreement, the Tramiels couldn't get good games.

Howard Lincoln told the press he considered the suit "meritless, simply an attempt to excuse Atari's poor competitive performance in the marketplace," and added that he relished the opportunity to go to court with the case. "Our defense is really simple," he said. "We are going to put Sam Tramiel on the stand and he is going to explain how, in 1985, he had 100 percent of the market for home video games, and [that] the home video-game business was synonymous with Atari and no one had ever heard of Nintendo. And then we're going to demonstrate how, through his own ineptness and idiocy and mismanagement, he took that franchise and shot it in the foot and killed it. We'll show how he was quite successful in doing it and literally went from owning 100 percent of the market to no market share. I think we will do fine."

Atari Corp.'s contention was essentially the same as that of Atari Games, only in reverse. The two-year limitation hurt Atari Corp. because Nintendo licensees' best games were tied up, and it hurt the licensees because it limited their market. Those were two years that software companies lost the ability to profit by selling their games for other systems.

The implications of the two suits were enormous. Dan Van Elderen said, "If Nintendo loses on that key argument, they have a huge liability—almost overwhelming. All the licensees could have been unfairly restricted by Nintendo." The Atari Corp. suit, the first to come to trial, could cost Nintendo almost half a billion dollars (damages are tripled in this kind of antitrust case). Other companies could also jump on the bandwagon and sue. Atari Games' case would be even costlier, and the nine-hundred-pound gorilla could be brought to its knees.

Arakawa was called by Atari Corp. lawyers into Judge Smith's San Francisco courthouse. Wearing a double-breasted navy-blue suit, he leaned forward on the stand, listening carefully to the questions. "Isn't it a fact that if Atari or Sega was also being carried, the salesman would go in and say they won't be able to carry Nintendo?" the attorney asked.

In a voice decibels lower than the lawyer's, Arakawa answered, "Definitely not."

The lawyer asked, "Did Nintendo ever tell any licensee that they could only make games for Nintendo?"

"No."

"Did Nintendo ever tell them, the licensees, that if they put games on any other system they would be penalized?"

"No."

"That Nintendo would reduce their allocation of chips during the chip shortage?"

"No."

"Cancel trade-show space?"

"No."

"Any threats to prevent them from making games for other home video-game systems?"

"No."

The court cases slogged along. Nintendo's business was "video games and litigation," said an employee. In the middle of a deadline for a new issue of *Nintendo Power,* Gail Tilden was handed another weighty stack of papers by a member of Howard Lincoln's staff of lawyers. She shrugged. "We spend a lot of time doing depositions these days." Arakawa noted that Nintendo's legal bills ($20 million a year by 1990), although a significant amount of money, were nothing when compared to total sales. "Nintendo made a billion dollars this year," Hide Nakajima said in 1991. "They can spend all the time and money it takes to destroy us." On the other hand, Dennis Wood claimed that the Nintendo cases could have broken Atari Games, and Nakajima concurred. In the meantime Atari Games pleaded with the judge for a stay of her ruling that forbade it to sell its games. The ruling was destroying the company. On April 11, 1991, the judge agreed to allow Tengen to sell its games pending the outcome of an appeal. A year later, in September 1992, Tengen lost the appeal and had to recall all its games that played on the NES. The judge meanwhile set a date for the trial of Nintendo's litigation against Atari Games in May 1993.

Howard Lincoln never believed Nakajima's contention that he was fighting Nintendo on his own. "There's no question in my mind who's running the litigation against us," he alleged. "It's not Nakajima at Atari Games, and it isn't Jack and Sam Tramiel at Atari Corp., who are just on the bandwagon like vultures looking for scraps. It's got to be Steve Ross at Time Warner. He wants to be back in the video-game business and he sees this as a tremendous opportunity to make some money. He has nothing to lose and everything to gain. We keep beating the hell out of them, and they keep hemorrhaging money. Why are they doing this?"

Time Warner's large interest in both Ataris lends weight to Lincoln's assertion, though Ross refuses to comment. An Atari Games spokesman repeatedly denied that deep pockets were fueling the lawsuit, but Manny Gerard, Ross's old Warner buddy, quoted a senior Warner executive as saying, "The biggest asset to come out of the Atari experience for Warner may lie in the Atari Games–Nintendo lawsuit." Gerard says, "For Steve, a significant percentage of a jillion dollars would be a nice consolation prize. At least he would have gotten something for the Atari nightmare."

The two Atari cases inched along. Months of discovery turned into years, with motion after motion, countless status conferences, and tons—literally—of depositions. Meanwhile, the clock ticked on Nintendo's invulnerability. It had made too many enemies to survive all these attacks unscathed.

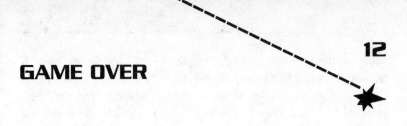

12

GAME OVER

WARNING: Your child may be addicted to a product pushed by a criminal racketeering enterprise. Kids as young as five have even given up TV for their habit, hooked by assertive characters named Mario and Luigi.

The alleged criminal enterprise is Nintendo. . . . If the legal accusation of toymaker as mobster is surprising, perhaps all is explained by the fact that Nintendo is a Japanese company whose success comes at the expense of an American former market leader.

—L. Gordon Crovitz,
The Wall Street Journal

When Nintendo's competitors failed in the marketplace, they not only turned to the courts, but also went to legislators with their accusations that the company was perpetuating an illegal monopoly that not only deprived American companies of their right to do business but, in the process, added billions of dollars to the existing trade deficit with Japan.

Atari Games' Hide Nakajima and Dennis Wood had found Nintendo's Achilles' heel, and they took aim at it: Nintendo was a Japanese company enjoying extraordinary success in America at a time when Japanese companies were the object of American distrust and hostility. Soon members of Congress, the Justice Department, and the Federal Trade Commission would all have their sights set on Nintendo. Nakajima, himself Japanese, may not have set out to place Nintendo in the front line of the trade war, but he did. Apparently all was fair in the video-game wars.

Slowly Americans began to notice that the game was over. The

Japanese had *already* landed. Disguised as a video-game system, the invaders had been carried into America's living rooms by children. The growing video-game business—worth in the neighborhood of $5.5 billion by 1992—was owned by Nintendo and a few smaller competitors, all of them Japanese. In the United States, people were finally beginning to ask, "Where are the American companies?"

As legislators realized that Nintendo was expanding its operations in America and seeking broader markets—markets that corporations such as Apple and IBM had assumed were theirs—Capitol Hill's interest began to heat up. Congressmen, meeting in closed-door sessions (which they followed with melodramatic press conferences) would soon portray the Nintendo system, this time without flattery, as a modern-day Trojan Horse—a computer hidden inside a toy with greater dependence on Japan and a larger trade deficit lurking within. In the media poised to jump on the bandwagon, Nintendo would become one of the worst perpetrators of the Japanese devastation of the American economy. The repercussions were manifold.

In early 1992, Hiroshi Yamauchi, previously unknown to the American public despite Nintendo's prominence, would land on the front pages of newspapers when it was announced that he was buying Seattle's major league baseball team, the Mariners. Unfortunately, Yamauchi's offer was announced in the same week that a Japanese legislator publicly made the indelicate suggestion that American workers were lazy. On the heels of this highly publicized slur, the Yamauchi-Mariners deal was the final straw: "They" had the audacity to attempt to buy a piece of American baseball, far dearer in the hearts of its citizens than Rockefeller Center.

Nintendo would become a tennis ball in the election-year face-off between the United States and Japan, with epithets volleying back and forth. Meanwhile, baseball commissioner Fay Vincent would argue that we had to keep *something* sacred, out of the grips of Japan—all this despite the fact that Seattle city fathers had solicited Yamauchi's bid in a desperate attempt to keep the team in town. Nintendo hardly needed the attention. It was already a target on Capitol Hill, in the offices of state attorneys general, and in the press. The inquiries that had begun into its business practices would, at the least, cost Nintendo some money. They might also, in the words of one investigator, "make it very unpopular to be associated with Nintendo in this country."

This counteroffensive against Nintendo began in 1989, when the Atari Games executives went to Washington and were greeted by the sympathetic ear of the chairman of a congressional subcommittee on antitrust, Representative Dennis Eckart, a Democrat from Ohio. Dan Van Elderen insists that the investigation on Nintendo was under way before Atari Games got involved, and Eckart corroborates the claim, saying that it was a response to complaints from consumers about Nintendo's high prices and short supplies, as well as his own observations about the video-game market. Howard Lincoln counters that Atari Games appealed to its local congressman, Tom Campbell, who happened to be on Eckart's subcommittee. A Campbell staffer confirms that Atari Games approached Campbell. In any event, Eckart set out to build a case against Nintendo.

Atari Games executives Van Elderen and Wood were star witnesses in the congressional investigation. Eckart also questioned dozens of others, many of them off-the-record informants who feared reprisals by Nintendo. If the congressman hadn't been convinced by Atari, the investigation confirmed for him that Nintendo was a huge, monopolistic demon.

Eckart accused Nintendo of antitrust activities related to the security system and the licensing agreements with software developers—what Atari Games called "anticompetitive conduct in the retail market." Eckart further charged that it was illegal (as tested in an IBM case from 1969) for a hardware company to restrict software companies from creating products for its machine. He attacked Nintendo's practice of "bundling" hardware with the software needed to make it work, and he claimed that its powerful position vis-à-vis retailers and competitors had enabled it to hike up prices by 20 to 30 percent, particularly during the preceding Christmas shopping season. Therefore the public was paying excessive prices because of NOA's monopoly.

Howard Lincoln first heard about the investigation when it was nearly over. He was at a cocktail party in Toronto in October 1989 when the head of the Software Publishers Association asked if he knew that his company was under investigation by Congress.

Two weeks later Lincoln was in Washington to meet with Nintendo's lobbyist, Don Massey, senior vice-president in the government-relations division of the firm Hill & Knowlton. (Nintendo had hired the lobbyist to push legislation through the Senate that would outlaw the rental of video games.) Lincoln asked Massey to find out what was up.

The following week Massey reported that there hadn't yet been hearings, but that boxes full of evidence had been gathered, and as many as sixty-five interviews had been conducted. It was ominous that Nintendo hadn't been contacted; it suggested that a lynching was in the making.

Lincoln asked Massey to request that Nintendo be given the opportunity to be heard. Hearing dates were set up and canceled twice, and Lincoln publicly charged that Nintendo was being blocked from testifying. Eckart responded by scheduling a hearing for December 4. However, that hearing was canceled too, and a few days later, Massey called Lincoln to report that the committee was issuing a report without a hearing, and that a press conference had been called for the next day.

"What?" Lincoln barked. "How can they have an investigation without hearing our side of the story?" He told Massey he would call the committee's counsel himself.

Lincoln could hardly restrain his fury. "Look, I don't know what is going on here, but this is not fair. At least hear our side of the story before you go and make an announcement."

The committee's attorney told him that the investigation was completed, and that there was nothing to be done; Nintendo's position had been represented by Massey. Lincoln snapped, "My lobbyist does not know my business! No one talked to *me.*" He demanded to talk to Eckart.

When the congressman got on the line, he coolly repeated that the investigation was over.

"Come on," Lincoln said, beginning to boil over. "Give me a break here. You haven't heard our side of the story; you never received any materials from us." When Eckart refused to budge,

Lincoln lit into him. "You have this press conference set up for tomorrow; you know what that date is as well as I do."

Eckart said he didn't.

"Tomorrow is December the seventh, Congressman."

Eckart tried to calm him. "It's going to be a real low-key press conference," he said.

Lincoln stopped trying to contain himself. "I wasn't born falling off a hay truck," he fumed. "You're trying to screw us."

Eckart said, "I'm sorry you feel that way."

Lincoln angrily tried again. "Come on," he said, "let's not have a circus."

"But the tent was on its way up," wrote L. Gordon Crovitz in a biting piece in *The Wall Street Journal*. Eckart held the news conference, as scheduled, on December 7, Pearl Harbor Day.

"Coincidence?" Crovitz asked. He quoted Howard Lincoln: " 'A lot of strange things happen back there in Washington, but at least they usually give the appearance of fairness.' " Crovitz continued: "[Lincoln] called Eckart a liar for saying Nintendo had its chance to be heard and he dismissed the congressman's antitrust allegations as 'grandstanding.' "

Crovitz's piece recounted the issues in the Atari Games–Nintendo lawsuit. "About fifty software firms, many of them American, have made fortunes under this system. But Atari Games and its subsidiary, Tengen, which was a Nintendo licensee, were not satisfied with this arrangement." In conclusion, Crovitz charged, "There is something wrong with a legal and political system whose vague laws strongly encourage competitors to seek market share outside the marketplace. . . . Nintendo competitors could better spend the time they devoted to huddling with lawyers and congressional staffers to dreaming up the next Mario and Luigi instead."

With Christmas a couple of weeks away and Congress out of session, there was little else for Capitol Hill reporters to do but flock to Eckart's press conference, at which the congressman announced that he was recommending an investigation of Nintendo by the Justice Department. When Lincoln got a taped copy of the press conference, he exploded. "Holy Christ! This guy had more microphones in front of him then Roosevelt did when he declared war on Japan. He went on about everything from resale price maintenance to our treatment of Tengen. Basically it was a story Tengen would have written. It was a disaster."

Tengen, on the other hand, celebrated. "Nintendo's behavior in the United States is so atrocious that it requires the action of the Department of Justice to restore free competition to the market," Dennis Wood said in a press release. "Since Tengen first introduced its line of independently manufactured Nintendo-compatible cartridges in December 1988, Nintendo has attempted to force Tengen out of the home video-game market through a deliberate campaign of distortion, intimidation and coercion. . . . We applaud Representative Eckart's recognition that Nintendo's monopolistic business practices transcend the legal disputes between the two parties, and fully support his recommendation for a full-scale investigation into Nintendo's unlawful domination of the U.S. home video-game market." The next day Atari Corp. also released a statement: "Nintendo has demonstrated its disregard for free and fair competition in America," said its president, Sam Tramiel.

Tengen further attempted to explain the issues in a white paper issued a few days later. "The Nintendo monopoly, like all monopolies, preempts the laws of supply and demand. By absolutely foreclosing competition and by strictly controlling not only the volume of the games but the product mix, Nintendo, not the consumer, is in the position of determining which games are the most popular. Further, through its allocation practices, Nintendo regulates the amount of revenue its licensees and retailers can generate, not the marketplace. Therefore, manufacturing their own products and filing the lawsuit became the only way that Tengen and Atari Games could meet the actual demand for their products and take control of their own destinies."

In Eckart's detailed, eleven-page letter to the Justice Department's antitrust chief, James Rill, the congressman said that Nintendo had stifled competition by blocking competitors' software cartridges from working, thus allowing Nintendo to make restrictive licensing agreements with software makers. "The net result is that there is only one game in town," Eckart wrote. He concluded, "You have indicated in a recent public address your concern regarding the 'growing public perception that antitrust has lost its purpose and its potency.' I applaud your commitment to a more active enforcement of our antitrust laws and urge your careful attention to this very important and visible area of high technology and intellectual property."

Responding to press inquiries, a Justice Department spokeswoman said that the agency would begin an investigation immediately.

Months later, on May 3, 1990, Eckart was called to address another subcommittee, this one chaired by Jack Brooks, a congressman from Texas. Brooks's committee was looking into violations of U.S. antitrust laws by foreign governments, specifically Japan, and had studied the many-tiered distribution system there, a closed club that would be illegal in America.

Brooks ceremoniously called his fellow congressman to the stand to discuss the results of his review of Nintendo. Eckart elucidated the antitrust issues from his investigation and sounded alarms about the high stakes. At the time there were 18 million homes with Nintendo systems. He asked, histrionically, "How did they get in American homes?" Pause. "They enticed their way in through our children's hearts . . . [and now] you can turn that video game into a low-level home computer. . . . [In Japan] they pose a direct threat to the low-end computer industry, and because of their 80 percent market concentration, they are in the most unique position imaginable to make that similar move here in the United States. Is that bad inherently? No. Unless, of course, they control the programming through the lock-out chip, through this artificial physical barrier."

Eckart responded to questions about the lock-out chip and bundling. "Many years ago we said no thank you to IBM when they wanted to bundle their programming with their hardware," he said. "In fact, now every ad you see always says, 'IBM-compatible,' 'Apple-compatible.' They brag that you can put this programming in anybody's hardware system. Not so in video games.

"However, if we allow them to retain this limited hardware gateway, this chip, what does that do to the principle that we established here with IBM? That was in the 1960s. We said the same thing to AT&T, you'll recall . . . there should not be hardware limitations to . . . access to information. So it has been the policy of this government through several presidents, through your predecessors, Mr. Chairman, and now your position in the chair to say, 'No. We want access. We want opportunity.'

"Mr. Chairman," Eckart continued, "if you can turn a toy into a computer, what's the next step?"

Before concluding, he noted that U.S. companies were held to a different standard than foreign companies—so much so that foreign companies had little to fear from the current application of America's antitrust laws—and he implored Congress to take action against Nintendo.

When Dan Van Elderen testified before the same committee, he attempted to avoid any appearance of Japan-bashing by pointing out that "our problem is not with any specific country. Our president is Hideyuki Nakajima, a Japanese citizen. . . . Our problem is with a specific company—Nintendo." He asserted that "Nintendo has been wildly successful, but we believe its success is the result of an illegal monopoly that is operating at the expense of American consumers and its video-game competitors. Nintendo has virtually eliminated competition and the concept of open markets in the home video-game industry."

Van Elderen cited Bill White, who was quoted in the *Los Angeles Times* as saying, "American companies don't play hardball like this. There's more of a sharing of the pie by American companies. In Japan, it's different: winners win big and losers lose." It was a restatement of Hiroshi Yamauchi's philosophy of one strong and the rest weak.

On the bundling issue, Van Elderen quoted from an article in *Computer Lawyer:* "Ever since IBM 'unbundled' hardware and software in 1969, it has been the accepted dogma in the industry that hardware and software are two separate domains. It has long been considered taboo for a hardware manufacturer to enforce restraints, by patent, copyright, or otherwise, on what software can be run on its system or who can manufacture or sell that software. Apart from the home video-game business [controlled by Nintendo], virtually every hardware segment has a dominant supplier that would be capable, if given the opportunity, of using its market position and/or financial muscle to suppress competition in its sector. . . .

"Nintendo, because of its total control of the manufacturing of the software, is in the sole position of determining which products get produced, how many titles and units of specific software are available on the market at a given time, and how many total units are available to consumers." Van Elderen also offered the complaint of an anonymous software company CEO: "We are at their mercy. They can make or break any of us overnight."

"It is," Van Elderen stated, "as if Ford Motor Company only allowed Ford gasoline to power its vehicles, or if Sony only allowed videotaped movies that it produced in Japan to be played on Sony VCRs."

The congressional hearings continued as the Justice Department ruled that the Nintendo case should be handled by the Federal Trade Commission. In collaboration with the attorneys general of several states, the FTC began concurrent investigations into the price-fixing charges and the implications of the lock-out technology. The investigation dragged on for more than a year.

There was sweating over the potential outcomes at NOA in Redmond since they threatened Nintendo's continued dominance in the United States. Hiroshi Yamauchi, in Japan, saw things differently. For him the FTC and American antitrust laws were "an inconvenience" that had to be worked around; they went with the territory. Yamauchi didn't ignore the potential disaster, however. It prompted him and Arakawa to look hard at markets that could replace America if the worst happened. Nintendo had already planned to intensify its push in Europe, but the specter of trouble in the United States caused them to expedite the European invasion. Nintendo would be poised there if any portion of the American gold mine was to be denied them.

On October 22, 1990, Greg Zachary, in *The Wall Street Journal,* revealed Nintendo's decision to ease its licensing restrictions. The decision had been announced in a memo, quietly circulated among licensees at the beginning of the month.

Zachary observed that the timing of the changes—which finally allowed some licensees to manufacture their own games—was related to the ongoing FTC investigation. The word on the street was that Nintendo was in trouble. "Indeed, Nintendo's business records and the records of roughly a half-dozen of its largest licensees have been seized by investigators over the past six weeks," Zachary reported.

Nintendo confirmed that it had loosened its restrictive policy. Zachary concluded, "It denied it took action, which it hasn't publicly announced, to ease criticism from both government officials and competitors, who charge that Nintendo locks others out of its business to keep prices—and profits—artificially high."

The fine print revealed that only a few small licensees could manufacture games on their own, and that they could do so only after purchasing proprietary chips from Nintendo. NOA would be paid for the chips and would also receive a high royalty. Howard Lincoln later admitted that although the loosening of the restriction was a trade-off in terms of Nintendo's profits per game—since the licensees still had to buy security chips from Nintendo and pay a hefty royalty (reportedly, 20 percent)—Nintendo made no less money. However, the eased restrictions did mean that licensees had a slightly greater degree of control of their own fates.

Nintendo also announced that it was rescinding all the exclusivity provisions in the licensing contracts. It was all right to release games on the Nintendo system and, say, the Sega Genesis system without waiting for two years. The reason for the changes, Nintendo claimed, was that quality had been successfully kept under control and *Nintendo Power* provided a de facto regulator; companies that continued to make high-quality games would get the attention of consumers through the magazine. One retailer privately challenged this claim, saying that by now an effective and legal intimidation factor had come into play. (A licensee said, "I'm not going to make games for competing systems because we know that Nintendo would get even, one way or another.") Van Elderen says the changes proved that Nintendo was guilty, and he suspects that the FTC forced the new rules. Nintendo consistently has denied it, and the FTC has refused comment.

Nintendo's announcement may indeed have resulted from a deal made with the FTC, or it may simply have been a way to cut its losses in case either one of the Ataris won in court. If its licensing agreement was illegal, presumably every licensee could sue NOA for restraint of trade and other damages, and the exposure would have been in the billions of dollars.

The easing up of restrictions may also have been a message to Judge Smith, who was hearing both the Atari Games and Atari Corp. cases. Nintendo would thus be saying, "Look, we are easing up." In reality, however, nothing much had changed. The companies that were allowed to manufacture their own games were the biggest licensees, the ones that had the best relationships with Nintendo in the first place. They were in so deep with Nintendo that there was no risk of their going against NOA's wishes on anything from pricing to releasing too many games. Nintendo still exercised enormous clout by rating games and choosing to feature them in—or exclude them from—*Nintendo Power* and other advertising. Nintendo could also influence retailers and, if worst came to worst, create a short supply of the essential chips.

The first FTC investigation, made public in April 1991, focused on allegations of price-fixing. The investigation, conducted in cooperation with attorneys general in New York and Maryland, sought to determine whether or not Nintendo maintained artificially high prices for its hardware and software by enforcing uniform pricing policies and punishing companies that offered discounts.

There were no price wars throughout Nintendo's first years in the marketplace because, it was alleged, Nintendo simply forbade them. Stores tried to discount the NES and the games, but Nintendo pressured them to stop. One chain reportedly lowered the price of the NES by a matter of cents and advertised it in Sunday newspapers, and a competitor called Nintendo, which immediately froze shipments to the company offering the lowered prices. The competitor who turned them in allegedly called Peter Main back to ask, "Do I get their allocation now?"

Hundreds of retailers were interviewed for the FTC investigation, as were representatives of licensees and distributors. Outside the protection of anonymity that came with the FTC investigation, it was more difficult to find vocal critics of Nintendo. The manager of a chain of toy stores who spoke to the FTC says, "The deal with

Nintendo was straightforward: play their way and they were all charm and good cheer; cross them and they would rip your lungs out."

The survival of many retailers depended on regular shipments from Nintendo, and they couldn't afford to start a price war if Nintendo would retaliate by short-shipping them. "Sales guys aren't antitrust lawyers. What were they going to do?" Atari Games' Dennis Wood asked.

When, in February 1990, the FTC and the attorneys general responsible for enforcing the antitrust laws in the fifty states served Nintendo with a civil investigative demand for resale price maintenance, Lincoln says that he cooperated fully. "They wanted to see everything. They wanted to understand our business."

A year later, some of Nintendo's outside attorneys reported back. "You have some problems," one reportedly said. "Your sales guys may have gone a little too far." John Kirby counseled Howard Lincoln to schedule a conference with the FTC.

At the meeting, Lincoln was told that the FTC was prepared to bring an action unless Nintendo signed a settlement agreement called a consent decree. He, Kirby, and Arakawa decided to sign it if it would "wrap up everything," including the investigations by the states. In subsequent meetings with representatives of the states of New York and Maryland, it was suggested that the resolution could be a redemption coupon offered to consumers, essentially to pay those who had overpaid for Nintendo products.

Under the agreement, Nintendo promised to refrain from price-fixing. It would not reduce the supply of Nintendo products to dealers, impose different credit terms on them, or terminate those who failed to adhere to minimum suggested prices. It also agreed not to ask dealers to report others who offered Nintendo products below resale prices suggested or established by NOA. Further, it had to send a letter to its dealers advising them that they could henceforth advertise and sell its products at any price without any adverse action by Nintendo.

Concurrent with the press conference held in Washington, D.C., on April 10, 1991, the New York State Attorney General's office released a strongly worded statement. "For the first time in more than a decade, the FTC has rejoined the battle against vertical price-fixing by manufacturers and retailers," Attorney General Robert Abrams said. "Nintendo sales representatives kept track of retail sales to make sure dealers towed the line on prices. Retailers who resisted Nintendo's pressure were threatened with a slow-down of shipments or a reduction in the number of consoles delivered for sale. . . . Nintendo was not satisfied with being the major supplier and seller of the biggest electronics game for kids in the country. It coerced some of the nation's biggest retailers into keeping the prices of its basic video-game system at $99.99." Abrams stated that Nintendo had threatened retailers who lowered the price of the system by as little as six cents, wishing "to extract every last ounce of profit." He said he was sending a message that was "not only reaching the boardrooms in America, but the boardrooms in Japan and in all other countries that do business in America."

The unbelievable part was the penalty Nintendo would "pay." Abrams announced that, anyone who had bought an NES system from June 1, 1988, through December 31, 1990, would get a coupon good for a $5 discount on Nintendo merchandise. NOA had to redeem a minimum of $5 million and a maximum of $25 million worth of coupons. What it boiled down to was that Nintendo was forced to offer a merchandising deal that Peter Main might have cooked up on a good day. It was, pure and simple, a promotion to encourage people to buy millions of Nintendo games. It was also an indication that the government's case against Nintendo was fragile.

In an editorial in *Barron's* in December 1991, the FTC, not Nintendo, was taken to task. The settlement, the piece read, "was declared a victory. Abrams said it sent 'the most powerful possible message across this country that we will not tolerate this kind of pernicious practice which takes millions of dollars out of the pockets of consumers.' Indeed, it was a victory for Abrams, who arranged for Nintendo to mail out the coupon [in New York] with a self-congratulatory letter from him, just as his campaign for the U.S. Senate got under way. In contrast, consumers had to spend $20 to $70 on Nintendo products in order to enjoy the $5 rebate. . . . Robert Abrams and the legions of trustbusting lawyers would be far more productively occupied playing 'Super Mario Bros. 3' than bringing cases of this kind."

Despite the fact that settlement was relatively painless for Nintendo, in a press release Atari Games' Dennis Wood indicated that the FTC decision was a vindication. "The FTC concluded what we've believed for a long time: that Nintendo has built its business on illegal activities." In *his* press release, however, Howard Lincoln emphasized that the settlement was agreed to in order to avoid a lengthy court battle and to foster consumer goodwill. "We were concerned with how Nintendo game players might view these allegations," he said. He did not acknowledge that the settlement was a boon to business. "We have still suffered the embarrassment," he says. "Even without admitting liability, it has hurt to be called price-fixers."

After the settlement was announced, Nintendo did its required penitence, sending out letters and indicating that it would include a disclaimer on any promotional materials in which it suggested a retail price to remind dealers that they were free to set their own prices. Nintendo also educated its salesmen and marketers. "We made it crystal clear to every employee that if anybody violated the antitrust laws, they were history," Lincoln says.

But the FTC wasn't finished with Nintendo. The investigations of licensing policies and the lock-out chip were ongoing. An insider suggested that the government was awaiting the results of the Atari antimonopoly cases to determine if Nintendo was culpable. If it was, the FTC would go after Nintendo with guns drawn. If not, the investigation would likely fade away.

Despite the findings of the FTC or in the lawsuits, some overriding questions persisted. One was whether competition should be legislated or litigated outside the marketplace. Another was whether prosecution would be part of America's counterattack on successful Japanese companies. "No one complained until we became so big," Arakawa says. Peter Main is adamant that Nintendo was a victim of the trade war. "If Nintendo were an American company, no one would have said a word." Acclaim's Greg Fischbach agrees: "It's more Japan-bashing." Non-Japanese companies were also targeted by the FTC (Nintendo's Redmond neighbor,

Microsoft, was under investigation), but the specific charge of price-fixing had been leveled exclusively at a series of Japanese companies, including Panasonic, Mitsubishi, and Minolta, all of which were found guilty.

Novelist Michael Crichton exploited U.S.-Japanese trade tensions in his best-selling 1992 thriller *Rising Sun,* a pop piece of propaganda that nonetheless cited facts. Crichton's fictional conglomerate, Nakamoto, is about to be investigated by journalists. "We're doing a big series on taxes," says a reporter. "The government is finally noticing that Japanese corporations do a lot of business here, but they don't pay much tax in America. Some of them pay none, which is ridiculous. They control their profits by overpricing the Japanese subcomponents that their American assembly plants import. It's outrageous, but of course, the American government has never been too swift about penalizing Japan before. And the Japanese spend half a billion a year in Washington to keep everybody calmed down."

"But you're going to do a tax story?" Crichton's protagonist, a police special services agent, asks.

"Yeah. And we're looking at Nakamoto. My sources keep telling me Nakamoto's going to get hit with a price-fixing suit. Price-fixing is the name of the game for Japanese companies. I pulled a list of who's settled lawsuits. Nintendo in 1991, price-fixing games. Mitsubishi that year, price-fixing TVs. Panasonic in 1989. Minolta in 1987. And you know that's just the tip of the iceberg."

Later, Crichton's cop gets another perspective when he discusses the issue with his partner. "But price-fixing is illegal," the special services agent says.

"In America," his partner responds. "Yes. But it's normal procedure in Japan. . . . Collusive agreements are the way things are done. . . . Americans get moralistic about collusion, instead of just seeing it as a different way of doing business."

Speaking anonymously, a game developer who worked with Nintendo said, "Nintendo exerts monopoly control over development, over the retailers, over manufacturing. . . . They only get away with it because there is no competition. If there were, developers would not agree to let someone else control the manufacturing, control the delivering of product, control everything. They would not allow Nintendo to charge for everything in advance."

Nintendo may have had justifications for manipulating the product released for its machine, but such manipulation was illegal, regardless of intentions. Part of the reason it limited the number of titles companies could release was because it didn't want to see any other company gain significant power with retailers—just as companies attempted to stop labor unions from forming. It didn't want coalitions to form. It forced companies to develop titles exclusively for its system to force them to choose—and what company could afford *not* to choose Nintendo? Of course the vast majority chose it over the competition despite its disproportionately high royalty, so prices remained high.

Nintendo also tied up developers with Howard Lincoln's lawsuit machine. "Once you signed up with them, they had you," the game developer says, because that signature legally acknowledged the validity of Nintendo's patents as well as receipt of confidential information required to develop games. Nintendo prohibited licensees from abandoning ship once they signed up. It could go after straying companies in court—just as it went after Atari Games—by arguing that they were able to compete only by exploiting proprietary information supplied by Nintendo.

For years Nintendo had been trying to stop the burgeoning industry of video-game rentals. It took the largest video-game rental company, Blockbuster, to court and worked with Hill & Knowlton's Massey to attempt to convince legislators to make it illegal. At the same time, Nintendo also attempted to stop video-game rentals by coercion and threats.

Since Nintendo wouldn't sell directly to video-rental stores, representatives of the stores bought games from retailers—just like anyone else, but in greater volume. Kids could then go to a place like Blockbuster and rent "Super Mario Bros. 3" or "Dr. Mario" for a few dollars instead of paying $40 or $50 to buy the game. To stop the rental companies from obtaining games from retailers, some of the largest Nintendo dealers issued a policy forbidding multiple sales of any game. But in at least one documented case, someone from Nintendo itself threatened retaliation.

Try Soft of America owned kiosks that were set up in malls in the Seattle area and sold nothing but Nintendo games. A memorandum was sent out on June 15, 1989, to Try Soft store managers from Brent Weaver, NOA's senior sales and marketing manager. "Nintendo of America is very concerned about video stores undermining our business by renting tapes [*sic*] to customers. They feel renting games decreases the perceived value by the customer and reduces our sales. During a recent meeting with top executives at NOA, Try Soft has promised NOA to follow a strict quantity limitation policy in all stores and mail order.

"Henceforth we will limit each customer to no more than two pieces of any NOA product, including software and hardware. There will be no exceptions to this policy, including multiple invoicing. Do not sell more than two pieces of any NOA product to any customer or employees of video stores. . . . if two employees of a video store come to your store together, *do not* sell a total of four pieces of any one title to them. We are essentially cutting off product to video stores. . . . Our policy concerning product manufactured by third-party licensees is a bit more relaxed. Your store will be given monthly a list of products which you may not sell more than two pieces of to a customer. All other product will be limited to a maximum of 10 pieces and must be multiple-invoiced.

"Nintendo has begun to watch our sales very closely," Weaver wrote, "and if we do not adhere to this policy, our allocations from them will be severely reduced. Failure of your store to follow these guidelines will result in a reprimand. . . . This policy is effective immediately."

Although Howard Lincoln denied the accuracy of the memo and fired off a response to Try Soft saying that it was all a misunderstanding, it was part of a growing body of evidence indicating that Nintendo had overstepped its bounds. Something more than price-fixing seemed to be going on. The question was whether, or in what instances, Nintendo was actually breaking laws. The Try Soft memo revealed that at least some people at Nintendo sought to protect their own games and were less concerned about licensees' games. Also, the controls on the licensees' games were apparently inconsistent; a list would determine what games could and could not be sold in bulk. The assertion by Atari Games that licensees were not all treated alike—in spite of what Howard Lincoln, Minoru Arakawa, and Peter Main claimed—was concretely supported by the memo.

The implications of Nintendo's tight controls, legal or illegal, were hotly debated. Some in the industry agreed that they were anti-competition and downright un-American, while others agreed with the Nintendo party line: NOA was exercising legal controls to manage an industry that had once self-destructed. Publicly, licensees agreed and nonlicensees disagreed. One of the latter, it was observed, originally disagreed but changed its mind once it became a licensee.

Jim Mackonochie, who had run Robert Maxwell's software company, Mirrorsoft, before joining Commodore International, believed that Nintendo's monopolistic practices could stunt the larger industry. "It's as if a writer could not publish unless he owned the printing press," he says. "How much poorer would our civilization be? In the computer industry, the most creative things often come when small groups of people, not necessarily associated with big companies that could get a license, get together to create. The best games—'Tetris,' 'Bamboozle,' 'Dungeon Master' —all came from individuals or small development groups. Nintendo's policies could block out the most creative people. It could kill the industry."

More likely, Nintendo's policies could kill the competitors. In the opinion of an industry analyst, "The lawsuits and government don't seem as if they will have much effect. The fact is, if they haven't been able to stop Nintendo in the marketplace, this is a feeble try. No one seems to be able to stop Nintendo." Another industry veteran says he gives Arakawa, Lincoln, and Main all the credit in the world. "They broke every fucking antitrust and racketeering law we have and got away with it. America hasn't seen anyone like them."

Nintendo remained consistently unrepentant, and the price-fixing settlement allowed it to deny guilt. Likewise, it denied charges about monopoly control and restraint of trade, and it never backed down on its commitment to the restrictive security system. As Hiroshi Imanishi said, "We made a choice and it turns out that our choice was the correct one." Arakawa added, "The chip has done exactly what we intended it to do. It is why we have been able to maintain such a strong list of games, and why consumers can trust the Nintendo quality seal." Howard Lincoln says the tactics described in the Try Soft memo may show that individuals associated with Nintendo went too far, but the company itself was guilty of nothing more than covering all bases in order to make as much money as it possibly could. "What's more American than that?" he asks.

Another Nintendo licensee believes that much of the anti-Nintendo sentiment was sour grapes. "It's ironic that the only people crying are the companies who are hurt by Nintendo's success," he says. "Nakajima and Wood sicced the feds on Arakawa even though his only crime was that he did the best he could for the parent company and the NCL stockholders. It's not lost on many of us that Atari, the loudest critic, is the company that destroyed video games the first time around. Nintendo was stronger and smarter. *That's* the bottom line."

It is also partly why the Japan-versus-U.S. battle worsened, with Nintendo in the forefront. For several weeks in early 1992, when Yamauchi's attempt to buy the Seattle Mariners was making headlines, the company was watched on both sides of the Pacific. In Tokyo and Osaka, the "attack" on Nintendo was an illustration of America's hostility toward Japan. Articles in the American press indicated that the United States was finally drawing the line. In a twist on Akio Morita and Shintaro Ishihara's *The Japan That Can Say No*, America was saying no to Japan: no to a Japanese manufacturer that had won a $122 million contract to build rail cars for the Los Angeles County Transportation Commission (the contract later went to an Idaho-based company); no to Japanese-made tractors in favor of John Deeres; no even to Kristi Yamaguchi, the Olympic figure skater, who received few of the sort of endorsement offers that Olympic gold-medalists normally get because, according to a National Public Radio report, she was "the wrong ethnicity"; and no to Hiroshi Yamauchi in his bid to buy into the American pastime.

In an interview with *The New York Times*, the Nintendo chairman attempted to show that his interest in the Mariners was his way of returning something to the country that had made him so rich. He explained that Seattle businessmen, and even the governor of Washington State and one of its U.S. senators, had solicited the bid. (The current owner was planning to move the Mariners to St. Petersburg, Florida.) Yet Yamauchi's patronizing "benevolence" was the last thing Americans wanted; they didn't need Japanese charity.

Still, local public opinion in the Pacific Northwest favored the sale to Yamauchi. Letters to the baseball commission called the decision to ban the sale racist and unfair. Major league baseball wanted local ownership, and Arakawa lived in Seattle. In any case, baseball was not exclusively American, since Canadians had been allowed to own teams since the 1960s.

On the other hand, one letter writer, in an attempt to prove that Yamauchi would make an undesirable owner of an American team, charged that Nintendo was a racist company that discriminated particularly against black Americans. A Nintendo employee named Carey Wiggins, an African-American with a degree in marketing, was the first to level the charge. Wiggins had moved to Seattle and found a job at Nintendo as a temporary warehouse worker. Nine months later he was still a temporary employee, with no benefits and no commitment from the company to advance him, in spite of the fact that less qualified workers who had worked for shorter periods of time had been hired as permanent, full-time employees. Wiggins claimed that he was passed over for jobs on a number of occasions, each time in favor of less experienced, younger, white employees.

In April 1990, Tim Healy of the *Seattle Times* did a series of articles on charges of discrimination at Nintendo, and reported Howard Lincoln's claim that Nintendo's explosive growth made it difficult to find enough qualified employees regardless of racial or ethnic background. Wiggins, however, didn't accept the excuse. As part of a group of twenty-five African-American Nintendo employees, he accused NOA of racial discrimination in the hiring and promotion of blacks. The employees met with a watchdog organization, the Seattle Core Group, which mounted an investigation. The Core Group determined that there were only thirty to thirty-five black employees out of the total of 1,500 to 1,600 in the company, and of these, only one was in management.

Healy reported that Nintendo's record was not atypical for American subsidiaries of Japanese companies, and cited a study that found a pattern of Japanese auto manufacturers locating plants in areas of the United States with low black populations. Lawsuits alleging discrimination had been brought against more than half a dozen American subsidiaries of Japanese companies. Honda and Nissan paid $600,000 and $6 million, respectively, to settle investigations by the Equal Employment Opportunity Commission (EEOC). Other companies accused of racism in the United States included Toyota, Nikko Securities, and Sumitomo Bank. It was true that in the video-game industry—Nintendo's industry—many games imported from Japan had to be modified so that they would be acceptable in America; the villains in these games were frequently black or otherwise dark-skinned.

There were also reports of anti-black jokes among both white and Japanese workers at Nintendo. According to the director of the Core Group, African-Americans were "given messages that they were not welcome at Nintendo."

The Core Group requested a meeting with Arakawa. After the state employment agency got involved, Arakawa agreed to a meeting with the Core Group and representatives of other Seattle-based groups concerned about discrimination. He gave a compassionate speech—he had been the victim of racism himself, he said—and pledged that Nintendo had begun training its employees in ways to avoid workplace discrimination, and promised to initiate an affirmative-action plan.

Meanwhile, the state of Washington suspended its practice of referring job seekers to Nintendo even though Phil Rogers contested the Core Group's statistics. He insisted that Nintendo employed 110 ethnic minorities, a number representing 14.3 percent of the permanent work force. (The percentage included all minorities, including Japanese.) He also said that three, not one, of Nintendo's 147 managers were African-American.

In 1991 Nintendo unveiled its plan for an affirmative-action program, but it was criticized as inadequate by the Core Group. Howard Lincoln defended it and pointed to steps Nintendo had taken to recruit blacks and other minorities. An EEOC spokesman in Seattle agreed that Nintendo seemed to have corrected the problems; there had been no new complaints. The state resumed refer-ring applicants to Nintendo.

Nonetheless, when Yamauchi made his offer to buy the Mari-ners, the head of the Core Group, Oscar Eason, brought the issue to the attention of baseball commissioner, Fay Vincent. Eason wrote: "This letter is to express our objection to the proposed purchase of the Seattle Mariners baseball franchise by a group of investors that would include Nintendo of America. Approximately two years ago over 95 percent of all African-Americans employed by Nintendo appealed to [us] for assistance to help improve the discriminatory conditions that existed at NOA's Redmond, Washington, facility. . . . Please understand; this appeal is totally unrelated to any discussion pertaining to racism toward Asians, 'Japan-bashing' or prejudice aimed at foreigners."

Nintendo's alleged racism, by then a hollow and unsupported charge, was not the primary issue. Howard Lincoln says, "There is a lot of Japan-bashing back in Washington, D.C. Nintendo doesn't have a host of friends, primarily because we're a Japanese com-pany. Nintendo is a well-known company with a large market share and it has been accused of all of these terrible things. . . . We do have the reputation of being big and tough, and I understand some of the sentiment against Japan in Washington. We do have trade problems. I object like any other American that there are barriers to Japan's market. But that is not what we are doing. I'm not causing those barriers."

What Nintendo did do was to aggressively protect its turf, meet-ing any perceived threat to its continued growth, no matter how small, head-on, sometimes with bigger guns than necessary. The idea was to nip any threat in the bud lethally. Howard Lincoln never denied the company's reputation for litigiousness. Rather, he exploited it. "Lincoln's motto was 'Fuck with us and we will destroy you,' " says a fellow attorney. "Otherwise he's a really nice guy."

To supplement their business in an increasingly competitive mar-ket, a few video-rental outfits began renting Nintendo games in 1987 and 1988, when demand for cartridges was enormous. For some of those stores, Nintendo rentals represented 30 to 40 percent of their business. For most it was a lower but still significant 10 to 15 percent. The largest rental chain, Blockbuster, had revenues of $1.5 billion in 1990, and perhaps $150 million of this was from Nintendo game rentals (Blockbuster would not disclose the exact figure).

The video-rental industry insisted that it was good for Nintendo; kids could try games out before buying them. Nintendo disagreed; kids could try games out *instead* of buying them. Howard Lincoln said that video-game rental was "nothing less than commercial rape. I can spend thousands of hours and millions of dollars creating a game," he says. "I expect, therefore, to be compensated every time the thing sells. All of a sudden, out of the blue, comes a system that distributes my game to thousands of people and I get no royalty. The video-rental companies exploit the thing—renting it out over and over again, hundreds and even thousands of times —and I get nothing. The guy who developed the game and Nintendo get screwed. What does the guy who's renting the cartridge contribute? What does he pay in terms of a royalty for the commercial exploitation of copyrighted work? Zip."

Movies, which enjoy a huge rental market, are protected. Studios decide if and when a movie will be made available to rental stores, commonly six months to a year after its initial release. The studios have milked as much money from the market as possible before they go for the gravy, the videocassette market. A similar system may be reasonable for video games. However, the same day a hot game is released to the public in toy and electronics stores, representatives from video-rental stores go to retailers and buy all the copies they can get.

Nintendo not only refused to sell to the chains and individual stores that rented movies and games and pressured its retailers to do the same. In addition, it attempted to address the problem in court and in Congress. Legislation was proposed in the Senate in May 1989, spearheaded by software companies such as WordPerfect Corporation and Microsoft and the Software Publishers Association, that would prohibit the rental of all computer software, including video games.

The trade association of video-rental dealers, the Video Software Dealers Association (VSDA), promised to crush the bill as long as video games were part of it; video-game rentals were too lucrative to give up. The computer-software companies and the SPA succumbed, according to Howard Lincoln, and cut a deal with the VSDA, agreeing to exclude video games from the bill. Since the revision of the proposed legislation required some justification, it was argued that video games, unlike floppy disk–based software, were not copiable. It was easy to rent computer software and make a permanent copy at home, but it was nearly impossible to make a copy of "Super Mario Bros." (although one Taiwanese company, Baelih, advertised a game-copying device).

Before 1989, Nintendo had never been represented in Washington and had relied on the SPA. However, Howard Lincoln charged, the association "sold us down the river," as did Microsoft, their neighbor in Redmond. (Howard Lincoln was rebuffed when he went to the company to ask for its support.)

Don Massey, a lobbyist who represented such clients as the government of Turkey and Gerber Products Co., was retained to attempt to influence the rental legislation. Nintendo's position was supported only halfheartedly by Washington State's congressional delegation, which was caught in the middle; their other constituent in Redmond was Microsoft, which wanted the bill to go through whether or not Nintendo liked it. With Massey's help, Nintendo fought the change vigorously, but lost in July 1969.

NOA then tried to push for a compromise bill that would prohibit the rental of a video game for a year after its release. This was proposed in Congress by Representative Joe Barton, from Texas, a friend of Byron Cook, the president of a Nintendo licensee called Trade West. Nintendo lost that one too. There was one more try, an attempt to pass a bill similar to one sponsored by the record industry that sought to ban the rental of records, but it never made it through the House Judiciary Committee. "We got killed on Capitol Hill," Lincoln admits.

There were other avenues Lincoln took in his attempt to stop the rental companies. He considered suing them but concluded that Nintendo wouldn't prevail; a guiding principle in copyright law called the "first-sale doctrine" said that the buyer of a piece of property could do just about anything with it. Since a ban on rentals by the courts or Congress seemed unlikely, Nintendo tried another tack. The video-rental companies were packaging game cartridges with photocopies of Nintendo's instruction manuals (original manuals rented with games were routinely lost or torn). Since these were copyrighted, Lincoln sued over the manuals.

NOA went after the biggest company, Blockbuster, filing a suit on the eve of a VSDA convention in August 1989, and eventually Blockbuster was hit with a preliminary injunction that prohibited it from copying the manuals. Nintendo followed this minor victory with an announcement that it would sue any other company that copied manuals, even though the action didn't hurt the rental business in the long run; companies simply came up with common instructions to go with games.

Nintendo also pursued video-rental outlets that named their video-game departments "Nintendo Centers" and retailers that used "Nintendo" as a generic term—i.e., "Come in and buy a Nintendo game." NOA felt it had to pursue all offenders if *Nintendo* was not going to enter the language as the word for video games. Of course Lincoln didn't complain when General Norman Schwarzkopf, in his press briefings during the war in the Persian Gulf, referred to Desert Storm as "the first Nintendo war," but few others were allowed to take the company's name in vain.

Lincoln's in-house and outside counsel also continued to pursue a host of counterfeiters, smugglers, and purveyors of illegal games through 1992. NOA spent millions of dollars going after counterfeiters, occasionally with the cooperation of the American government. In mid-1991, in conjunction with U.S. Customs authorities, Nintendo infiltrated a huge counterfeiting ring. Individuals associated with prominent Taiwanese companies that manufactured and exported counterfeit games were arrested in Chicago, San Jose, Los Angeles, and Miami. Among those arrested were executives of United Microelectronics, a multimillion-dollar Taiwanese manufacturer of integrated circuits. United Microelectronics was Taiwan's largest semiconductor manufacturer, and the bust had the potential to precipitate an international incident, particularly when it was revealed that the Taiwanese government held a 30 percent interest in the company. It made not only counterfeit games but also counterfeit memory chips, which were sold to dozens of other Taiwanese counterfeiters.

"We finally hit pay dirt," Lincoln says. "It's like stopping a drug ring. You find the little guys but keep missing the big guys. We finally hit the big guys." The big guys turned out to be a government that sanctioned counterfeiters operating in huge, ultramodern plants, churning out nothing but counterfeit product. U.S. Customs said that 70 percent of counterfeit computer and electronic products seized in the United States between October 1990 and March 1992 was from Taiwan. Nintendo was the first company to tie the illegal operation to the government itself.

Taiwanese game cartridges worked on the NES by "zapping" the security chip. A specially designed chip shot a negative voltage spike into one of the input pins of the security chip in the NES base unit. This surge, or "zap," knocked the security chip senseless; its program got lost in a repeating loop in "the weeds," as a techie explained. "It was lost in space."

Packed with up to a hundred games on one cartridge, the counterfeit games sold throughout the world for from fifty to several hundred dollars. They were available in major cities at electronics or video-game shops, mostly small ones.

In early 1992, Nintendo asked the U.S. Trade Representative to take action against the Taiwanese government. In 1990 alone, Nintendo claimed to have lost more than $1 billion in wholesale sales. Up to 100 million counterfeit games were sold that year in the United States, and most of the ROM chips in them were manufactured by United Microelectronics.

Nintendo also sued retailers that carried the illegal games and, when possible, pursued the manufacturers and importers; there were three hundred suits in Canada and the United States. The games were also sold through mail order, newspaper ads, and flea markets. Nintendo went after any violator it could find.

The counterfeit business was so enormous that even if Nintendo stopped it in the United States, Canada, and Europe, it couldn't touch the mammoth industry in the People's Republic of China, where 100 percent of the growing Nintendo business was counterfeit, both hardware and software, almost all from Taiwan. The counterfeiters actually set up a company in China called Nintendo, according to Howard Lincoln. Nintendo tried to bring the issue to the attention of the Chinese government, but to no avail. In 1992 the U.S. Trade Representative agreed to intercede on Nintendo's behalf.

Closer to home, Nintendo went after a relatively small San Francisco–based toy company, Lewis Galoob, not for counterfeiting, but for a peripheral device. Galoob manufactured a product (under license from a British firm) that allowed players to modify Nintendo games. Mario could jump higher, become invincible, and have a hundred or even an infinite number of lives—whatever the player chose. Galoob sold the device in the United States under the name Game Genie.

Nintendo claims that Galoob approached them about a license for Game Genie. Galoob denies this. Howard Lincoln insists that Nintendo considered it, but decided that the invention would destroy the experience of many Nintendo games. "It creates derivative works," he said. "It not only alters Nintendo games, infringing on copyrights, but can make them less fun, too easy to play."

Nintendo sued in June 1990 when Galoob announced its plans to release Game Genie on its own. Nintendo's argument was that millions of dollars went into designing games, that decisions about the number of lives a player was allowed and the capabilities of characters were essential to game play, and that altering these key components could destroy a good game. Howard Lincoln argued, "If you spend enormous time and effort to come up with 'Super Mario 3' and the player can get all the way through it, basically by cheating, then the challenge is gone, and challenge is what our business is about."

The changes to a Nintendo game, Galoob countered, were temporary—they did not create new versions of the games—and consumers had the "freedom and right . . . to play their Nintendo games in ways they see fit."

In June 1990, Nintendo got a preliminary injunction that barred Galoob from producing or marketing the Game Genie. Galoob appealed to the Ninth Circuit, which upheld the injunction. In a nonjury trial in April 1991, with Judge Fern Smith, the same judge hearing the Atari cases, presiding, witnesses included Sigeru Miyamoto, flown over from Kyoto, who testified for a full day.

But Judge Smith ruled in Galoob's favor in July 1991, deciding that Game Genie did not create derivative works, and that even if it did, Galoob was protected by the fair-use doctrine. Galoob was given the green light to bring Game Genie to market, and its stock shot up 20 percent the day the court's decision was made public, despite Nintendo's announcement that it planned to appeal. Analysts predicted a boost in Galoob's sales by $30 million a year, 20 percent more than its pre–Game Genie totals. There were orders for half a million units immediately after the decision was announced, even before a massive ad campaign was launched for the Christmas season in 1991. Eight hundred thousand units, all Galoob could make, were sold by the time of the Toy Fair in February 1992. The company released a Game Genie for Sega, too, and planned to launch both versions in Europe, Australia, and, in late 1992, Japan.

Reporters jumped on Judge Smith's ruling as "the first crack in the fortress that Nintendo has established in the video-game market," as the *San Francisco Chronicle* put it. "The congratulatory calls began pouring in as soon as the decision was announced," reported the *Examiner.* Howard Lincoln admits, "They all loved it. Godzilla in Toyland had gotten beaten. It was like a dragon had been slayed. . . . If we win on the appeal," he adds, "I'll write the press release myself." (Nintendo lost the appeal in September 1992.)

Nintendo was not only in the business of suing. Fighting against Nintendo's ability to lock them out, some renegade companies, including American Video Entertainment (AVE), Camerica, and Color Dreams, made Nintendo-compatible games that defused the security system with "zapper technology." In 1991 American Video Entertainment discovered that their games didn't work on the newer NES systems; the circuitry inside them had been revised. Richard Frick, AVE's president, believed that Nintendo had intentionally reworked its machine to reject his cartridges. In a letter to Nintendo, he charged that the company was trying to put him out of business.

Howard Lincoln says that the NES hardware revision that "happened" to lock out Frick's games was part of Nintendo's ongoing war against counterfeiters. NCL had already updated the NES on fourteen or fifteen different occasions, and the latest version was designed to stop illegal Taiwanese multi-game cartridges from working on the system. "We were trying to stay one step ahead of the counterfeiters," Lincoln insists. "American Video Entertainment accused us of deliberately doing it to them. I didn't respond, but the answer is: '*Tough!* We can make our product any way we want. The change was not to drive you out of business; we didn't even know you were *in* business before we got your letter.' "

Frick joined the list of companies suing Nintendo with a $105 million action in January 1991, charging that the company had altered the NES machines in a deliberate attempt to render his company's cartridges incompatible, and, once again, that Nintendo was guilty of monopolistic practices designed to kill competition. The suit also claimed that Nintendo failed to inform buyers that only NES-approved games would work on the machine, even though, in the course of the Atari Games litigation, it had stated that outside cartridges *could* work on the NES. One thing *was* proved by AVE: Nintendo used its licensing agreement to keep prices high. Frick demonstrated that he could make a profitable game retailing for $20 because he didn't have to pay Nintendo's manufacturing fees and royalties. He said he had made a sizable profit on sales of 60,000 "F-15 City War" games.

Howard Lincoln dismissed AVE's claims out of hand. Even Dan Van Elderen of Atari Games indicated that Nintendo would likely get away with the system revision that blocked the zapper technology, even if this had been its sole purpose. The technology, Van Elderen concedes, actually could harm the NES. "The revision was a sound engineering decision that made up for a deficiency in the system. . . . The NES was never designed to withstand that kind of high-voltage charge."

Frick said he planned to go after Nintendo with everything he had. In addition to fighting Nintendo in court, he decided to go to the public, urging consumers to return "deficient" NOA units to retailers and to demand refunds or machines that played *all* games. Consumers, however, did not respond to his call.

Nintendo couldn't legally pursue companies like AVE or Color Dreams, which never had licensing agreements and whose games didn't infringe the security-system patent, and had to fight those companies in other ways. There were new charges of distributor and retailer intimidation, with Nintendo allegedly threatening companies that carried Color Dreams or AVE games. In 1992 Richard Frick claimed that a Nintendo representative warned the head of the electronics departments of Wal-Mart that if customers used games employing the zapper technology on an NES they would thereby invalidate the Nintendo warranty. It was one more variation on an old theme: Sell only authorized Nintendo product or else. Wal-Mart would not confirm the accusation, Frick says, because, "You tell on Nintendo and you lose."

Howard Lincoln insists the charges are untrue. "We're not idiots," he says. "We would never threaten retailers. It's wrong, and most retailers would consider it inappropriate." The fact was that Nintendo was probably able to repress the upstart companies without any overt action because of its control of distribution. Retailers dealt with the biggest companies, all licensees, because it was easier. Also, the best games tended to come from the major players, who had the resources to invest in game development. Nintendo controlled sales by promoting only approved games in *Nintendo Power,* and retailers were unlikely to waste valuable shelf space on games that didn't appear in the magazine. The retailers also felt more comfortable selling games backed by Nintendo's and licensees' service and support networks. The small companies simply could not compete.

AVE did find customers in the video-rental industry. Blockbuster led the rental chains with an order for 2,050 copies of one AVE game with new technology that circumvented the Nintendo system revision that had locked out earlier AVE games. The game, "Wally Bear and the No Gang" (it was originally ". . . the Just Say No Gang," but Nancy Reagan had trademarked that phrase), was endorsed by the American Medical Association for its message: in spite of peer pressure to drink and do drugs, Wally Bear said no. The cartridge worked on all NES units, although there was no way of knowing whether Nintendo would come up with another system revision to block "Wally Bear" and AVE's technology.

Howard Lincoln looked for other ways to stop the companies that got around the security system, and said he was investigating other patents or copyrights the companies might have infringed. If he found grounds, he would sue. There was, however, at least one other maker of nonlicensed games that Lincoln was unlikely to go after. The company was called Wisdom Tree and it made the Bible Adventure Series of video games which included "Noah's Ark," "Save Baby Moses," and "David Versus Goliath." The possible headlines—for instance, NINTENDO SUES BABY MOSES CREATOR— would have given even polished Nintendo PR chief Bill White nightmares.

13

FROM RUSSIA
WITH LOVE

As Nintendo grew and its target audience, the classic audience for video games (young boys), was nearly saturated, the company went after new groups of customers. Peter Main says it was a strategic decision from early on. "We wanted to break out from the historic video-game user, boys eight to thirteen, because the thirteen-year-old boy will soon be fourteen years old and pass from our grip. Our objective from day one was to move beyond that narrow base."

Girls were an increasingly important market for Nintendo in mid to late 1991, when more powerful systems by competitors had distracted some of Nintendo's formerly loyal boys. A survey Peter Main commissioned revealed that girls from six to fourteen were the primary players in many households. Their level of satisfaction was "intensifying," according to the market researcher.

In Japan, adults had never been considered part of Nintendo's market. Inherent in the Japanese culture was *jyukyu,* a system of ethics dating from the seventeenth century, which held that the most important values were hard work, thriftiness, and seriousness, and that luxury and leisure were a waste of time. *Jyukyu* was behind Japan's phenomenal productivity, the high savings rate of its people, and, ultimately, its obsession with education and work. Vestiges of *jyukyu* made it unthinkable for many adults to sit in front of video-game machines.

Not in America. Adults in the West comprised many of the computer- and video-game players, and increasingly they wrestled controllers away from their children in order to play Nintendo.

NOA encouraged them. Arakawa had made a point of selling Nintendo products in electronics as well as toy stores. Peter Main says, "We wanted to be in electronics outlets, department stores, and discount mass merchants in addition to toy stores because we knew there were varying signals that emanated from a product depending on where it was distributed." Since the initial test in New York City, Arakawa had fought to get into Circuit City and Crazy Eddie—among the stereos and VCRs—as well as Toys "R" Us.

A growing number of licensees came up with games for adults, alternatives to shoot-'em-ups and karate games, like "Jeopardy," "Trivial Pursuit," and sophisticated sports games. "Bases Loaded," made by a licensee named Jaleco, was particularly ingenious, even beyond its impressive graphics and sound, and serious baseball fans could obsess over its trivia and decision-making options. In "Bases Loaded 2," pitchers were guided by an ERA and batters by individual statistics. Programmed into the game were what Jaleco called "a special player-performance system" that programmed players to have streaks and slumps. You decided if a player was having a bad day and if you ought to bench him. More than 4 million "Bases Loaded" games were sold.

Adults played against each other, against their children, and in secret after the rest of the family had gone to bed. *Nintendo Power* featured celebrity adults who played, and the game counselors heard from parents who had devoted entire nights to practicing so they could trounce their kids. Still, nothing compared to the adults who flocked to Nintendo because of a new hardware product launched in 1989.

Gunpei Yokoi and his forty-five-man R&D 1 team of designers, programmers, and engineers came up with a device that married the NES and Game & Watch. Like the NES, it was a video-game system that played interchangeable cartridges, and like Game & Watch, it was quite small. Here was a video-game system that could be played anywhere—in planes, trains, and automobiles, or in the quiet of a bedroom. Inside its pretty package, Game Boy combined portability, miniaturization, and entertainment, three of the most important attributes of emerging technologies, according to Greg Zachary of *The Wall Street Journal.* Game Boy was so slick that a Sony engineering team was reportedly chastised by managers and executives in Japan for being beaten by Nintendo. "This Game Boy should have been a Sony product," a manager is said to have told the group, which worked on portable entertainment devices. Some of the engineers were shuffled to new departments, and one was so shamed that he left the company.

Game Boy, decidedly playful (Sony probably would have called it Game*man*), was housed in a gray plastic case the size of a transistor radio. On the face were buttons and a palm-sized, black-and-sallow-green liquid crystal display (manufactured by Sharp). Game Boy was released at a lower price than Yamauchi had aimed for, $100. He predicted Game Boy sales of 25 million units within three years.

Once again, hardware sales were only the beginning. Game Boy was a tiny computer that could, if successful, create a viable software market all its own. Game Boy games, the size of a saltine cracker, would sell for $20 to $25, so a sizable sub-industry was possible.

When Game Boy was released, it was criticized because it didn't have a color screen. Yamauchi had decided to forgo color for cost and efficiency. A color display would have required the power of four or eight AA batteries instead of Game Boy's two. Color would also have drained the batteries too quickly. Low-priced color screens had other problems. They were difficult to see in bright light (Game Boy was difficult enough as it was). Competitors Sega,

Atari Corp., and NEC released expensive hand-held systems with color screens that sold a small fraction of the number of Game Boys. Two hundred thousand Game Boys sold out in two weeks in Japan, and 40,000 units were sold the first day in the United States. NCL sent 1.1 million Game Boys to the United States in its initial shipment, and Toys "R" Us tried—unsuccessfully—to get them all.

Of course lots of kids played Game Boy—two could be connected for competitive play—but a remarkable aspect of its success was the number of adults who played it. Game Boys were frequently seen in first-class compartments on cross-country flights, in corporate lunchrooms, and in desk drawers and briefcases. President Bush, in the hospital in May 1991, was pictured in newspapers commander-in-chiefing a Game Boy.

Nintendo unleashed Game Boy marketing campaigns that targeted adults. "This Father's Day treat Dad like a kid," read one of Bill White's ads. Another ad ran in an airline magazine to attract business travelers. "If you're reading this ad, you're very bored," the copy read. "You've mastered the safety instructions in every language, and the flight attendant won't give you any more almonds. Now what?" The choices: "Travel to another galaxy, golf . . . [all with Game Boy]. Game Boy won't ask you for your dessert, and fits just as neatly into the mouth of that screaming child beside you as it does into your briefcase. . . ."

Minoru Arakawa first saw a prototype of Game Boy in Kyoto in 1987 and promptly topped Yamauchi's predictions. Before it was over, 100 million Game Boys would sell throughout the world, he said. To reach such astronomical levels, however, it would need a monster game, and Arakawa saw it at an arcade-industry trade show in June 1988, where Randy Broweleit of Atari Games was showing off a prototype of an arcade version of a game called "Tetris." At the convention with Howard Lincoln, Arakawa stood in front of the console and played, transfixed. Broweleit told Lincoln that Atari Games had the coin-op rights and that Atari's Tengen had the rights to produce a home video-game version. Broweleit said that "Tetris" rights had also been sublicensed in Japan—coin-op rights to Sega and home-game rights to Henk Rogers's company, Bullet-Proof Software. Arakawa said he was pleased that "Tetris" would be coming out on the NES. He did not say he was also pleased that there was no mention of the hand-held rights, which might still be available.

Back in Redmond, Arakawa's engineers created a test version of "Tetris" that worked with a prototype Game Boy. When Arakawa plugged the tiny "Tetris" into Game Boy and played it, it was as if they had been made for each other. The "Tetris" pieces were large enough to be seen easily on the small screen, and the game was simple and hypnotic. Arakawa believed its appeal would cut across all age groups, as well as across the gender line, and said, "We must have this game."

Arakawa instructed Lynn Hvalsoe, the company's general counsel under Howard Lincoln, to track down the hand-held "Tetris" rights, but she couldn't determine who held them. Apparently Atari Games had bought its rights from Mirrorsoft, the London-based software group owned by Maxwell Communications, and Mirrorsoft claimed it had secured all "Tetris" rights from the creators in the Soviet Union. "They were giving us all sorts of bullshit," says Howard Lincoln. "We weren't getting anywhere." The reaction from Mirrorsoft made Hvalsoe suspicious, and she reported to Arakawa that Mirrorsoft's ownership of the rights appeared dubious. This led Arakawa to send an emissary to the Soviet Union to attempt to contact the inventor of "Tetris."

Alexey Pajitnov had the build of a medium-sized bear. His face was framed by auburn hair and a harshly clipped beard. He had grown up in Moscow, where his father was an art and theater critic, and his mother wrote for newspapers and for a weekly cinema magazine. As a child, Pajitnov's passions were mathematics and movies. As a rare privilege of his mother's occupation, he was given passes to the yearly Moscow Film Festival, where he watched five movies a day, fifty in ten days. "It was the only window to the outside," he says. His other love, math, was expressed in playing games that involved geometry, algebra, and other systems of logic.

When Alexey was eleven his parents divorced. He lived with his mother in a one-room state-owned apartment. He was seventeen when they were able to acquire a private apartment in a modern, fourteen-story building at 49 Gersten Street, in a fancy neighborhood of embassies and hotels not far from the Arbat, Moscow's far humbler Champs-Élysées.

Apartment 106 was, by Moscow's bleak standards, spacious and airy, with two bedrooms, a fair-size living room, and a small kitchen with a view of St. Basil's Cathedral in Red Square. Bookshelf-lined walls were crowded with scientific books, computer manuals, classic novels in Russian, French, and English, and books on art and film. On other walls were framed prints of paintings by Monet, Renoir, Matisse, and Modigliani.

Alexey was a good student, particularly in math; when he was fourteen, he was a finalist in a citywide mathematics competition, and he spent the last three years of school in a specialized math program. During that time, in the summer he turned fifteen, he sat down in front of a computer for the first time and created his first program, a number game.

"Mathematicians are usually very strange people," Pajitnov says. "They are fully in their abstract worlds, and I was there with them." But he had other interests too. With friends he played cards, drank beer or vodka, and headed to the countryside for camping expeditions. He hung out with girls and was in every way, he says, a "normal schoolboy."

After his university studies, Pajitnov took a position at the Moscow Institute of Aviation, a technical university, in the department of math applications. He enjoyed teaching there, but one day he abruptly quit. His passion for math had been replaced by something new: by the universe he discovered in computers, where the exploration of numbers, games, programming languages, and mathematical logic had no limits. Thereafter computers consumed him. "It doesn't matter to a hacker what he is working on—it could be a game or an abstract math problem, but if a computer is involved, he is a god and can do whatever he wants inside that world."

This new interest led to Pajitnov's next job, at the Computer Center of the Moscow Academy of Science, one of the Soviet government's preeminent R&D labs. His office at the Computer Center was in a wood-paneled room crowded with a dozen metal desks separated by chest-high partitions. There he hunkered over the keyboard of an archaic Soviet microcomputer, an Electronica 60, spending long days and many nights drinking black coffee, smoking unfiltered cigarettes, and exploring artificial intelligence and the capability of the computer to recognize the human voice. Inevitably, he also created games and puzzles.

To most of us, puzzles are a diversion, but for Alexey Pajitnov they are metaphors and mirrors that reflect nature, emotion, and patterns of thought. The young mathematician had turned to computers with the belief that they could model consciousness. Where better did electronics and humanity collide than in computer games? At their best, games were sublime examples of the intersection of logic and humanity. They worked because of logic and mathematics, but also because of psychology and emotion. The best games held in them a challenge, but also a reward and certain elemental experiences: discovery, recognition, frustration, and completion.

Inspiration for Pajitnov's games came at unexpected moments. Visiting an aquarium one day, he wandered past tanks of eels and seahorses, a shallow pool of starfish and anemones, and a large pool of rays, salmon, and sharks. He stopped in front of a tank of flatfishes and became transfixed there, staring at the liquid world of the flounder, plaice, and sole. The fish were barely distinguishable from the rocks, sand, and sea grasses on which they rested. When a plaice glided over white gravel, it metamorphosed before his eyes, as if its orange spots had been bleached white. A brown-speckled flounder floated over a bed of kelp and turned a soft green. Pajitnov was mesmerized as he contemplated using nature's splendid invention in a puzzle; he envisioned chameleon pieces that hid by altering their colors or shapes.

Another time, walking along a quiet boulevard, Pajitnov stopped to peruse a sidewalk display of imported knickknacks—porcelain dolls, paper umbrellas, and brass incense burners. From a clay pot he withdrew a Chinese fan. As he unfolded the pleats, what had been obscured was revealed: a crimson crane surrounded by golden flames. Laughing aloud, he imagined how fantastic it would be to re-create in a game the essential emotion of the experience—recognition.

Pajitnov had read about Pentominoes, geometric puzzles designed by an American mathematician named Solomon Golomb. The puzzle pieces were formed out of different configurations of five squares (a line, a "T," an "L" shape, and so on). The pieces could be fitted together into a perfect rectangle.

In a small toy shop, Pajitnov found a Pentomino puzzle. When he removed the dozen pieces from the rectangular case and mixed them up, he discovered that "it was a big problem" to put them back. He imagined a computerized version of the game in which randomly generated pieces would appear one at a time and with intensifying rapidity. An electronic version of Pentomino would require very quick thinking. He envisioned the puzzle pieces plummeting down from computer heaven and the frantic attempt to arrange them.

Sitting in front of his computer, Pajitnov experimented with permutations of Pentomino and finally settled on a simpler version, with each piece made out of four instead of five squares. From *tetra,* a form of the Greek word for four, he named the game "Tetris."

There is a theory in psychology that humans can process seven things (plus or minus two) at once: seven digits, seven shapes, seven concepts. It is the reason most people can remember seven-digit phone numbers but have difficulty beyond that. It so happened that seven different configurations of the four squares were possible. Seven "Tetris" shapes, Pajitnov reasoned, could be memorized and instantly recognized, and the reaction to any one of them could be almost visceral, reflexive.

Since the Electronica 60 had no graphics capabilities, Pajitnov's puzzle pieces were actually spaces outlined by brackets. Generated by the computer and sent onto the screen, they would descend slowly at the easiest levels, fast and furiously at the most difficult. The player had until they reached the screen's bottom to turn them or move them so that they would fit snugly into a solid row when they landed. If the pieces did fit perfectly together and an unbroken row was formed, it disintegrated: success! If any spaces were left unfilled, however, that row became the beginning of a wall that would grow until it overtook the screen.

Pajitnov realized his game would be more fun if the computer code were translated into real-time graphics—that is, if the brackets delineating the "Tetris" pieces were replaced by the real, movable shapes they represented. A young hacker named Vadim Gerasimov set out to create a color version of "Tetris" that would play on IBM-compatible computers.

Gerasimov was sixteen years old at the time and was still attending high school, but he was so far ahead of his peers that his teachers allowed him to drop in for classes a couple of times a semester. Raised by his mother, a nuclear physicist, Gerasimov had a revelation when he got his hands on a computer for the first time. "He saw the computer and forgot about the other world," Alexey Pajitnov says.

The wiry-haired Gerasimov, with enormous blue eyes behind thick-lensed glasses, was bean-thin and tall, with a slight stoop, and often wore the same shapeless gray wool sweater. Another programmer named Dmitri Pevlovsky introduced him to Pajitnov, who put the young boy to work. Gerasimov had a knack for finding glitches in programs, and had technical skills that neither Pevlovsky nor Pajitnov possessed. He had taught himself to program with Microsoft's DOS operating system from the West. He knew the BASIC and PASCAL languages, and how to perform miscellaneous feats on computers, breaking supposedly unbreakable copy protections and ferreting out viruses. Computer Center scientists twice his age asked for his help on programs, occasionally slipping the boy a few rubles.

Gerasimov worked with Pajitnov for two months to convert "Tetris" to work on an IBM-compatible computer. In the end, the "Tetris" pieces lit up in solid colors. Pevlovsky added a table that tracked high scores. When the program was bug-free, he copied it and distributed disks throughout the Computer Center. His colleagues congratulated him on his brilliant and addictive program. A friend who worked at a psychology institute gave the game to his staff, but soon realized that not much work was getting done because of it. One night, after everyone had gone home, the man went from desk to desk collecting the "Tetris" disks and destroyed them all.

All across Moscow the game was catching on in computer circles —"like a wildfire," says Pajitnov. In a computer-game competition held in Zelenodolsk, in November 1985, "Tetris" took second prize.

Pajitnov worked on other programs, including one called Biographer, a sort of therapist in a box. Information about someone's life was fed into the program, which revealed behavioral patterns and, in a crude way, drew conclusions. The program was a simple implementation of the idea that a computer could be more objective and more patient than a human psychologist. Through early 1986, Pajitnov continued to work on Biographer and other explorations of simple artificial intelligence. In the meantime, he suggested to one of his superiors that something should be done with "Tetris" outside the U.S.S.R. "We had no copyright laws at all," he says. "Certainly, by the spirit of our law we had no right to sell anything to anyone. We could do nothing for personal gain." It would, however, be a significant accomplishment to have a program published.

Victor Brjabrin, who oversaw twenty researchers at the Computer Center, was particularly enthusiastic about "Tetris," and he sent an evaluation copy of the game to SZKI, the Institute of Computer Science in Budapest. Hungarian-born Robert Stein, who ran Andromeda, a software company in London, happened to be visiting SZKI at the time.

Born in 1934, Stein arrived in Britain in 1956 as a political refugee. He first got work as an instrument- and toolmaker in London. Later he worked for Olivetti, selling adding machines, then left that business to sell mechanical check-writing machines while studying marketing at night. In business school he discovered that he was as good at training as he was at selling, and he got a job teaching engineers how to communicate with clients. He consulted in this field for a while, serving such major clients as Texas Instruments. After that he set up a company to sell TI calculators, and eventually sold digital watches and the first television game, Atari's "Pong." When that company went bankrupt, Stein founded a new company to sell chess computers to Harrods and other depart-

ment stores. The business grew when he started selling Commodore's Vic 20 computer. He soon realized the computer
sold in direct proportion to the amount of software available
for it, which led him to the software business.

When Commodore was preparing to introduce their C-64, a
more powerful computer, they asked Stein to return to Hungary in
search of innovative software. In 1982 he helped establish a software company there with Hungarian engineers. To sell their games
and business programs, Stein founded Andromeda, which kept 25
percent of whatever the Hungarian programs brought in.

From his London office, Stein sold the rights to Hungarian
products to Commodore and software companies in England such
as Mirrorsoft, and he made a number of deals with Jim Mackonochie, the man who ran Mirrorsoft for the Maxwells.

In June 1986, Stein was at SZKI in Budapest to see Hungarian
programs when, on a nearby computer, he noticed "Tetris." He sat
down to try the game and couldn't stop playing. "I was not a game
player," he said, "so if *I* liked it, it must be a very good game." He
asked the director of the institute where the game had come from,
and was told that it had been sent by a friend at the Computer
Center of the Academy of Science in Moscow.

The same day, Stein claims, he was shown another "Tetris," this
one on a Commodore 64 and Apple II. It was the same game, Stein
says he was told, adapted by Hungarian programmers. Although
they had obviously converted the Russian program to the other
machines, Stein says he told the Hungarians he would license the
original PC game from the Russians and the Commodore and
Apple versions from the Hungarians.

Back in London, Stein sent a cable to the Moscow Computer
Center indicating that he was interested in "Tetris." Victor
Brjabrin saw the telex and delivered it to Pajitnov. "It looks like
someone is interested in your game," he said.

Since there was no one else to do it, Pajitnov began the negotiations himself. "We had no idea what to do," Pajitnov says. "It was
an absolutely new experience for us."

Answering the telex was comically difficult. Pajitnov's English
was a little sketchy, so he composed an answer in Russian and
showed it to the chief of the computer center, Professor Ju. G.

Evtushenko, who had to sign an approval to have it translated. He then had to figure out how to send it.

Pajitnov had no access to a telex machine, but he learned that there was one in another division of the Academy of Science that he might be able to use if he could gather the required authorizations. His requisition had to be approved by his supervisors and half a dozen others at the Academy. It was weeks before he could send back the simple answer: "Yes, we are interested. We would like to have this deal."

Stein was already shopping the game around. He had showed it to representatives of British and American software companies. His plan was to determine if there was interest, and then, if it seemed worth his while, to secure the rights; he assumed it would be easy to convince the Soviets to make a deal. As it happened it took more than a year finally to secure the rights to Tetris. On reflection, Mr. Stein thought that it would have been easier to use the Commodore 64 version produced by the Hungarians as a master source.

Stein pitched "Tetris" to Jim Mackonochie at Mirrorsoft and to an American software company, Broderbund, but neither company showed much interest in it. Mackonochie wasn't convinced the game would sell, so he showed it to the two men who ran Mirrorsoft's sister company in America, Spectrum Holobyte, telling them he would license "Tetris" for Europe if they would do so for the United States and Japan.

Phil Adam and Gilman Louie ran the California-based Spectrum Holobyte, 80 percent of which was owned by Robert Maxwell's Pergamon Foundation. Adam, with impeccably trimmed hair and fingernails and perfectly matched casual clothing (wool sweaters, khaki slacks, and oxford-cloth shirts), was a smart and well-liked manager of Spectrum Holobyte, which was known for its flight simulators. "Falcon," created by Gilman Louie, a computer whiz, was one of the best flight-simulation games available for PCs, and it sold more than half a million copies. Spectrum Holobyte went on to create a host of other computer games, in addition to simulators for the military.

His tortoiseshell glasses pushed up on his thin nose, Adam sat down at three o'clock one afternoon in front of a computer on which "Tetris" had been booted. At seven, dinner companions who had been waiting for him for an hour came and literally had to unplug the computer in order to drag him away.

Louie, tall and slender, his jet-black hair untamed, had thick-framed glasses and, behind them, a curious, decisive expression. He too played "Tetris" and loved it. After conferring with Adam, he told Mackonochie, "Put it in a red box and get the rights."

According to Robert Stein, Mirrorsoft and Spectrum Holobyte bought all the "Tetris" rights except for the arcade and hand-held versions. From these companies, Stein got a small advance—about 3,000 pounds—against a fluctuating royalty of between 7.5 and 15 percent.

In his next telex to the Soviets, sent on November 5, 1986, Stein offered a firm deal: the Russians would get 75 percent of whatever he collected for "Tetris." This sum would be a percentage of gross sales. He also offered a $10,000 advance.

Pajitnov responded favorably to the offer in a telex on November 13. Signed by the Computer Center's Evtushenko, it said that the Academy of Science was ready to transfer the copyright to Andromeda. In another telex, Stein had offered to pay the Computer Center, in part, with Commodore computers, and the Soviets agreed to this as well. They also noted that the deal was for the IBM-compatible version of "Tetris" only; they would consider non-IBM versions of "Tetris" in the future.

Alexey Pajitnov now claims he indicated only that the deal sounded good. He did not mean to give Stein a firm go-ahead. "I had no idea that this kind of polite telex can be a document," Pajitnov says. "I think of a document as something very serious which needs to be signed, changed, and signed again; then you shake hands and drink champagne."

Stein paved the way for the signing of a contract in telexes to Evtushenko. The Academy of Science's licensing group, Licensnauka Prasolov/AcademySoft, took over negotiations and sent Stein a cable in late December, inviting him to come to Moscow.

When, much later, Stein did visit Moscow, he arrived with plans to talk to the people at the Computer Center about establishing a relationship similar to the one he had in Hungary: he would act as agent to sell Soviet-created software in the West. His immediate concern was to leave with a signed, legal contract that confirmed his rights to "Tetris." He met with a group of the Soviets in a hall at the Academy, a cavernous, poorly heated, ill-lit room, as grim as anti-Soviet propaganda in the West would have painted it. Down the center of the room was a massive wooden slab, a table that could have seated fifty, across which Stein faced six Russians. One of them was Alexey Pajitnov, the chain-smoking creator of the game.

Trying to ally himself with Pajitnov, Stein said, "Gentlemen, in our country the most important person is the one who designs the game. I'm here to listen to his wishes, because if we don't sign a contract, it is he who will suffer."

Stein also tried, unsuccessfully, to hide his eagerness. The contract he had drawn up was on the table in front of him, ready for the Russians to sign, so he was unprepared to be met with suspicion.

The Soviets were swimming in uncharted waters—software licensing was baffling—yet they made up for their naïveté with caution and obstinacy. If Stein offered 75 percent, they pushed for 80; if he offered $10,000, they held out for $25,000. They asked for protections and limitations. The details were hammered out over several days of meetings, but Stein, although he caved in to their demands for a higher advance and royalty, left Moscow without a signed contract.

The Russians had tried Stein's patience, and he wondered whether the Mirrorsoft and Spectrum Holobyte people should abandon the PC version and use the Commodore 64 version created by the Hungarian programmers, which they had licensed simultaneously with the PC version from the Russians. Stein later said that the biggest mistake he made was that this did not happen.

But it was too late for that. Mirrorsoft and Spectrum Holobyte had seen great value in the fact that "Tetris" was the first game to come from behind the Iron Curtain, which was still intact at the time. As Gilman Louie had suggested, they slickly packaged "Tetris" in a red box, emphasizing that it was from Russia with love. Atop an illustration of St. Basil's Cathedral, "Tetris" was written in Cyrillic, the final character taking the form of a hammer and sickle. Programmers at Spectrum Holobyte in the United States added battle scenes as background pictures and a simple animation that played at the start of the game: a Cessna flew across the screen and landed in Red Square. It was a homage to Matthias Rust, the young West German pilot who had flown his small plane from Helsinki to Moscow, past all the Soviet radar and air defenses, and landed in Red Square, embarrassing the Central Committee. Rust had been arrested, tried, and imprisoned.

Other modifications to the program were made. One was the addition of a "boss button" to some versions of "Tetris"; when a certain key was pushed, the game instantly disappeared and the computer would appear to be running a serious accounting program. Also added were high-quality graphics, and, for computers with sound-generation capabilities, music.

In April 1987, Stein informed the Soviets that the rights to "Tetris" had been sold to Mirrorsoft, the Maxwell Communications software company, and to its American counterpart, Spectrum Holobyte. He specifically noted that the rights were only for the IBM PC (and compatible) versions of "Tetris." By then he had given up plans to license separate Apple and Commodore versions of the game from Hungary, and he noted that there would be separate advances for those versions when they were launched. He pressed for a contract to be signed.

In June, Stein signed the contract he had negotiated among Andromeda, Mirrorsoft, and Spectrum Holobyte. The contract designated that he was selling the "Tetris" rights for the IBM PC, although the rights included "any other computer system." In addition to warranting that the work had no obscene or indecent material, Stein confirmed that he was the rightful owner of the copyright, free to grant the license, although there was still no contract with the Soviets.

In a letter sent to the Academy of Science in December, Stein pleaded with the Soviets to confirm his rights to "Tetris," and offered "to travel anywhere and meet anybody" in order to get a signed contract. If nothing else, he wrote, "we need a simple letter stating that you approve the terms under which we have signed this contract with Mirrorsoft."

Mirrorsoft and Spectrum Holobyte, unaware of Stein's troubles, released "Tetris" for personal computers in Europe and America in January 1988. It sold well in computer stores and received rave reviews.

A reviewer for *The Times Educational Supplement* quoted from the Mirrorsoft press material and wrote, "Another 'remarkably simple, addictive, abstract puzzle game' doesn't fill me with excitement, but the provenance of this one is interesting. It was invented by a 30-year-old researcher at the USSR Academy of Sciences. Coding was done on an IBM personal computer by a teenage student at Moscow University." Many British computer journals heaped on praise. "I warn you. The addictive power of 'Tetris' is frightening," a reviewer wrote. "It sounds so simple that you cannot believe your abysmal score, so you try again and again; only a manic lunge at the reset switch can save you." Another reviewer called it, "a stonking good game," reporting that it had been banned from the PCs at Mirrorsoft's central office, "and no doubt it will be banned elsewhere." Yet another reviewer concluded, "There are very few games that are so addictive you find yourself playing them in your sleep, but 'Tetris' did exactly that for me. . . . After a poor night watching shapes float down past my closed eyelids, I had to be firm with myself and refuse to play it again for at least 48 hours. I cannot give a higher recommendation."

In Moscow, meanwhile, Pajitnov was busy with his day-to-day tasks at the Computer Center. A version of Biographer was working well; he wondered if it could be sold as educational software. In discussions about it, he met with a newly created Soviet organization called Elorg—short for Electronorgtechnica, the ministry for the import and export of hardware and software. In his meeting with an Elorg director named Alexander (Sasha) Alexinko, Pajitnov mentioned the difficulties he was having licensing "Tetris." Alexinko interrupted him. Licensnauka and the Academy of Science shouldn't be negotiating at all, he said. They were academic institutions, which were forbidden to indulge in commerce. This was Elorg's domain, and Alexinko said he would take over the "Tetris" negotiations.

Examining the communications between Moscow and London, Alexinko concluded that Pajitnov had blundered in his negotiations, and that his telexes could have been misinterpreted. Now the inventor became the target of blame. "You allowed this game to be published without our approval," he was told. "We must stop it now."

Stein, further and further out on a limb, having sold the rights to a game he didn't own, now heard from Elorg. It was taking over "Tetris" negotiations, and his deal was off. The Soviets were going to take over the international sale of "Tetris" directly.

Backed into a corner, Stein composed a carefully worded memo to Moscow. He threatened to create a scandal for the Soviets. It would look very bad, he said, if the Soviet state stopped a commercial deal at this point; it would be extremely embarrassing in the international community. On the other hand, here was a chance, he pointed out, to begin an alliance that could be politically and economically significant. A deal had to be consummated.

Stein and Elorg tested one another with tentativeness, straining over minor points, but finally agreed that a deal could be made. Stein flew to Moscow to meet with Elorg's Alexinko in late February 1988. On February 24, a written contract was proposed that included a stipulation that Elorg had to approve any versions of "Tetris" created in the West. It also gave Andromeda the right to adapt "Tetris" to "different types of computers."

Stein and the Soviets negotiated for four days to finalize a contract. Even after this there was long-distance fine-tuning for several months until finally, after almost two years' worth of draft agreements, one was signed in May 1988. Stein breathed an enormous sigh of relief: he had successfully confirmed his exclusive rights to sell "Tetris" for computers. In a memo, he told Mirrorsoft that the contract included "TV games," but excluded coin-op (arcade), hand-held games, and other concepts "which we did not dream about yet."

In the meantime, "Tetris" had become the best-selling game in England and the United States, publicized by word of mouth and on computer networks. In the United States, in 1988 "Tetris" received the Software Publishers Association award in two categories: Best Original Game Achievement and Best Entertainment Program.

An article about "Tetris" from a London computer magazine reached Elorg in Moscow. It described the version of "Tetris" that was being sold in the West. It mentioned the graphics—the battle scenes and the image of Matthias Rust flying his plane across the Russian sky and into Red Square. Alexinko showed the article to Pajitnov, who was amused by the reference to Rust. He was not amused, however, by the battle scenes. Pajitnov had come to view "Tetris" as a small but meaningful bridge between cultures at a time when the Cold War was thawing. "Tetris" was a game of the intellect, completely nonviolent. He informed Stein that he wanted "Tetris" to be "a peaceful game heralding a new era in the relationship between superpowers and their attitude toward world peace." The bureaucrats in Moscow were far more concerned about the reference to Rust. The Central Committee viewed the young pilot as a terrorist and did not consider the invasion of its air space a practical joke. Broadcast around the world, Rust's "raid" had been a great humiliation.

The Soviets expressed their "grave concerns" in their next meeting with Stein, who immediately contacted Jim Mackonochie. "It would be useful if all battle scenes, as background pictures, would disappear," he wrote, "and planes flying across the title screen . . . would be omitted." As a result, Mackonochie and Gilman Louie revised their games again.

Stein felt the Soviets had to be placated as much as possible if he was going to be able to obtain other "Tetris" rights that Mirrorsoft wanted, particularly the coin-operated and hand-held rights. Atari Games was already preparing to release its own "Tetris" arcade game in America, and it had sold the rights to Sega to release the arcade game in Japan. (Mirrorsoft had gone ahead and sold Atari Games the right to do this based on Stein's assurance that they were forthcoming.) Stein, however, was in the midst of another sluggish negotiation. He had made a firm offer for the coin-operated rights in July 1988—a guarantee of $30,000 as an advance against royalties—but there had been no response.

Stein telexed a month later, noting that he was being "pressurised" and needed an agreement by mid-August.

Alexinko met with Stein in Paris on July 5. While Stein pushed for more rights, especially the coin-op rights, the Russian had another agenda. He intended to register his dissatisfaction with the results of the deal that had already been signed because no checks were coming in from Andromeda.

Stein explained that it took time for money to be dispersed from company to company, but promised he would do what he could to expedite the royalty payment process. Alexinko wanted a penalty for late payments to be agreed upon, and threatened to withhold any new rights until the situation was remedied.

More telexes were fired back and forth between Stein and Alexinko after they had returned to their home bases. Stein implored the Russians to grant him the coin-op deal, while Alexinko, responding in a tersely worded telex, indicated that the existing deal was not satisfactory "because no payment is effective yet," even though "Tetris" had been "more than six months on the market."

In another telex he asked Stein to add a clause to the original agreement that stated that a 5 percent fee would be charged monthly for late payments. Alexinko insisted that this clause was important "to expedite the positive decision" regarding the "Tetris" coin-op rights.

Stein sent word that he agreed, and once again begged Alexinko to sign a coin-op contract because "someone will steal the product from under our noses."

Typically, a company that licenses a game will exploit those rights as much as possible with sublicenses to other companies for other markets. Spectrum Holobyte and Mirrorsoft had a hit game on their hands, and it was not surprising that they set about to sell all the sub-rights they could. The chart of "Tetris" licenses and sublicenses was beginning to look like a tangled family tree.

Henk Rogers, who had seen Spectrum Holobyte's computer game at an electronics trade show in January 1988, went after the rights to release the game in Japan on computers, video-game systems, and coin-op arcade machines. Since Spectrum Holobyte had been granted the rights for Japan in the deal with Stein, Rogers negotiated with Gilman Louie. Two separate deals—one for computer-game (floppy-disk) rights, another for video-game rights

—were agreed upon. Rogers also wanted the coin-op game rights, but discussions for these were put on hold.

The day after Rogers signed a deal with Spectrum Holobyte for computer-game "Tetris" rights for Japan, which was the day *before* he was to sign a deal for the Japanese home video-game rights, Gilman Louie called Jim Mackonochie in England to report on the deal. Mackonochie "had a fit," Louie remembers. The Mirrorsoft chief told Louie it was impossible for Spectrum to go ahead; he had already sold those rights to Atari Games. Hide Nakajima was getting "Tetris" rights to North America and Japan in exchange for worldwide rights to an Atari game called "Blastroids."

Gilman Louie hit the roof. "What are you talking about?" he shouted. "Those are *my rights*!"

Louie argued that not only were all the Japanese and American "Tetris" rights his, not Mirrorsoft's, but Mackonochie was making a terrible deal; "Blastroids" was an atrocious game. Besides, he said, he had made a terrific deal with Henk Rogers.

Mackonochie explained that there was nothing Louie could do. The Maxwells had financial control of both Mirrorsoft and Spectrum Holobyte, but both men knew that the family had far greater interest in Mirrorsoft, which was personally overseen by Robert Maxwell's son Kevin. Kevin Maxwell supported Mackonochie's deal, so Louie was out of luck.

Louie insisted that he at least had to honor the deal he had already signed; Henk Rogers had to be able to have the sublicense for the floppy-disk game in Japan. Since it was the least valuable of the rights under discussion, Mackonochie agreed.

Atari Games' agreement with Mirrorsoft was not actually signed until May 30, 1988, only two weeks after Stein's initial deal with the Russians was signed. Hide Nakajima planned to exploit the "Tetris" rights in as many ways as possible. In the United States he planned to release both a coin-op "Tetris" and an NES version under the Tengen label, and he planned to sublicense these rights in Japan.

Gilman Louie called Henk Rogers to apologize; he explained that unbeknownst to him, Atari Games had been awarded the "Tetris" video-game rights for Japan and the United States. Rogers's floppy-disk rights were secure, but if he wanted the other Japanese rights, he would have to negotiate with Atari Games.

Rogers tried, contacting Randy Broweleit and Hide Nakajima. Broweleit said the coin-op rights for Japan were gone; they had been sold to Sega. Dejected, Rogers said he would at least like to secure the rights for "Tetris" on Japanese home video-game systems, including Nintendo's Famicom. What, he asked, would he have to do for these?

Broweleit was noncommittal, so Rogers went directly to Nakajima. The two men had dinner on August 16, 1988. It took until October to hammer out a deal. As far as he knew, Rogers had finally sewn up the rights to sell "Tetris" in Japan, not only for computers but also for the far bigger market, Nintendo's Famicom and other home video-game systems.

Now Rogers headed to London for a meeting at Mirrorsoft. He brought videotape copies of versions of "Tetris" for the Russians to approve. After the meeting, Rogers returned to Japan, where he received a faxed go-ahead to produce his game from a Mirrorsoft attorney.

Rogers's "Tetris" for PCs was launched in Japan in November 1988. A month later he shipped his version for the Famicom. Tetrismania quickly swept Japan, just as it had the United States. Two million copies of the game were sold there.

By now Minoru Arakawa had decided that he wanted "Tetris" for Game Boy. His lawyers had figured out that Mirrorsoft was probably tap-dancing around the fact that it didn't have the rights; indeed, it was possible that no one owned the hand-held rights. That is when Arakawa decided it might be beneficial to try other avenues to get the game. In a meeting with Henk Rogers, Arakawa made an offer. If Rogers could get the hand-held rights, NOA would take out a sublicense from him. Rogers was let in on the still-secret reason that Arakawa wanted "Tetris": he was shown a prototype of Game Boy.

For all Rogers's impressive insight into game play, he was sharp about the business side of video games as well. When Arakawa challenged him to get the hand-held rights to "Tetris," the younger man viewed it for what it was: a potential fortune. "If you've met

Rogers, you know that he is capable of finding his way in the middle of any storm," Howard Lincoln says. "Telling him that we were ready to license from him was like showing red meat to a hungry lion."

Henk Rogers faxed Robert Stein in London on November 15, 1988. He wanted to bid for the worldwide hand-held rights to "Tetris." Stein responded that he was trying to get the rights from Elorg, but the agency had not made a decision about them. He was close to pinning them down, however, and would get back to him.

Stein immediately wrote Jim Mackonochie. "We must go for those rights immediately," he said. But Stein had new worries. In late 1988 he received a telex that informed him that Sasha Alexinko had been replaced at Elorg. The reasons for his departure were unclear—he ended up leaving Elorg to found his own trading company, and there was speculation that he used his government work to set up the company—but there was nothing ambiguous about the character of his replacement, the vice-director of Elorg, Evgeni Nikolaevich Belikov. Belikov, large-framed, slightly balding, and red-cheeked, was described by some who knew him as vicious, cutthroat, and bullheaded, and by others as savvy, amiable, and razor-sharp. All, however, agreed that he was a worthy adversary in any negotiation.

In late 1988 Belikov became the man to charm, outmaneuver, or otherwise win over in all future "Tetris" deals. It was easier said than done. As Alexey Pajitnov discovered, "He is an excellent actor." Stein found him "instantly dislikable; a creep."

Rogers continued to try to push Stein to get him the hand-held "Tetris" rights. Many calls, letters, and faxes later, however, he came to the conclusion that Stein was useless. His only chance to get the rights was to head directly to Moscow, which he did, in February. But he wasn't the only one to do so.

Kevin Maxwell helped run his father's empire, then thought to be the tenth-largest media concern in the world. In addition to newspapers with circulations that totaled in the tens of millions, he looked over such companies as Marquis, Thomas Cook Travel, Berlitz language schools, and MTV Europe. He also oversaw Maxwell's electronic-media companies—that is, on-line networks and the computer software company Mirrorsoft. This made him Jim Mackonochie's boss.

Kevin, a graduate of Balliol College at Oxford and a serious, morose workaholic, had spent his entire life working in the family companies. His brother Ian Maxwell had once lost his job with the Maxwell organization because he chose to meet a girlfriend in Paris rather than show up at the airport for a meeting with his father. That was a choice Kevin Maxwell would never have thought to make.

Like his father, Kevin tended to be inaccessible when his top executives needed his input, though he would meddle on a whim. And, like his father, he was short-fused and inclined to tantrums, tending to jump when it might have been prudent to think.

From the beginning, Kevin Maxwell had kept tabs on the "Tetris" negotiations in consultation with Jim Mackonochie. When negotiations got so bogged down that Mackonochie decided to go to Moscow himself, Maxwell stepped in. He was going to Moscow anyway, he said, and he could easily straighten things out with the Soviets.

Robert Stein, frustrated because he couldn't nail down the rights to "Tetris" for hand-held or coin-operated games, worried because there were already coin-operated versions being sold in the United States and Japan, and under pressure from Henk Rogers, realized that he also had to head to Moscow again. Unbeknownst to one another, the three men flew to the U.S.S.R. at exactly the same time.

THE "TETRIS"
SONG

Alexey Pajitnov could tell instantly that Henk Rogers was a man after his own heart. Of all of those he had dealt with, from Stein to the bureaucrats at Elorg, Rogers was the only one who seemed to truly love "Tetris." He was a hacker, he spoke the language of games, and he understood the pure beauty of the "Tetris" design.

Elorg and the Academy both hoped to establish a relationship with someone from the West other than Robert Stein. Of course they also wished to make as much money as they could, so when Rogers appeared in Moscow, after tracking down Pajitnov through chess players and computer hackers, the Russians had a second bidder and leverage against Stein.

For his part, Rogers was surprised by how naive the Soviets were about licensing deals. "Whoever they had been doing business with obviously didn't explain what the world looked like," he says. The jargon was unfamiliar to them, and he could have snowed them. At their first meeting at Elorg, he accepted the coffee the Soviets offered. Sounds echoed in the unadorned, chilly room. Rogers did most of the talking. Before he knew it, he was walking the Soviets through the video-game business as if he was teaching a course.

When the meeting ended, Pajitnov and Rogers struck up a conversation and ended up having dinner together at a restaurant. Then Pajitnov invited Rogers back to his apartment to show off other software he had been working on. It was an evening of frank discussion and good cheer. Pajitnov, suspicious of all the Westerners trying to get pieces of "Tetris," found that "the most important thing about Henk was that he didn't ask for protection in the deal. He offered me nothing and asked for nothing."

The following day, at Elorg, Rogers presented an offer for the hand-held rights to "Tetris." The deal was ironed out within a few days, and signatures confirmed it on February 21.

Delighted, Rogers promised an advance check right away. He also assured the Russians that the royalties would be significant; hand-held "Tetris" would bring them a substantial amount of money. In a celebratory mood, he mentioned that he had brought with him a copy of the home video-game version of "Tetris" he was selling in Japan. The Russians looked at one another dumbfounded as he produced from his briefcase the brightly packaged Nintendo game cartridge and proudly passed it around the room.

Sitting on the edge of his seat, Nikolai Belikov spoke first. "What is this game?" he asked.

Rogers explained that it was his "Tetris" game for the Famicom, the video-game system made by Nintendo in Japan.

The Russians had never heard of Nintendo.

Rogers reminded the group that they had seen and approved a videotape of the game.

Belikov shook his head. "We have approved nothing. We never licensed anybody to make this!" he snapped.

The mirth evaporated, and Rogers realized he was in trouble. Stammering, he told the Soviets, "But I bought those rights from Tengen. I paid lots of money for them!"

Belikov's fists rested on the table. "We don't know this company Tengen," he said. "We do not understand."

It dawned on Rogers that the license he had bought from Atari

Games/Tengen to market "Tetris" for the Famicom was, as he said later, "a sham."

Rogers now recounted to the Soviets his negotiations for the home video-game rights that began with Spectrum Holobyte and ended when he was told that Atari held the rights. He said he had negotiated with Atari over the course of more than six months, and reported that Atari had announced that it was releasing a home video-game version of "Tetris" for the Nintendo system in America. But there was more: Atari had also sold a company called Sega the rights to create an arcade version of "Tetris" in Japan.

The Soviets were speechless. "An arcade game has been sold?" Belikov asked. Rogers nodded.

Belikov finally spoke in Russian, uttering instructions to an assistant, who disappeared for a few minutes while Belikov told Rogers, "We have not licensed anyone to make 'Tetris' on home video-game systems or coin-operated games. I will show you."

The assistant returned with a tall stack of documents, which he set on the table. Belikov flipped through the papers until he came to a copy of Elorg's contract with Robert Stein. He pored over the pages for a few minutes, and when he found what he had been looking for, set the document on the table and pointed with his forefinger to the paragraph explicitly stating that the rights granted to Robert Stein's Andromeda Software were for the IBM PC and other computer systems.

Rogers was as shocked as the Soviets were, but his immediate concern was to hold on to the video-game rights that he had just secured. He said he must have been lied to, and that he wanted to make things right with the Russians; pending negotiations, he would pay them directly for all the games he had sold so far.

Belikov indicated that this offer was acceptable, but that they would adjourn for the day. Rogers was to return in the morning, and an arrangement would be negotiated then.

When Rogers returned the next day, he came with exact calculations of the number of Famicom cartridges he had sold and offered to pay what amounted to a second royalty on the 130,000 cartridges. He immediately wrote a check for $40,712, representing a portion of that.

Meetings continued for the next few days, during which Rogers examined the Soviets' documents on the deal with Andromeda. He was convinced that the Russians had never—intentionally, at least —sold the video-game rights. The upshot of the meetings was an offer from Belikov: Rogers had three weeks to determine whether he wanted to make a proposal directly to them for *all* the home video-game rights to "Tetris."

Rogers warned, "There will be trouble." The companies that were selling "Tetris" rights with abandon—Mirrorsoft and Atari Games—were heavyweights. He said he had a plan: he would return with a partner who not only had an enormous amount of money but clout enough to fight them. This partner, the company that had control of the largest market for video games in the world, was Nintendo.

Robert Stein had raced to the Elorg office that morning in a taxi from the Hotel Kosmos. Exhausted from travel and from the prospect of renewed negotiations, he waited obediently in a small room, where a rickety table held only a pitcher of water and a glass.

Finally Belikov, Elorg's new negotiator, stormed into the room. He refused to engage in small talk, but simply threw a document in front of Stein and told him to sign it.

Stein asked what it was. "We already have a contract," he pointed out.

Belikov explained that this was an amendment to the contract. Confused, Stein said he didn't understand; he was in Moscow to negotiate for the hand-held and coin-op rights to "Tetris," not to sign a new contract.

Belikov held firm; he would continue negotiations only if Stein signed the paper.

Stein examined it. It was an addendum that noted that it and the original contract were to be read as one. Concentrating on the payment schedules and fluctuating percentages, Stein's eye skipped over one line that defined computers, as referred to in the original contract, as "PC computers which consist of a processor, monitor, disk drive(s), keyboard and operation system." He was told that the one-page document—alteration no. 1—was to amend the contract signed the previous May, which was why it had been backdated to May 10, 1988; it would be effective as of that date.

Stein returned to his hotel to study the amendment. It seemed to him that the most important item was about the assessment of penalties for late payments of royalties. He knew the Russians were concerned that money had not arrived frequently enough. The amendment was, in these circumstances, understandable. Though he read and reread the paper, he was unconcerned about the innocuous line clarifying the definition of computers. Later he deduced that everything in the document *other* than this line was "a smoke screen."

"Henk Rogers orchestrated it for Nintendo," Stein believes. "He advised the Russians." The charade was all designed to take away most of his rights and offer them to Nintendo, which swooped in and snatched them.

Stein returned to the Elorg office the following day. He had no problem with the amendment, he announced, but said that he would not sign it unless the deals for the other rights were made. It was the only leverage he had. He had drawn up by hand an offer for the coin-operated and hand-held "Tetris" rights, and he presented it to Belikov. In the document he included minimum guaranteed sales, advances, and percentages of royalties, but it was, Stein says, "a mockery. They knew what they were going to do before I arrived." He was told he could not secure hand-held rights at this moment, but he could have the coin-operated rights. He would have to pay dearly for them; within six weeks he had to come up with an advance of $150,000 or the deal would be off. He signed the contract and the amendment two days later, on February 24, 1989.

Oozing charm and self-importance, Kevin Maxwell was also at Elorg on February 22, meeting with the Soviets in a small room. He, Stein, and Rogers might well have run into one another in an Elorg hallway that day.

After small talk, Maxwell asked Belikov why the deal for coin-operated and hand-held "Tetris" rights was taking so long to settle. But Belikov had another agenda. Reaching into a sack, he withdrew, as a magician might reveal a rabbit from a hat, a video-game cartridge and placed it on the table. "What is this, Mr. Maxwell?" he asked.

Maxwell reached for the cartridge—Henk Rogers's Japanese video-game version of "Tetris"—and examined it. He had no idea that his own company had licensed the game via a sublicense to Atari Games, so he shrugged. Then the Russian asked him to look at the copyright notice on the cartridge. It read "Elorg, Mirrorsoft, and Tengen."

Maxwell said that Mirrorsoft had not licensed home video-game rights to "Tetris," so the cartridge must be a pirated game.

It was a crucial error. As a result, the Russians decided once and for all that they could maintain that Mirrorsoft had no right to the home video-game version of "Tetris."

Unaware of the gravity of the situation, Maxwell returned to his own agenda, the coin-op and hand-held "Tetris" rights. Belikov excused himself and disappeared for a while. When he returned, he told Maxwell that he would sign a "protocol agreement" promising Mirrorsoft the right of first refusal on ancillary rights to "Tetris"—including coin-op, hand-held, and merchandising rights —contingent on Maxwell's assurance that he would make an offer for the video-game rights within one week. "We must clear up the matter of this pirated cartridge," Belikov said. "Therefore we must have a deal within a week."

A protocol agreement was signed guaranteeing Mirrorsoft the right to bid on all remaining "Tetris" rights, even though the coin-op and hand-held rights were simultaneously being granted to Henk Rogers and Robert Stein. In exchange Elorg would get the rights to publish in the Soviet Union Maxwell Communications properties, such as *Collier's Encyclopedia* and other reference books. The Soviets may not have known much about the video-game business, but they demonstrated their uniquely effective method of negotiating. Their juggling of Rogers, Stein, and Maxwell had put them back in a position of complete control.

The upshot of Belikov's busy week was a deal for the hand-held rights for "Tetris" with Henk Rogers, plus a chance for a new deal within three weeks on the home video-game rights with Nintendo. He had a signed deal for coin-operated "Tetris" with Stein and the promise of a check for $150,000. Kevin Maxwell had given Elorg more than the rights to publish Maxwell Communications reference books: he had characterized the "Tetris" game his company had licensed as a pirated game instead of insisting that Mirrorsoft had licensed the games from rights it held—thus supporting Elorg's position that it had never sold the video-game rights in the first place. If Mirrorsoft wanted those rights now, it would have to outbid Nintendo. It was not a bad week's work.

A letter was drafted to Henk Rogers on February 24 confirming that Elorg had not granted anyone the license "to make, have made, duplicate, market, distribute, sell or in any way use" "Tetris" for use on "video games or TV games or game consoles, which are defined as computers which have no keyboard."

Looking back on the week in Moscow after the fact, Robert Stein said that his only satisfaction was knowing he emerged from the meetings with the coin-op rights while Kevin Maxwell emerged with nothing but worthless paper. Otherwise, he said, the week was a disaster, the product of lying, cheating, and back-stabbing. It was impossible, he said, for the Russians not to know about the BPS Nintendo game before Rogers showed it to them; he claimed he had provided Elorg with a copy of Rogers's videotape back in December 1988. He concluded that he had been set up, and that the amendment, the lies, and Kevin Maxwell's admission about the Japanese "Tetris" cartridge all cleared a path for the Russians to double-deal him. "I was set up because of stupidity," he says, "but also because I was under tremendous pressure to walk away with a contract for coin-operated games because Atari Games and Sega were already selling them all over the world without a contract." He insists that Jim Mackonochie and Mirrorsoft had put him in the position of "signing my life away" by knowingly selling rights they didn't own, and that he was forced to cover Mackonochie's tracks.

Bitterly, Stein says, "I will never know if Jim Mackonochie is a good friend or if he knew he was screwing me. Maybe he was just a corporate animal. The fact is, he went right behind my back. I was fighting for my bloody life and he screwed me, acting as if the coin-op license was already theirs, so that if I didn't get it for them I would be sued out of existence. Then"—he shakes his head wearily—"he concealed from me the fact that Kevin Maxwell was in the next room burying me, burying us all." Now Stein finally understood why Belikov kept disappearing in the middle of their meetings.

Henk Rogers tied matters up in Moscow and quickly returned home, where he called Minoru Arakawa. His news was better than anyone could have hoped for. First, he had the hand-held rights to "Tetris," and Arakawa's deal with him gave Nintendo the rights to "Tetris" for Game Boy. (He retained the rights to sell "Tetris" on other electronic hand-held machines, such as the Sharp Wizard.) Reportedly he would receive $1 for every "Tetris" sold with a Game Boy system, and more for games sold separately. This deal was worth between $5 million and $10 million to Rogers.

Best for Nintendo, however, was the news that the Russians claimed never to have sold the "Tetris" home video-game rights to anyone. The rights that Robert Stein had sold to Mirrorsoft and that Mirrorsoft had sold to Atari Games and BPS were, according to the Russians, bogus.

Rogers was covered whatever happened—he had Japanese "Tetris" rights from Atari Games or Nintendo, whoever ended up with them—but the beauty of the new development was that Nintendo could almost certainly have "Tetris" for the rest of the world's home game machines. Potentially this represented tens of millions of dollars and, for Arakawa and Lincoln, something even more delicious: the pleasure of depriving their old friend Hide Nakajima—the friend who, they felt, had betrayed them—of those potential millions. Revenge, Lincoln admits, was a prime motivator. There were no second thoughts: he and Arakawa would do whatever it took to get the home video-game rights.

Lincoln and Arakawa decided to send Rogers back to Moscow, this time with a lawyer. Lincoln called around and learned about John Huhs, a New York attorney who had worked in the Soviet Union. Huhs, who had been part of the Nixon White House, spoke fluent Russian, and, although he knew nothing about video games, was a talented international lawyer. Lincoln gave him a crash course on the phone before Huhs set out to meet Rogers, who was already en route back to Moscow.

After an initial meeting at Elorg, Rogers and Huhs made an

offer for the home video-game rights to "Tetris" on behalf of Nintendo. Included in the offer was an astronomical guarantee. The tough Soviet negotiators betrayed amazement over the figure. Arakawa wanted to be sure of clinching the deal; no one—not Atari Games, not even Robert Maxwell—was likely to touch the number he had authorized Huhs and Rogers to offer.

The same day, March 15, Elorg sent a telex to Mirrorsoft, noting that Mirrorsoft had promised to give Elorg its proposal for home video-game rights to "Tetris" within a week of the meeting with Kevin Maxwell in Moscow. As it was well past that week and Elorg had a competing proposal valid only until March 16, Mirrorsoft had one day to make an offer.

By design there was no chance for a Mirrorsoft bid. Elorg had elicited it in such a way as to address Maxwell's right of first refusal because it had been guaranteed in the protocol agreement. As planned, there was no response within that day, and so the path was cleared for Nintendo. Rogers and Huhs placed a call to Redmond. Huhs was convinced the deal could be signed and sealed if Lincoln and Arakawa could come to Moscow by Monday. He said they should head first to Washington, where their visas would be waiting at the Soviet consulate.

Arakawa and Lincoln told only Peter Main and Phil Rogers where they were going, for they were concerned that Atari Games would discover their destination. Everyone else at Nintendo thought they were going to Japan. To reach the consulate before it closed for the weekend, they had to fly to Los Angeles from Seattle and then take a red-eye to Washington. Arriving there at dawn, they checked into a hotel, showered, dressed, and hopped a cab for the office of the Russian consul general. There, in the visa office, the agent in charge had never heard of an Arakawa, or a Lincoln, or Nintendo. There had been no communication from Moscow, and without it a visa could not be given. All the Nintendo chiefs could do was wait. At 4:00 P.M. a telex finally arrived authorizing the issuance of visas.

There was just enough time to dash to the airport and catch the next flight to London, another red-eye. They passed out on the plane and awoke as it touched down at Heathrow, where it was now late afternoon. There was an evening to kill in London before the next flight to Moscow.

The two retired after dinner. Lincoln said he would ring Arakawa's room at 7:00 A.M. so they could catch the nine-thirty flight. But he slept through his wake-up call, stirring at about eight. When he realized the time, he frantically called Arakawa. The two threw their clothes in their bags and raced to the airport, where they sprinted through the terminal and tried to blast through the security checkpoint.

A bevy of security guards approached them. It was the first time they had paused to look at each other. They were unshaven, their hair was tousled, and Lincoln was wearing his pajama top under his suit jacket.

Fast talking got them past the security checkpoint and they got to the plane just as the door was about to be closed. The two looked as if they were on the lam from a madhouse.

They slept on the flight to Moscow that Sunday, March 19, 1989. Rogers and Huhs met them at the airport. Rogers, who had rented a black Mercedes 190, maneuvered the car through Moscow while Arakawa and Lincoln stared at the passing scenery, the heavy-coated pedestrians, and the surprisingly European architecture. It seemed like a film of New York in the 1940s.

Rogers told Arakawa and Lincoln that he had managed to find a Japanese exporter who had a fax machine they could use. He also had a portable computer and printer set up in his hotel room. He said he and Huhs would meet them the next morning at Elorg—he supplied the address on a piece of paper—and then dropped off his exhausted passengers.

At the front desk, Arakawa was informed that rooms were unavailable but that he and Lincoln needn't worry; the hotel would put them in an apartment in an adjacent building. It had a disconnected stove, a refrigerator missing a door, a ragged sofa, and, down a musty hallway, a small bedroom. They looked at the single bed and, without uttering a word, pulled out their wallets, slid out dollar bills, and played liar's poker. Arakawa lost and got the couch.

Despite their fatigue, the two men left the apartment in order to stock up on supplies and soon found what they were looking for: a liquor store with the crucial provisions. Their arms loaded with Heineken and cognac, they returned to their room, where they drank until Lincoln retired to the bedroom and Arakawa passed out on the couch.

In the morning, after rendezvousing with Huhs and Rogers, Lincoln and Arakawa were escorted into a conference room with high ceilings and shaded windows at Elorg's office. There they were introduced to the game's designer, Alexey Pajitnov, the Elorg chief, Nikolai Belikov, and some of his associates. Pajitnov attempted to size up Arakawa and Lincoln, but they were "like people from another planet." Lincoln was stoic, immune to Pajitnov's joking and teasing; Arakawa was introverted and inaccessible. But because of their alliance with Henk Rogers, Pajitnov was inclined to trust them and was their ally throughout the first meeting.

Although he had tried to talk fishing with Belikov and one of his assistants, once the major issues were on the table, Howard Lincoln was neither chatty nor amiable. Money was not the most important issue for the Nintendo team. They were more concerned with Stein's and Mirrorsoft's claims to the home video-game rights. Lincoln needed assurance that the Russians had never meant to sell those rights to Andromeda or anyone else. Satisfied with the answers, he said emphatically several times that he had to be absolutely certain that the Soviets would commit to the deal and stick with it to the end, and that the Russians had to be prepared for a variety of counterattacks from Andromeda, Mirrorsoft, and Atari Games. Arakawa sat by patiently, his hands folded on the table.

The Soviets had come prepared with copies of all the letters, telexes, and contract proposals, in addition to the signed contracts with Stein and Andromeda. Lincoln examined them. One document was worth all the others: the amendment Robert Stein had signed that defined a computer. The NES was without a "monitor, disk drive(s), keyboard and operation system." It was not a computer.

The Russians eagerly broached new business with Nintendo. Glasnost had begun, and partnerships with the West in the form of joint-venture companies were being encouraged by the government. Belikov said that they wished to form a partnership with Nintendo, a joint venture that would provide NOA with terrific new video games like "Tetris."

Lincoln suggested that they complete one negotiation at a time.

Next the Soviets asked why they couldn't manufacture the "Tetris" cartridges themselves.

"That's not what we had in mind," said Lincoln. "Nintendo manufactures all the cartridges."

The Russians then said they wanted to make Nintendo systems themselves and sell them in the U.S.S.R.

"We make them in Japan, thank you."

To impress the men from Nintendo with Russia's engineering prowess, an Elorg representative brought out a small box. "Our people made this," he said.

Arakawa opened it. Inside was a Game & Watch that played "Donkey Kong." There was no sign of Nintendo's name or trademark on the watch. Arakawa politely complimented the Soviets on the product.

After the initial meeting, Arakawa and Lincoln took Huhs, Rogers, and Alexey Pajitnov out for dinner at the only Japanese restaurant in Moscow. It had no liquor license, so a waitress was dispatched to a store and returned with several large bottles of beer.

When a plate of sushi arrived, Pajitnov's first, he tried a small bite. Arakawa explained that sushi should be eaten a whole piece at a time. Bravely Pajitnov tried some *toro,* the fatty tuna belly set on a tiny brick of rice, carefully balancing it on his chopsticks and maneuvering it into his mouth. The taste was surprisingly pleasant, he found. He got more proficient with the utensils as he tried yellowtail, eel, crab, and *tamago,* a miniature omelette set on rice. Then he attacked a gluey green ball on his plate and popped it into his mouth as his companions, in a chorus, shrieked, *"No!"* They were too late; Pajitnov had downed a mound of *wasabi,* the wildly hot horseradish meant to be used in small amounts to spice the soy sauce for dipping. He felt a stinging explosion in his nostrils and behind his eyeballs, which vibrated as if they were ready to launch from his head. He tried to drown out the burning with beer, but in vain.

Arakawa couldn't hold back his laughter. Rogers joined him, and soon the others did too while Pajitnov wiped the tears that streamed from his eyes. He had better luck managing the courses that followed: *shabu shabu,* ginger fish, and vinegared seaweed, all washed down with beer.

After dinner, the party shifted to Pajitnov's apartment to see a new game he had created which he planned to sell through a joint-venture company in Moscow. Lincoln was concerned that the game, "Welltris," might be derivative of "Tetris," and that perhaps he should sew up the rights to it as well. Arakawa was more interested in seeing a Russian apartment; he had come equipped with a Game Boy for Alexey's children.

When they all entered the Gersten Street building and got into the elevator, Arakawa and Lincoln exchanged glances as it began its creaking, unstable ascent. Below them, the elevator shaft could be seen between the floorboards, and Lincoln pressed himself against a wall in an effort to keep his weight off the elevator floor. When they reached their destination, the elevator doors opened three or so precarious feet below the landing, and everyone had to climb up to reach the floor.

Inside the cheery apartment, Pajitnov's wife, Nina, served icy vodka and Russian brandy. Both Arakawa and Lincoln had questions to ask about life in Moscow, and the Pajitnovs were happy to answer them while their eldest child, Peter, played his father's game on Game Boy. Peter was told he was the only child in the Soviet Union with a Nintendo system.

The following day, the Nintendo representatives returned to Elorg ready to iron out the deal. Howard Lincoln outlined his plan to prove that the video-game rights to "Tetris" had never been sold, then confirmed Nintendo's offer. The negotiations continued for three days. Lincoln was determined not to leave Moscow without a signed contract, and with Henk Rogers on the word processor, they knocked one out, paragraph by paragraph, in Rogers's hotel room. In morning meetings with the Soviets, the details were spelled out, and in the afternoon the changes were typed out and printed.

In one meeting, Lincoln said he wanted it to be clear what the game's author's rights were. The Elorg people answered that since

Pajitnov worked for the Computer Center and had created the game on company time, the copyright was owned by the Academy of Science, and that as the trade organization, Elorg was authorized by the Academy to license "Tetris." Pajitnov nodded his confirmation. He was resigned to a small degree of glory, perhaps some opportunities for the future, but no money.

Lincoln insisted on a clause in the contract that committed the Soviets to cooperate in any litigation that might ensue; they would have to come to the United States to testify if it became necessary. Then, at the last minute, the Soviets began to squabble over the royalty Nintendo had offered, but Lincoln said the deal was no longer negotiable. Nintendo was going to be responsible for the legal expenses, which would probably be sizable because many people would be upset about the agreement they were about to sign. These included, he noted, Stein, Atari Games, the Mirrorsoft people, and Mirrorsoft's owner, Robert Maxwell. The mention of Maxwell dampened the discord, particularly since Belikov was all too aware of the latest telex from Mirrorsoft that had arrived that morning: Jim Mackonochie, responding to the last Soviet communiqué, had written to insist that Mirrorsoft didn't have to offer anything for the video-game "Tetris" rights since it already owned them.

By way of reply, Elorg sent a telex informing Mackonochie that neither Andromeda, Mirrorsoft, nor Tengen had been authorized to distribute "Tetris" on home video-game systems, and that the rights were no longer available since they had been granted to Nintendo of America. The telex was sent on March 22, the same day the Nintendo contract was finalized. Arakawa signed for Nintendo, Pajitnov signed as the author, and Belikov signed for Elorg.

The signing ceremony was attended by representatives of Nintendo and Elorg, as well as two senior officials of the Soviet government, Edward A. Maksakov, deputy chairman of the State Committee for Computer Systems and Informatics, and Dr. Stanislov I. Gusev, head of the Department of Scientific-Technical Information at the Computer Center of the Academy of Sciences. The advance guarantee was kept confidential, but rumors had it at $3 to $5 million. Lincoln says it was less, but will not divulge the amount.

Dennis Wood, Atari Game's counsel, says the amount "would entice anybody to double-license."

An attempt to derail the Nintendo deal arrived on March 23, addressed to Belikov. Kevin Maxwell wrote: "I give you formal notice that you are now in grave breach twice over of our agreements with you." He added that the matter would be raised during the forthcoming visit to London of President Gorbachev and stated flatly, "We already hold the worldwide rights to 'Tetris' on the Nintendo family computer. Indeed, we have been marketing it accordingly, both directly and through Tengen in the United States and Bullet-Proof Software in Japan since January 1989. . . ."

Maxwell said he was coming to Moscow and was willing, "in the spirit of reconciliation," to meet Belikov to "hear how you intend to remedy your double breach of our agreement." He concluded by threatening that if the Russians didn't make good, he would carry the matter to the highest legal and political levels.

It was too late. That evening, Arakawa, Lincoln, Rogers, and Huhs celebrated at the Japanese restaurant with Alexey Pajitnov. Sitting at the Teppan Yaki bar, they asked the Japanese waitress if she would go to the liquor store for beer.

"Finnish beer?" they heard her say.

Fine, they responded, and she scurried off.

Arakawa tried to contain his elation, but his smile was enormous. "We've got it," he said exultantly to Lincoln. Spirits were so high that the table almost levitated.

The waitress hadn't returned with the beer and it was time for a toast. Lincoln saw the waitress and called her over. "Where is the beer?" he asked her.

"Finish beer," she repeated. "Finish beer."

"Fine," Lincoln said, "but . . ."

With her delicate hands, the woman made an X. *"Finish beer!"* she said firmly.

Roaring with laughter, they toasted instead with soft drinks. (Later Henk Rogers sent Howard Lincoln a case of Finnish beer as a present.)

They all exchanged warm hugs after dinner and said good night.

Rogers was particularly happy. Not only did he have the hand-held rights to "Tetris," sublicensed to Nintendo for Game Boy, but as a reward he had also been given a sublicense to distribute "Tetris" for home video-game systems in Japan, the rights he thought he had bought from Atari. This time, though, he got the game at Nintendo's cost. This meant his profit would be $5 to $8 more per cartridge than what other licensees made on games manufactured by Nintendo. Combined, he was about to make perhaps $30 to $40 million on "Tetris."

That difference was pocket change for Arakawa and Lincoln, who, back at their squalid apartment, finished off the warm Heineken they had bought earlier. They were too excited to sleep and sat up all night talking.

Heading to the airport the next morning to fly home, Arakawa said, "I'm never coming back to this place again."

"Not so fast!" Lincoln interrupted. "We promised we'd be back with a bunch of Game Boys for their hospitals and orphanages."

Arakawa shook his head and grinned. "We did," he said, "and they'll be very grateful for them when *you* deliver them."

"We knew we had those bastards by the balls," Howard Lincoln says. "We knew we were going to make a fortune on this product and they, in turn, were going to get kicked in the head." He worried only about what Robert Maxwell would do when he heard that Nintendo had snatched "Tetris."

In late March, Belikov sent telexes to Mirrorsoft and to Stein at his London office. Hand-held "Tetris" rights were no longer available. The telex stated, "It is a pity that we were forced to conclude the contract concerning 'Tetris' for hand-held with another firm."

Stein sublicensed his hard-won coin-op rights to Mirrorsoft, but Maxwell's company was unable to secure the most valuable rights, in particular the home video-game rights that it had already sublicensed to Atari Games. Mirrorsoft was in trouble because Atari Games had already invested millions in the Tengen version of "Tetris."

Back in Redmond, Lincoln relished the moment he sent a fax on March 31 to Hide Nakajima and Atari Games in California. It informed Nakajima and company that they must "cease and desist from any further manufacture, advertisement, promotion, offer for sale or sale of 'Tetris' for the NES or any other home system" because the rights belonged to Nintendo.

An attorney from Dennis Wood's office ripped the fax from the machine and read it quickly, stared at the paper, shocked, and rushed to Dennis Wood's office. Wood read and reread the fax before walking, stony-faced, into Hide Nakajima's office.

Tengen quickly called Mirrorsoft to find out what was going on. The initial response from Mirrorsoft was not to worry; the rights were theirs. Whatever Nintendo and the Russians were up to would not work, Dennis Wood was told.

It took until April 7 for Tengen to respond to Nintendo. "We are in receipt of your letter . . . and quite frankly are quite confused. As Nintendo has known since last year, Tengen received all NES rights to the game 'Tetris' in early 1988. These rights are, in Tengen's view, clear and unequivocal. . . ."

Howard Lincoln offered to discuss things further, but by then, on April 13, Atari Games had filed an application for a copyright on the "audiovisual work, the underlying computer code and the soundtrack" for "Tetris" for the Nintendo system. Atari did not inform the Copyright Office that its version of "Tetris" was simply a spruced-up version of Alexey Pajitnov's game, or that Nintendo had informed Atari that it held the exclusive rights to the game.

In a conference in London, Jim Mackonochie told Kevin Maxwell about Nintendo's frontal attack. Maxwell decided it was time to inform his father, and the senior Maxwell "went berserk," as an associate put it. "He went ape shit."

When told that the Soviets had broken the protocol agreement, Robert Maxwell had Kevin explain the details of the "Tetris" negotiations. A protocol agreement is not a legal document, but it is, the elder Maxwell insisted, the equivalent of a gentlemen's agreement. To break it was, according to Robert Maxwell, tantamount to a slap in the face. The ultimatums Mirrorsoft had been given by the Russians were clearly for show; the legal right of first refusal had been mocked.

At that time, Robert Maxwell was steadfastly building a global media empire that would span more territory than Her Majesty's empire ever did. "Information is growing at 20 percent a year,"

Maxwell said in the early 1960s. "Communications is where oil was ten years ago. There will be seven to ten global communications corporations. My ambition is to be one of them." He had pursued this ambition tenaciously, gobbling up or founding communications-related companies from Britain to China, the Soviet Union, and Brazil.

Maxwell not only had a formidable world presence as a businessman, but used his position to gain remarkable influence in world politics. He was a trusted adviser of leaders in Israel and Canada and a powerful force in opposition to the Conservative governments of Margaret Thatcher and John Major in Britain. He spoke nine languages fluently, and his phone rang incessantly with calls from world leaders. When a secretary told him that the prime minister was on the phone, he asked, "Which one?"

Maxwell was trusted by Soviet president Mikhail Gorbachev, but he had been a familiar face in the Kremlin even earlier. He had known and published books by four former Soviet leaders—Brezhnev, Andropov, Gromyko, and Khrushchev—so there was every reason to believe his boasts about his influence in the U.S.S.R.

Although Maxwell's son Kevin was in charge of Mirrorsoft over Mackonochie, the elder Maxwell had a twenty-four-hour-a-day watch on all aspects of his parent companies, Maxwell Communications Corporation and the Mirror Group. It was a way to be certain that no one, not even his sons, knew exactly what he was up to. He was a general who kept his commanding officers in the dark on most important operations, informing them on a need-to-know basis and playing them off against each other. He staged surprise troop inspections to keep his top brass on their toes.

Kevin Maxwell tried to avoid going to his father for anything, but big guns were required in the "Tetris" deal. When he was told that the Soviets had double-dealt them, Robert Maxwell punished his desk with his fist. "They won't get away with it," he bellowed. "Rest assured of that." He promptly wrote letters to his friends in the Kremlin, including the minister of foreign economic relations, who catered to him when he visited Moscow. "We attach high importance to our excellent commercial relations with the Soviet government and many leading agencies in the fields of information, communications, publishing and, indeed, pulp and paper production," Maxwell wrote. "We face the prospect of all this being jeopardized by the unilateral action of one particular agency."

That agency, Elorg, was concerned when there were rumblings from above. However, this was perestroika and, as Jim Mackonochie put it, the Elorg bureaucrats were "feeling their oats." Still, when the foreign economic relations minister began to pry into the agency's affairs, Belikov realized trouble was brewing.

Next Maxwell contacted his own government and asked Lord Young, secretary of state for trade and industry in Britain, to intervene; he wanted "Tetris" to be discussed between the heads of state during a forthcoming visit by Gorbachev.

Word filtered back to the Moscow Academy of Science that Maxwell was throwing his substantial weight around, and people there and at Elorg worried that their authority might be undermined. At the same time, they were also delighted. In a strategy meeting with Belikov, the Academy chiefs debated how to respond to the Central Committee of the Communist Party, which was certain to react to an inquiry from the secretary general.

Belikov felt justified in having made the deal with Nintendo, and he planned to stand up for it. For all his dinner parties with the Gorbachevs, Maxwell had offered the Academy, via their dealings with Elorg, only a fraction of what Nintendo would bring to the country's coffers. In addition, Mirrorsoft was continually behind in its payments. Most of all, Belikov was convinced that Mirrorsoft had willfully stolen the Russians' game, and that Gorbachev himself would understand that Elorg had made the correct decision. Elorg, Belikov decided, would defend its decision whatever the pressures.

The infighting was epic as Elorg and the factions of the Party loyal to Maxwell exchanged urgent messages. There were threats of prosecution, and that the KGB would be used against individuals who refused to cooperate. The pressure on the Russians peaked when Robert Maxwell flew to Moscow to meet Gorbachev directly. He was prepared to discuss his planned printing ventures and the newspapers he wanted to launch, but first on his list was "Tetris."

Maxwell arrived in Moscow on his private jet and was whisked to the Octoberskaya, the government's elite hotel, by police motorcade. The meeting that afternoon was friendly, and he brought "Tetris" up only after initial small talk and joking. Maxwell later claimed that after the discussion, Gorbachev promised him the matter would be resolved to his satisfaction; "He said I should no longer worry about the Japanese company."

Lincoln returned to Moscow in late April and was joined by his New York lawyers, Huhs and John Kirby, as well as one of Kirby's associates, Bob Gunther. For the New York attorneys, the trip began with a comedy of errors. Gunther dropped and broke a printer he had lugged from New York, and then left his wallet, which contained $1,000, in a Moscow taxi. Kirby had all his shirts stolen at Kennedy Airport, and as a result he had to shop for clothing in Moscow before he could begin the series of meetings. He found a stack of pitiful polyester shirts at a concession stand on a street near his hotel.

The Nintendo team showed up for a meeting in the main Elorg conference room, where Belikov, Pajitnov, and half a dozen other Russians were waiting, visibly shaken. They were not unfriendly, just burdened; increasing pressure was descending on them from the top. The meeting progressed with no mention that there was anything wrong, but Howard Lincoln sensed trouble. "What is it?" he asked Belikov. "What has happened?"

Belikov shook his head vigorously. "Nothing has changed," he said, but during a break in the meeting he pulled Lincoln aside. "You do not understand," Belikov said under his breath. "We have done the right thing with you, but the Maxwells have threatened us. We have said, 'No, we will not be threatened by you. A contract is a contract and we will honor it and Nintendo is our licensee.'" Whispering, he continued, "But I must tell you, Mr. Lincoln, we are getting calls from the Kremlin, calls from people who never before knew we existed. Many of them have shown up to examine our records and to question us on this deal. We have told them we have done the right thing. We have stood up to them, but we do not know what will happen."

The meeting resumed with a hint of counterespionage in the air. There were worries about spies in Elorg and KGB surveillance, not only of the meetings but everywhere twenty-four hours a day—tapped hotel telephones, monitored strolls, and bugged restaurant tables.

Preparing for the worst, the Nintendo attorneys interviewed Pajitnov the next day, as well as people at the Academy of Science, the Computer Center, and Elorg, and examined every scrap of paper that dealt with "Tetris." Belikov also wrote a lengthy letter to John Kirby recounting his version of the "Tetris" history, and this was later included in the court record as Belikov's declaration.

In the meantime, on his jet winging its way from London to Jerusalem for a meeting with Israel's Prime Minister Yitzhak Shamir and Defense Minister Moshe Arens, Robert Maxwell was asked by a reporter why his intervention with Gorbachev had apparently been futile. He snapped, "How do you know about that deal? How do you know about the meeting?" Then he shrugged it off as if it had been inconsequential. "So much money was involved, his people convinced Gorbachev to work with the Japanese company," he said. "I did what I could." He blamed his losing the fight to Gorbachev's tenuous hold on power. "He said other people in the government felt strongly that it should go the other way, so we were stopped." Maxwell insisted that the principle was what had goaded him. "I am an honorable man and I expect honorable treatment, but you take your lumps along the way." It was not the last lump Maxwell would take.

In the middle of the night, Howard Lincoln was awakened in his Moscow hotel room by the telephone. The operator said she had a call from America.

The time in Redmond, eleven hours behind Moscow, was two in the afternoon, and the caller was one of his associates at NOA; "Tengen has sued us," she said.

At Elorg the next morning, Lincoln announced this news with a bit of glee. The Soviets were in for a taste of the American legal system, as sluggish and inefficient as the leviathan Soviet bureaucracy.

To begin to prepare, Lincoln, together with Huhs, Kirby, and Gunther, continued to interrogate each of the principals involved in the "Tetris" negotiations, wanting to be certain their case was airtight. Before it was all over, Alexey Pajitnov would tell his story a few dozen times. When he was satisfied, Lincoln flew to Japan to confer with Yamauchi and Hiroshi Imanishi before heading home. Yamauchi was delighted with everything that had happened, unconcerned about the lawsuit. It was the kind of wheeling and dealing he admired. "You and Arakawa-san have done well," he said.

Back home, Lincoln filed a countersuit against Tengen, and lawyers on both sides girded for battle. Evidence was gathered and depositions were taken in the United States, England, and then, in June, in Moscow.

John Kirby's staff continued to investigate on Nintendo's behalf in the United States. Kirby found that Tengen had filed applications for trademark registration of "Tetris" in the United States, Japan, Australia, Canada, the United Kingdom, West Germany, Italy, and Spain. Pajitnov sat for still another interview, this one four hours long. Huhs had Pajitnov reconstruct, in ponderous detail, the story of "Tetris," from its inception to the first letter from Stein and up to the present.

Tengen shipped its first batch of "Tetris" cartridges in May 1989, despite the notice it had received from Nintendo and the pending litigation. Setting out to sell what the company felt would be Tengen's hottest game ever, Randy Broweleit and Dan Van Elderen placed a full-page ad in *USA Today* announcing "Tetris": "It's like Siberia, only harder," the ad read. "It's here, America . . . The nerve-wrackingest mind game since Russian Roulette. . . . So round up a few of your high-IQ pals, okay? You know, macho men with the first-strike capability to beat the Russian programmers who invented it. . . . But there's one little catch. If you can't make the pieces fit together an avalanche of blocks thunders down and buries you weaklings!" This was hardly in tune with Pajitnov's vision of "Tetris" as a peacemaker.

Tengen held a grand reception for retailers, trade reps, and the press at the Russian Tea Room in New York on May 17. The place was packed, and there was free-flowing vodka and Russian hors d'oeuvres amid the Tea Room's year-round Christmas motif. Russian music played in the background, and Tengen "Tetris" games were set up for play.

Beginning in June, the case was heard in San Francisco in the courtroom of Judge Fern Smith, who also was trying the Nintendo–Atari Games antitrust and breach-of-contract cases. Ultimately the "Tetris" suit hinged on personalities, semantics, and two lines buried in the pounds of documents. Stein's contract with the Russians stipulated that he was being given the rights for computers, and no one argued this point. But Atari Games contended that the Nintendo system was a computer, a microprocessor-based machine that ran software. To prove that there was no valid distinction between the NES and other computers, Atari Games' attorneys noted that Nintendo itself viewed its machine as a computer, with planned hookups that would connect to the expansion portal. The anticipated peripherals—such as a modem, keyboard, and, ultimately, a CD—were proof that the NES was a computer. In Japan, the NES was even called the Famicom, or Family Computer. As one Tengen spokesman observed, "In court Nintendo went to great lengths to say that the NES was a toy and its cartridges were the equivalent of Barbie's arms and legs, but at the same time they were signing up AT&T to use its machine for stock reports. There was a Nintendo computer network in Japan and one planned in the U.S. Sounds like a computer to me," he said.

The Atari contingent echoed the charges made by Robert Maxwell that the Soviets saw they could make a lot more money from Nintendo, so they found a loophole and pleaded ignorance. This was in spite of Alexey Pajitnov's insistence that the deal never was meant to include more than PCs; Pajitnov, the Atari contingent charged, was Nintendo's dupe, instructed on exactly what to tell the court.

Dan Van Elderen believes the Russians were less innocent than they pretended to be. "Whether the language was ambiguous or not, they knew they had sold all those rights until they figured out, counseled by Henk Rogers and Nintendo, that there was a loophole. They realized they could have gotten a lot more money, so they double-dealt us all."

Tengen's Randy Broweleit revealed in his deposition how much was on the line for Atari Games. In 1988 his company had devoted more than three personnel years to "Tetris," and more than $250,000. By January 1989, Tengen had committed to manufacture 300,000 "Tetris" cartridges and spent $3 million on them, plus millions on packaging, engineering, and marketing. One hundred thousand units had been shipped, and there were initial orders for 150,000 before the game was released in May 1989.

Hide Nakajima insisted that Nintendo had colluded to steal his game. "Something went on between the Russian author and Nintendo," he charged. "Nintendo knew we had the license, and it urged us to go forward with the game. Nintendo only cared once we filed the antitrust suit against them. They went after us. Howard Lincoln and Arakawa wanted to stop us. It was revenge." Howard Lincoln has affirmed this last point. "It *was* revenge," he says. "And you know what they say about how sweet revenge can be."

Nintendo's argument was straightforward: in spite of their innocence about international software licensing, the Soviets knew exactly what they were doing when they assigned the rights to Stein. Computers were computers. Just as the Soviets' contract with Stein excluded the rights to hand-held and arcade games, the Russians had had no intention of selling home video-game rights. Two lines in the contract proved it: the line stipulating computers in the main body of the final contract, and also a line in the amendment—alteration no. 1—which specified that a computer had a keyboard, monitor, and floppy-disk drives. The NES machine had none of those; ergo, it was excluded. Nintendo held fast to the position that it had bought the "Tetris" rights fair and square, giving the Russians a fair deal, while Tengen's rights were part of a faulty chain. The weak link was the original one with the West, Robert Stein's contract, which covered PCs clearly and explicitly. Assumptions made beyond that, whether by Stein, Mirrorsoft, or Tengen, were nothing short of thievery.

Nintendo and Tengen were trying to stop each other from selling "Tetris" with cross motions that they filed for preliminary injunctions to prohibit the other from selling the game. A hearing about this was held on June 15, 1989.

After reviewing the depositions and mass of documents, Judge

Smith decided that there was no evidence that Tengen (and the licensing chain that awarded it the rights) had ever been granted the video-game rights. She said she believed that Nintendo was likely to prevail in court, and therefore she granted Nintendo's request for a preliminary injunction. Tengen was enjoined and restrained from manufacturing and selling the home video-game "Tetris" as of June 21.

At this point, Hide Nakajima, Dennis Wood, and Dan Van Elderen (Randy Broweleit had left Atari Games to start an independent software licensing company) could only hope that the court ultimately would reverse this opinion, although it seemed unlikely (and no trial was scheduled as of late 1992). Tengen's production of "Tetris" cartridges ground to a halt. Although it claimed that its version of "Tetris" was superior to the one Nintendo released, Tengen had to lock its games away in a warehouse pending a final verdict, and its "Tetris" soon became a collector's item, selling for as much as $150.

Nintendo released its NES version of "Tetris"—slickly redesigned, with a score of Russian music—and it sold rapidly, remaining on the Nintendo top-ten most-popular game list (behind "Super Mario Bros. 3") for over a year. Pajitnov laughed when he heard that when millions of American children watched the evening news and saw a shot of St. Basil's Cathedral in Red Square, they shouted excitedly, "Look, the 'Tetris' towers!" Similarly, Tchaikovsky lost credit for his "Dance of the Sugarplum Fairy"; kids knew it only as "the 'Tetris' song." On a modest level, Pajitnov's dream that his game would be a bridge between cultures was realized. "Tetris" contest winners were awarded a ten-day tour of Kiev, Leningrad, and Moscow, "home of Alexey Pajitnov." *Nintendo Power* ran features about his homeland, and kids who played the game saw that something wonderful had come from the former "evil empire."

Grown-ups flocked to "Tetris" too. Arakawa had predicted correctly: feedback from its customers told NOA that a third to a half of the "Tetris" players were adults, and Nintendo's presence in the adult market increased to such a degree that almost half (46 percent) of the Game Boy players in the West were adults.

Arakawa was also right about another thing: "Tetris" sold millions of Game Boys. A total of 32 million of them sold worldwide through 1992, more than Hiroshi Yamauchi had predicted. A U.S. senator, a "Tetris" addict, joked that the game was a Soviet plot to distract and hypnotize Americans.

The game also did things for Nintendo that Arakawa hadn't anticipated. When the company was attacked by educators and psychologists for the mindless violence and lack of redeeming value of its games, Nintendo now had fodder for counterarguments. Some theories claimed that "Tetris" play increased intelligence scores (at least in the area of spatial relationships). Also, a study in Moscow showed that "Tetris" helped improve driving skills because it trained players to make decisions extremely quickly, shortening drivers' reaction time.

Kids were getting Tetrisized and played compulsively. After they stopped playing, however, they complained that "Tetris" shapes remained impaled somewhere in their consciousness. Grown-ups became Tetrisized as much as kids. A reader wrote in to a national women's magazine; "['Tetris'] led me to beg my coworkers not to leave me behind in the office when they left, for fear I'd stay [there] all night playing. I removed the game from my computer at home and threw it away, but I passed a Game Boy in a store and couldn't stop. I went in and bought it." A Russian cosmonaut even took one into space. (The Russian had been given the game by Howard Lincoln, who had returned to Moscow with his sixteen-year-old son, Brad, on what was, for the most part, a social visit. He brought with him the one hundred Game Boys that Arakawa had promised. Arakawa kept the other part of his promise by staying home.)

The journey of Alexey Pajitnov's program from Moscow to most places on the globe—and to space and back—left a number of casualties in its wake. Robert Stein says, " 'Tetris' made enemies out of friends and corrupted people left, right, and center." Andromeda, Mirrorsoft, and Atari Games, he says, felt that every penny Nintendo earned on "Tetris" should be theirs. "So why don't we all get together instead of fighting like lunatics?" he asks. But fight like lunatics they did, so infighting tied up most of the profits earned by the versions of the game not controlled by Nintendo and

BPS. Mirrorsoft saw modest profits on its floppy-disk "Tetris," but almost nothing from the licenses it sold to Atari Games, which refused to pay Mirrorsoft anything pending the outcome of the litigation with Nintendo.

Atari Games released a coin-op game and sold 15,000 to 20,000 units, according to Dan Van Elderen. It also earned a royalty on the arcade games Sega sold in Japan, but Atari's sublicense to Henk Rogers was useless and it would probably have to return Rogers's advance for the home video-game rights in the original deal.

Robert Stein admitted that over the years he had made about $200,000 on "Tetris" but said he could have made millions. Instead, he watched as the Soviets severed all of his ties to the game, citing nonpayment of royalties. Stein lost his rights to the computer versions of "Tetris" in 1990. Spectrum Holobyte had been paying royalties to Mirrorsoft, which refused to pay Stein. The Russians' 75 percent of Stein's nothing was nothing, so Elorg finally revoked his license. In order to retain the rights to "Tetris" on PCs (and to retain the rights to sell "Tetris 2"), Spectrum Holobyte had to negotiate a new deal directly with the Soviets. Gilman Louie found that the Soviets had learned a great deal from the "Tetris" experience, and he had to pay a far higher royalty than in his deal with Mirrorsoft for the license he already had.

At this point Stein still held the coin-operated "Tetris" rights, but he received nothing for them as long as Atari Games didn't pay Mirrorsoft. Since he didn't pay, Elorg announced in February 1992 that it was terminating the coin-op deal as well. Stein vowed to fight, but it would be an uphill battle. The man who had discovered "Tetris" for the West lost all his rights to the game.

The lawsuit remained unsettled well into 1992, although rumor had it that Atari Games would settle. If this happened (or if Nintendo prevailed in court), Atari Games would probably go after Andromeda, Mirrorsoft, and, ultimately, Maxwell. Maxwell and Stein had warranted that they owned the rights they had sold, and probably would be held responsible. Stein wouldn't be worth going after, but Mirrorsoft, with Maxwell's deep pockets, would have been—that is, until it turned out that those pockets were actually black holes. The upshot of the scandalous collapse of the Maxwell organization was the dissolution of Mirrorsoft (its meager assets were bought by Acclaim Entertainment) following Robert Maxwell's suspicious death.

Other "Tetris" players fared better, although Kevin Maxwell was left to suffer for his father's corrupt business practices. Not only was he left with no assets or income, but there was a good chance he would be indicted, despite the fact that he may have been kept in the dark about Maxwell senior's illegal maneuverings.

Jim Mackonochie had been forced out of Mirrorsoft well before it collapsed, back when Kevin Maxwell restructured the company in 1991. Mackonochie ended up working as a consultant in the industry before being hired to work on CDTV software by Commodore International in London.

As their country transformed, the Russians at Elorg and the Academy of Science scattered, although Nikolai Belikov remained at Elorg long after the Communist Party fell from power. A freer country meant increased opportunities for trade, and Belikov, no longer saddled by the pressures of the Party's interests, saw abundant possibilities for exporting Russian technological achievements. His first Yeltsin-era task was negotiating the tough deal with Gilman Louie for "Tetris 2"—designed along with Pajitnov and others.

The Academy's Sasha Alexinko wound up in Vienna, where he formed a trading company. Victor Brjabrin also left Russia and found challenging work in Western Europe with a nuclear-regulatory commission run by the United Nations. Young Vadim Gerasimov left Russia too. At only twenty, he moved to Tokyo, where he studied Japanese and worked with a software developer, who then advertised that the codeveloper of "Tetris" was on his staff.

In America, Phil Adam left Spectrum Holobyte in the hands of his partner, Louie, who took Spectrum on to new ventures, from new combat simulators to other Nintendo games. In 1992, Louie debuted a futuristic video-game system for arcades and shopping malls. Kids climbed into a slick pod or stood inside a device that looked like a gyroscope, strapped on binocular-like goggles, and entered computer-generated virtual realities. In one multicontestant game, players stalked each other in a surreal cybernetic environment of multicolored platforms and stairways. Armed with a missile-lobbing blaster, they "flew" around and attempted to nail enemies (who broke into pieces if hit) before being shot themselves, though occasionally a pterodactyl swooped down from the sky and carried them off.

In his modest office in London, Robert Stein continued to struggle to keep Andromeda afloat. He distributed Atari Corp.'s computers in England and attempted to take advantage of the post-Communist revolution in Eastern Europe, particularly in Hungary. He also kept trying to sell Hungarian games in the West; perhaps there was another "Tetris" out there. But he had learned his lesson: if he found a great game, he would sew up all the rights *before* selling it.

Henk Rogers made more from "Tetris" than any individual save, ultimately, Hiroshi Yamauchi. The Russian bureaucrats at Elorg and the Academy made almost nothing, although the Russian government made millions from the game, mostly from the Nintendo deal. They also took in roughly $150,000, all told, from Andromeda, plus advances and royalties directly from Spectrum Holobyte.

As always, Nintendo did best of all, though it is impossible to calculate exactly how much it made from "Tetris," since there is no way to measure accurately how much "Tetris" contributed to the success of Game Boy. Three million "Tetris" cartridges for the NES were sold, plus all those Game Boy units. Once a customer bought one, Nintendo could sell more games, an average of three a year, at $35 a pop. Not counting Game Boy, "Tetris" brought Nintendo at least $80 million. Counting Game Boy, the figure is in the billions of dollars (in both 1991 and 1992, Game Boy earned nearly $2 billion).

Alexey Pajitnov made very little money directly from "Tetris" royalties or advances. Elorg had made and then canceled a side deal that would have granted him the "Tetris" merchandising rights (Nintendo eventually got them, too), so he ended up getting nothing on the "Tetris" watches, clocks, board games, and the like.

Westerners criticized the Soviet system that robbed Pajitnov of a stake in the game that made so much money for so many people, but Belikov defended it. "If 'Tetris' had been made by a Boeing employee on Boeing time and Boeing sold the license, would the designer have received any more than Pajitnov?"

On the other hand, if Pajitnov had retained the "Tetris" rights and signed a deal typical of those in the United States, he would have earned up to 15 percent of net revenues. Pajitnov would have seen at least $3 million if he was earning this standard percentage of the Soviet government's share. If he had licensed it directly, the number would have been as high as $20 million, perhaps more. Instead, the Computer Center awarded Pajitnov his own personal computer, an IBM AT clone, for which he was grateful since it would have taken him sixteen years to be able to buy one on his Academy of Science salary.

Henk Rogers, who came out of the deal with a good relationship with Elorg, appealed to Belikov on Pajitnov's behalf in a letter. He wrote, "If someone plants an apple tree and it brings you many, many apples, you ought to give them some apples—it would encourage them to plant more trees."

There was no response. The Elorg team was not particularly sympathetic. Pajitnov's apartment was nicer than the homes of most of his Academy superiors and the Elorg bureaucrats, and in addition he had gained recognition throughout the world, far more than any Soviet citizen dared hope for.

It amazed Pajitnov that Americans couldn't believe he wasn't bitter. This, he came to realize, was one of the key differences between him and most of the people he met in the West, where financial reward was the measure of accomplishment. "For me, to have my game played everywhere is the greatest thing to know," he says. In 1989, he was called to the telephone at the Computer Center to talk to a reporter who was writing a story about "Tetris." Every question was slanted to make Pajitnov admit that he was resentful, but he told the reporter, "I will make my games and send them to you. *You* can fight over them."

The Soviet Union thawed, and as trade opportunities increased, Pajitnov was able to take advantage of the success of his creations by licensing games and other programs through several joint-venture trading companies that paid small advances and royalties on his designs. With the income that trickled in, he bought a car—his first, a used Jugoli, a Russian clone of an outdated Fiat. The Pajitnov family had something more: Peter and Dmitri, Alexey's sons, had one of the only two Nintendo Entertainment Systems in the U.S.S.R., which had been sent by Henk Rogers (the children of his friend Vladimir Pokhilko had the other).

One day in spring 1989, Pajitnov found himself stuffed into a tourist-class seat on an Aeroflot flight to Tokyo. Pajitnov, who had rarely been out of Moscow, stared blankly out the window. Feverish from a flu, he couldn't sleep. He watched the spotty clouds below him and, below that, the ice blue of the Arctic Ocean, and waited patiently to see land.

At Narita Airport, after claiming his one small suitcase and fumbling his way through customs, Pajitnov saw no one he knew and grew fearful. Perhaps, he thought, it was all a mistake. He spoke no Japanese and only shaky English. It was almost certain that no one in this airport spoke Russian, so he waited, staring at a large-screen television set up in the waiting area.

It was some time later that Pajitnov heard his name paged amid the Japanese. He found a telephone and shouted his name. Henk Rogers was on the phone: "Sit tight. I'm almost there," he said.

Finally, Pajitnov looked up through the crowd and saw Rogers's face behind his familiar thick black beard, and he rose and threw his husky arms around his friend.

Pajitnov wanted nothing more than to sleep off his flu, but Rogers wouldn't hear of it. Gunning the car, he headed into the megawatts of neon that gave daytime Tokyo a surreal pallor, then dragged the Russian into a building and up an elevator, to the lookout atop a Tokyo department store. The breathtaking view was lost on Alexey, though he was duly awed when Rogers stopped on the way home for groceries at a supermarket. There was more food than Pajitnov had ever seen in one place. "When you see it in movies, you think they put the stuff there just to make it look good, and that it's not really like this," he says. He couldn't believe that people passed over the incredible goods, picking some, examining a box, rejecting fruit because of a scratch or a bruise. His wife, Nina, had asked for pictures of Tokyo, so he snapped shots of the aisles full of food. Before he left, he bought jeans, a VCR, a small color TV, a CD player, a couple of Walkmen, and toys for his children with money Rogers had given him.

Then they drove through Tokyo to Rogers's Yokohama home, where Pajitnov slept off his illness before beginning his three-week Japanese adventure. He spent the first week accompanying Rogers to the Yokohama BPS offices, where he met with the staff and asked a nonstop string of questions about the video-game industry. He was amazed at the technical prowess of programmers and of their highly sophisticated development tools, and he devoured information about how games were marketed and distributed.

Then Rogers took Pajitnov to Kyoto on the bullet train. The Russian wore a necktie—for the first time since his wedding—to meet Nintendo executives. He was treated with reverence at headquarters, where he met with the general manager, Hiroshi Imanishi, as well as the company's marketing director and other NCL executives. He also met his Japanese peers—engineers like Gunpei Yokoi and his R&D 1 staff, and game designer Sigeru Miyamoto.

In the afternoon, Pajitnov adjusted his tie and patted his hair down before he was ushered into Hiroshi Yamauchi's office, where a Russian translator sat with them for a brief, stilted conversation. Yamauchi told Pajitnov he hoped he would make another "Tetris," and said they should have a long and fruitful relationship.

Pajitnov did some work at NCL, too. A slightly modified version of "Tetris" for Game Boy hadn't yet been approved by him (Nintendo provided the right of approval that Robert Stein had only promised). Testing it, Pajitnov found a programming glitch, and worked to correct it with a team of NCL programmers. He was taken through the company's development area, and it impressed him even more than what he had seen at BPS. Rows of automated game-testing machines filled the rooms. Nintendo tested game cartridges that had been manufactured and assembled by a dozen competing subcontractors. Eighty cartridges from each batch of a thousand were examined. Some were disassembled by engineers so the tooling could be inspected, and others were tested electronically. If even one of the eighty cartridges had a problem, the entire thousand were returned. "It was as rigorous as the military in Moscow," Pajitnov said.

Aside from business, the Russian was taken to restaurants, where he drank a good deal of sake and beer. For the first time in his life he had a gin and tonic, and he went to a karaoke bar but refused to sing.

Pajitnov first visited the United States in January 1990. The trip was sponsored by a joint-venture company he had teamed up with in Moscow. His first stop, after changing planes in New York and Chicago, was Las Vegas, where the Consumer Electronics Show was in progress. Direct from the breadlines of Moscow, he found himself a star attraction at the 1990 consumer show, where the only lines were for the $3.69 all-you-can-eat buffet at the hotel. Lighting a Kool cigarette with an I LOVE LAS VEGAS lighter, he stared, mouth agape, at the lobby of his hotel. "So this is a typical American city," he said over drinks with Gilman Louie.

After interviews and meetings in Las Vegas, Pajitnov flew to San Francisco, where Louie escorted him through a maze of dinners and parties. When he had his first taste of Kentucky Fried Chicken, it was a profound moment for him, and KFC became his favorite American food, a staple when his hosts weren't pushing fancy French meals and California cuisine on him. (At Stars, in San Francisco, he laughed at the nouvelle version of Russian blini with caviar, as pretty as a painting and almost microscopic on the massive plate.) He also tried tequila for the first time. "Very enjoyable," he says with a smile.

Pajitnov's schedule in the Bay Area was hectic. Spectrum Holobyte booked four or five interviews a day, but he didn't complain. "I have to take care of the royalty," he said. He was written about in dozens of computer magazines and daily newspapers. He also visited Seattle, where he dined with the Arakawas and Lincolns, and then traveled to the East Coast. In New York City, at the Modern Art and Metropolitan museums, he saw, for the first time in his life, the originals of some of his favorite paintings. It was, he says, as thrilling as anything that had ever happened to him. He was riveted by paintings he knew intimately from books in his Moscow home—Picassos, Braques, Légers.

In Boston, Pajitnov visited MIT's Media Lab, where he was invited to play with a NEXT computer. There were, of course, more interviews. After several, he met with a computer-magazine photographer at his hotel. When he was left alone with the photographer, the man asked him to change into lighter-colored pants. Pajitnov was bewildered; he had no extra clothing with him. Then the man handed him a twenty-pound VGA computer monitor and asked him to put it on his head. Pajitnov satisfied this unusual request; as instructed, he sat on top of a table near a window overlooking downtown Boston and balanced a computer monitor on his head as the photographer's strobe flashed and camera's auto-wind whirred.

His whirlwind American adventure almost over, Pajitnov flew to meet Henk Rogers in Oahu for a vacation, where they swam, kayaked, and drank lots of mai tais. Rogers asked Pajitnov if he would come to work full-time for BPS in Washington State. Would he consider leaving the Soviet Union?

Pajitnov became quiet and cast his eyes downward. "I do not have an answer for that question," he replied.

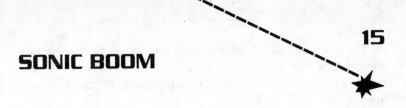

15

SONIC BOOM

Between them, Hiroshi Yamauchi and Minoru Arakawa had created a new mammoth industry and, with it, a field for competition. Seven American and Japanese companies were marketing video-game systems by 1988. But the contenders had little success in damaging Nintendo's share of the market, which was 85 to 90 percent on both sides of the Pacific. Atari sold a handful of its 5200s and 7800s, and Sega sold a total of 2 million Master Systems. Other companies sold too few to count.

Having failed to break Nintendo's lock on the NES generation of video-game systems, the would-be Davids attempted to topple Nintendo's Goliath in the next generation with more powerful hardware. They took aim at Nintendo's single vulnerability: its success. Nintendo was dominant, and such companies tend to stagnate by sticking with old technology. The problem for Nintendo, which was raking in a large part of its fortune from licensees, was that it had so much invested in the NES-Famicom technology. If the company planned to release a new system, the game-designing companies would worry that the NES was obsolete, and the shift could precipitate an early crash of its bread-and-butter NES business.

Nintendo also suffered from a malaise typical of industry leaders. Fat and happy, it had been lulled into a sense of invulnerability. Yamauchi and Arakawa felt they didn't have to react to competitors simply because they were Nintendo. This could have been a fatal mistake.

At first the competitors were like termites gnawing at the base of a giant sequoia: merely pests. Nintendo went about its business of selling millions of systems and tens of millions of games to a faint buzzing sound in the background.

If there was any threat, according to Yamauchi, it was when NEC, the Japanese computer and communications giant, entered the video-game business. If NEC was a termite, it was a voracious one. With $22 billion in net annual sales, the company was sound and well run. Each year it invested a hefty 16 percent of its net sales in R&D and engineering programs—$3.7 billion in 1988—more than any of Nintendo's annual net sales until 1992.

Yamauchi also viewed NEC as a potential threat because of its semiconductor business; it had a direct, inexpensive source of chips. Most important, NEC had a reputation for maintaining a long-term view. Backed by its size and substantial resources, it put up a prolonged and ruthless fight for any market it wanted. It had done so in personal computers and laser printers, with well-engineered products and perseverance. It was the tortoise in the races it entered, in for the long haul.

To push its first video-game system, NEC formed a home-entertainment group and released PC Engine in Japan in October 1987. A more expensive system (at $200) was released to the American market in 1989. The TurboGrafx-16 was an expandable system, with 16 bits of power.

For a long time, to the kids who comprised the primary video-game players, bits and bytes were only slightly more relevant than Latin. Then NEC fired the first shot in the battle of the bits. The TurboGrafx-16 had twice as many bits as Nintendo, and kids learned that more bits meant more realistic games, with more and brighter colors and awesome sound effects—arcade-quality games.

The essential computing chip that runs computers and video games, the microprocessor, functions like a traffic cop at a busy intersection. The processor directs a steady stream of information (from other integrated circuits and programs) at the busiest intersection in town to wherever it needs to go in order to make the computer and program function. The net result of all the high-speed traffic is, say, a high jump by Super Mario to the accompaniment of an electronic cymbal crash.

An 8-bit processor, such as the one at the heart of the NES and Famicom, can work with 64K (that is, 64,000) strings of information that are eight characters (or bits) long. Each bit is actually an electric impulse that is either turned on or off. A *one* means on; a *zero* means off. Each message is unique, depending on the specific configuration of the eight ones and zeros.

A 16-bit processor reads messages made of sixteen bits—that is, sixteen ones and zeros. It can, therefore, "understand" and process 250 times more messages—16 million. It means, simply, that a 16-bit machine can do a lot more, and do it a lot faster, than an 8-bit machine.

When NEC launched the TurboGrafx-16, video-game players were impressed with the meatier, more textured feeling of the first games. There was, however, a flaw in the more-is-better logic behind the new technology, and NEC learned the lesson the hard way. While its powerful system played games with better graphics and sound, NEC hadn't improved its video games. They were not as much fun. In the end, NEC's 16 bits could not compete with "Tetris," "Super Mario Bros.," "The Legend of Zelda," and hundreds of other great Nintendo games. In spite of its clout, less than 1 million TurboGrafx units were sold. For all their technical prowess, NEC's machines could be no better than the software that ran on them, and NEC had limited access to good games. "Bonk's Adventure," in which a bantam-weight caveman bounces his way through a Paleolithic paradise, was a good game that accounted for many TurboGrafx sales, but most of the credit-card-size cartridges NEC released were unexceptional. Since it had no experience making games, it depended on third-party developers to build a library. But the best entertainment-software companies were too busy making Nintendo games to bother making ones for TurboGrafx.

Still, NEC probably could have gotten the support of some software companies if techies in the industry had been truly excited by the TurboGrafx-16 technology. Some companies specialized in software for the state-of-the-art technology regardless of its limited market. What designers and programmers found, however, was that TurboGrafx was only incrementally better. NEC advertised its system as a 16-bit machine, but actually it had an 8-bit processor that was souped up to emulate 16 bits. "It's going to run out of gas," said one software engineer. "The technology has severe limitations. It is not a true 16-bit system." Since the designers weren't excited about the machine and the installed base was so low, for the most part NEC could get only bottom-of-the-barrel software. TurboGrafx turned out to be no threat and Yamauchi relaxed.

Sega never was a threat as far as Yamauchi was concerned. The $700 million Japanese company—founded, curiously enough, by an American—had a reasonably successful history in the video-arcade business in Japan and the United States, but it seemed too small and too specialized to make inroads into Nintendo's vast consumer business. Sega had released the Master System as a competitor to the Famicom and NES but never gained more than 5 percent of the market. Although it, unlike NEC, was an able software company, it never seemed to be playing in the same league as Nintendo.

Yamauchi underestimated Sega, whose executives understood the importance of software to drive hardware sales. This philosophy was built into the company's 16-bit contender, the Genesis, which it launched in 1989 in Japan (1990 in the United States). Genesis was the first dedicated video-game system powered by a true 16-bit processor; it had the same 68000 processor that ran the Macintosh computer. Sega had simply taken the design of its 16-bit arcade machines and adapted it for Genesis. It could therefore boast not only such 16-bit features as high-definition graphics and animation, a full spectrum of colors (more than 500), two independently scrolling backgrounds that created impressive depth-of-field, the illusion of three dimensions, and near CD-quality sound, but also a proven software catalogue: Sega's arcade hits. As a peripheral to the Genesis, Sega also released a unit called the Power Base Converter. For $35 it allowed Master System games to be played on the Genesis.

Sega got a reputable distributor in America (Tonka), spent $10 million in advertising, and set out to fell Goliath. The machine, priced originally at $199, was launched on the strength of software titles that kids who hung out in arcades already knew. One was "Altered Beast," a brutal game in which a hero metamorphoses into a werewolf, weredragon, and werebear, gaining the powers of the revolting creatures he kills. Sega attacked Nintendo head-on. "Sega Genesis does what Nintendon't," its slogan read.

Competition was exactly what Nintendo and the video-game industry needed, though Nintendo was unlikely to acknowledge this. The American automobile industry had floundered in the absence of competition. In the 1920s there were 181 American automobile companies. As the Big Three gobbled them up, replacing what had been a vibrant industry with a small club, competition declined, and with it went the American auto industry. In Japan there remained nine strong automobile manufacturers in a cutthroat, competitive environment. Competition kept innovation moving at a fast clip; there was no time to waste, no room for stagnation. Nintendo, however, stuck with its haughty above-the-fray posture. "We listen to our players," Bill White told the press. "They tell us they are extremely happy with the existing system and are totally involved with the games. We haven't maxed out our 8-bit system yet." This attitude left the company in the dust of the 16-bit war.

At first the market reinforced Nintendo's confidence. During the first two years Genesis was on the market, Nintendo sold 18 million 8-bit systems in America. Sega's arguments for 16 bits weren't supportable. The first Genesis games, even the knock-offs of the arcade hits, were not as much fun as the best Nintendo games. In many cases, the Sega programmers were so intent on exploiting the possibilities of detailed graphics and exciting sound that they forgot what made great video games.

The company spent multimillions to sign Michael Jackson, himself a video-game fanatic, to a contract to codevelop "Moonwalker," a game based on Jackson's 10-million-selling album, *Bad*. Jackson worked with Sega's Al Nilsen, who ran the marketing of the American home video-game operation, to come up with a story line for a game. Sega programmers created it, and Jackson helped fine-tune its features. The plot loosely re-created the *Moonwalker* video, in which Jackson did his trademark fancy dancing and, meanwhile, saved some of his young friends. At the end, he turned into a menacing robot.

The final product had remarkable visual and audio touches. Jackson's face and dance moves were digitized; there were electronic re-creations of half a dozen of the songs from the album and Jackson's digitized voice—incessant *whoop*s, *yeah*s, and *oooo*'s. When he danced on piano keys, the piano "played."

When the game was released, Sega sold a sizable number of Genesis systems because of the Jackson name and its state-of-the-art appearance. However, in the end, "Moonwalker" had a fatal flaw: it was repetitive and boring. Flash was no substitute for substance.

The Genesis continued to flounder through its first couple of years on the market, although Sega showed Sisyphean resolve. It sold machines to anyone it could, mostly to older boys who were diehard video-game buffs. They were the kids who insisted on having (and could afford to have) both the NES and the cutting-edge machine. Sega released some better games and sold more systems—a hundred thousand here, a hundred thousand there. The savvier kids extolled the virtues of 16 bits and scoffed at the dweebs still playing Nintendo. The result was that Sega began to embody cool. NOA commissioned a study that confirmed it: younger kids and girls liked Nintendo, but the trendsetters in the video-game world, young teenage boys, were talking Genesis.

Sega built on the growing buzz by learning from the Michael Jackson experience and making licensing deals that produced high-profile games that were also challenging and fun. Working with sports celebrities such as Arnold Palmer and Tommy Lasorda, the Sega designers figured out ways to take advantage of the Genesis's 16 bits for "deeper"—more complex—games. The resulting games were superior to the ones on the NES. Joe Montana signed on for a reported $8 million, and Sega released a great football game. Its sequel, "Joe Montana 2: Sports Talk Football," had running commentary by a semirealistic-sounding announcer who screams, *"Montana drops back. He's got a man open. . . . He passes. It's—no good. Incomplete."* Behind him is the sound of a wall of cheering fans.

Sports games and arcade knockoffs remained Sega's strong suits. Yet by and large, its designers came up with great-*looking* games—better than any that had been seen to date—but not great-*playing* games. Examples were those that came out of Sega's licensing agreement with Disney, such as "Fantasia" and "Castle of Illusion," both featuring Mickey Mouse. "Fantasia" had rousing classical music from the film and great-looking Sorcerer's Apprentice brooms that danced. In "Castle of Illusion," the expressiveness in Mickey's face set a new standard for video games. However, they weren't that much fun to play. Sega was missing what Hiroshi Yamauchi had long before acknowledged to be the most important single asset of a video-game company: "one true genius." It needed a Sigeru Miyamoto or an Alexey Pajitnov.

While Sega hoped for a "genius" to emerge, it got an enormous boost from its first third-party licensee. Although its base of 1 million systems paled in comparison to Nintendo's, Trip Hawkins of Electronic Arts calculated that Sega had created a promising market, one for which the price of admission was far less than Nintendo's. There was little competition—the software companies in the video-game business were all after Nintendo's 70 million players—and the million people who had invested in the Genesis were dying for good games; indeed, they were demanding more games than were owners of personal computers.

Hawkins met with the Sega chiefs, who planned to initiate a licensing agreement with similar restrictions and fees as Nintendo's. Hawkins refused—Sega wasn't big enough to throw that kind of weight around. EA engineers had succeeded in reverse-engineering the Genesis without using any confidential information, and Hawkins maintained that he would release games for Sega with or without a license.

Determined to maintain a stake in the Genesis software market, Sega said it would sue. To avoid "a pissing contest," Hawkins says, particularly since both companies had a common goal—to advance the Genesis marketplace—he agreed to enter into a licensing agreement under terms he found acceptable, far less restrictive (and costing less) than those for a Nintendo licensee.

EA's stock crept up as it shipped its first Genesis games in 1990. The next year it shipped nine more, four of which shot onto the list of the top-ten bestsellers. Having hedged his bets—he had also become a Nintendo licensee—Hawkins thrived in the Genesis market and made a killing. A quarter of EA's sales in 1990 were from Genesis games.

The Genesis's 16 bits meant that EA could convert some of its floppy-disk bestsellers to the Sega system. EA released "John Madden Football," which went up against Sega's "Joe Montana Football," and an impressive list of other games, from an award-winning martial-arts game called "Budokan" to a gory version of Will Harvey's "Immortal," in which 16-bit graphics allowed some particularly repulsive blood and guts.

Other software companies signed on (reportedly with less liberal licensing deals than EA's but still preferable to the Nintendo straitjacket), though many Nintendo licensees feared retribution by Nintendo. EA took the risk, gambling (successfully, it seemed) on the Sega technology, and some other software companies went to Sega because they had nothing to lose. Tengen, for one, out of the Nintendo business pending its litigation, jumped in with a series of Genesis games (licensed this time; it could afford no new lawsuits).

With each new convert Sega was more and more legitimized. Fueled by games released by licensees, the list of Genesis software grew—good software, to be sure, but no "Super Mario Bros." Sales of the Genesis hardware also grew, so that by mid 1991, there were well over 1 million in use. By then NOA had sold 31.7 million units in the United States, but Sega had established itself as the market leader of the next generation. Mighty Nintendo, which had announced that it would enter the 16-bit market only when it was good and ready, was in trouble.

Hiroshi Yamauchi had had a 16-bit system in the works for years. Masayuki Uemura, in charge of the top-secret project, had been experimenting·with a follow-up to the Famicom-NES since the late 1980s. Yamauchi left the technical specifications to the engineers, but he did insist that his company must be poised to jump into the 16-bit market by 1990. There was no perceived urgency, however; the NES was flying so high that Nintendo felt no pressure to rush.

One issue that Yamauchi made the engineers consider was the compatibility of a new system with the hundreds of millions of Nintendo games in circulation. New-generation hardware was always resisted if it made old software obsolete. Yamauchi correctly anticipated a backlash against Nintendo, particularly from parents, if its new machine couldn't play the libraries of games collected for its 8-bit technology.

Uemura accomplished many feats in his design of the new Super Family Computer (dubbed the Super NES, or SNES, in the West), but one of them was not compatibility. The cost would have been too high—at least $75 would have been added to the price of each unit. Uemura found that a major leap in technology was, almost by definition, *not* compatible with old technology.

Arakawa and Yamauchi discussed the problem and decided they could deal with backlash generated by the compatibility issue. After all, consumers were buying CD players even though the new technology wasn't compatible with their libraries of tapes and records. Arakawa insisted that video-game customers would do the same.

Uemura had better luck with other features of the new system. Around the central processor were special chips for sound and extremely high-resolution graphics and video. The new SNES could generate many more colors than the Genesis's 512—32,000 in fact, many of them barely distinguishable (especially on most televisions). This would be important, however, when video games began to incorporate real film footage. There was also a math coprocessor that allowed the hardware to do some of the work normally done by software and would make it easier to create games for the new system. Like the Genesis, the SNES could display several layers of backgrounds so that it could create the illusion of three dimensions. It also had the capability to generate (and move) large objects on the screen, and many more things could happen simultaneously in a game. It also had one other key feature: an improved version of the patented Nintendo security chip.

The Japanese version of the Super system was made to look somewhat like the original Famicom, but Arakawa had a different version designed for the United States. Don James and product designer Lance Barr sought a balance so that it would be sleek enough—Arakawa said it had to fit in next to a VCR—and accessible. The Genesis was housed in a black box with rounded edges. James and Barr came up with a far more elegant gray with right angles. Their colors represented the images of the respective machines. The Genesis, in black, was the outsider, the heavy metal of video-game machines. The SNES, sleekly styled in gray, was commercial and pop.

As "Super Mario" had sold the NES and "Tetris" had sold Game Boy, Yamauchi and Arakawa had to decide what game would be used to spark sales of the SNES. In the end there was no real debate. "Super Mario Bros. 3" had been the most successful video game in history. What would convince more buyers to cough up $200 for a new system than "Super Mario Bros. 4"? Sigeru Miyamoto was assigned to create it.

After the exhausting months of completing "SMB3," Miyamoto's team had taken fifteen months off to do nothing but conduct technology experiments to explore the outer reaches of the SNES. He and his thirty-person team were still scratching the surface of the machine's potentials when he was asked to press forward on a game that would show off the 16-bit bells and whistles and improve on "Super Mario Bros. 3." It was an onerous assignment, made more difficult by the undisguised pressure to succeed. "Super Mario Bros. 4," renamed "Super Mario World," was supposed to be so great that there would be no question when it came to choosing a 16-bit platform.

"There's no emotion in a game," Tony Harman says. "It's all ones and zeros; we're creating illusions. The magic comes from inspiration. . . . The best of the rest of the efforts can be no better than second best."

Miyamoto set to work to create sleight of hand beyond any he had managed before, and he did manage some great innovations. Still, when it was completed, "Super Mario World" was disappointingly similar to its antecedents. Mario had new skills, and the game offered some fascinating innovations. There were opportunities throughout it for players to hone new skills—first in nonlethal, then in semilethal, and finally in dangerous situations. To encourage players to enjoy Mario's ability to fly, for example, Miyamoto designed a world with no enemies where players could practice in a sky filled with coins. Every hundred coins were good for a free life, so there was plenty of incentive to wallow there, take a running start, pump the A button, and take off.

Miyamoto also made the game nonlinear—a player could return to different worlds at any time—and there were other touches that seemed as if they would lead to future developments. However, "Super Mario World" wasn't a sufficient departure from its predecessor. "People don't know how to write 16-bit software yet," Greg Fischbach said at the time. "It will be revolutionary, but it will take some time to understand." There would be more lifelike and emotion-filled games because of 16-bit processors. Miyamoto says, "Wait, and I will learn more about the limits of this machine." In the meantime, "Super Mario World" was a disappointment, particularly when it was compared to a new game that was released for Sega's 16-bit system.

An independent development team contracted to Sega came up with "Sonic the Hedgehog," a game featuring a cute creature who impatiently tapped his foot—er, paw—when the player took too long to act. Impatience was Sonic's essential characteristic: he had places to go—and quickly. He zipped along, collecting brass rings when he could find them, before rolling up into a ball and flying down slides with loops and underground tunnels. For the player, it was like a roller-coaster ride. Sega pronounced "Sonic" the fastest video game in history. It had finally found its Mario.

"Sonic" was not a great game, but it was novel, the character was likable, and it had good graphics and a bouncy musical score. Like so many games, it was plagued by repetition, but with not much competition at that time from Nintendo or any licensees, "Sonic" took off and sold hundreds of thousands of Genesis systems. Best of all for Sega, "Sonic" was there to mock Nintendo's launch of its Super system.

In spite of Sega's growth, Nintendo believed it would release its Super systems in Japan and America and easily do away with the Genesis, "Sonic" and all. Sega may have stockpiled cash ($400 million) and earned a cooler-than-Mario reputation, but this was only because Nintendo hadn't yet entered the field. Finally it announced plans to release its 16-bit machine. The announcement was, in large part, a warning to would-be Genesis buyers not to make a mistake they would regret.

The actual launch began a year later in Japan, where anticipation of the new system was rabid. In late October 1990, there were rumors that the Super Family Computer was coming, and stores were inundated with calls. As soon as Hiroshi Imanishi informed his sales team that a shipping date was forthcoming, some stores began taking orders. The Hankyu department store in Osaka announced that it would accept "reservations" beginning on November 3. A week later it had accepted more than it would be able to fill and stopped accepting orders. Other stores announced lotteries, and some had customers pay the 32,000 yen (about $200) for the system plus "Super Mario World" in advance.

Yamauchi and Imanishi jointly directed Operation Midnight Shipping, which commenced in the wee hours of November 20, 1990. Kenji Takahashi, in his recent book on Nintendo (*Light and Shadow of an Enterprise That Surpassed Matsushita and Sony*), described the secret transport: "On a night in the fall . . . with the wind blowing through the Kyoto Basin, adding a considerable chill, an unusually large number of over-sized ten-ton trucks congregated at a warehouse in the city. Workmen quietly loaded the trucks, which then disappeared, one by one, into the darkness of the sleeping town and onto the state highway. . . .

"At the same time, other trucks departed from warehouses throughout Japan. The last truck left the last warehouse as dawn was breaking."

All the secrecy was to head off thievery, Hiroshi Yamauchi subsequently revealed. A *yakuza* ring was rumored to be planning to hijack some of the trucks. Worth their weight in gold, the goods would have ended up in mob warehouses, from which they would have been parceled out to special customers. Fearing leaks, the Nintendo executives informed their staff about the details of Midnight Shipping only on a need-to-know basis. Similarly, only the key people expecting shipments were told when the trucks would arrive. When they did appear, carefully predetermined allotments of systems and software were parceled out.

The hundred trucks, each loaded with three thousand Super Family Computers and boxes of the first two Super Famicom games, "Super Mario World" and "F-Zero" (a racing game), had dropped off their secret cargo by the end of the business day on the twentieth. The next morning, store managers braced themselves before announcing that the next Nintendo generation had arrived. If high drama was part of the marketing plan, it worked. One department store closed down its toy department by 11:30 A.M. because it feared a riot. A small toy store on the main street near the Shakujii Koen train station in Tokyo received only six units. "It was not enough to meet even the request from our neighbors and the children of our friends," the elderly store owner reported. "As we would have been embarrassed to turn down our friends, we closed the store and posted a notice saying that we had taken a trip."

Three hundred thousand Super Famicoms were delivered that night, though the orders numbered 1.5 million. Four out of five customers were disappointed, including some who had paid in advance. It was a mess, although Hiroshi Yamauchi enjoyed hearing that the full supply was gone by the third day. A total of 2 million Famicoms were sold within six months, and over 4 million within a year.

The U.S. launch was trickier. The NES market had already peaked for a number of reasons, among them the fact that the United States was in an economic recession. But what was even more relevant was that Nintendo had reached relative saturation of its largest group of buyers, households with young boys. The arrival of the 16-bit generation of technology diverted some potential customers and deterred others who decided to wait and see what happened before choosing a system to buy. Another factor was that neither Nintendo nor any licensee had released any super-hot, must-have game since "Super Mario Bros. 3" and "Teenage Mutant Ninja Turtles" (although "The Simpsons," released by Acclaim, designed in collaboration with Matt Groening, was one of several big sellers during this period). For all these reasons, there was to be no Operation Midnight Shipping in the United States. Nintendo had its work cut out for it if customers, en masse, were to be convinced to shell out $200 for the new system.

Peter Main and Bill White stepped in with a marketing blitz. Twenty-five million dollars was spent on TV commercials for the September 1991 launch, and Gail Tilden at *Nintendo Power*—which then had about 1.2 million subscribers and 4 million readers —hyped the SNES shamelessly. There were also SNES promotions with Pepsi ($8 million worth), Kool-Aid, and other companies.

Parents, the media reported, resisted. "I have spent about $1,500 on Nintendo and Nintendo games," said one father, quoted in a San Francisco newspaper. "I said, 'No way!' " Compatibility reared its head as an issue, just as Minoru Arakawa had feared. "They made a new system that didn't play any of the old games so we would have to start all over," another parent grumbled, admitting that nonetheless he had succumbed.

The comic strip called "Foxtrot" by Bill Amend re-created a debate heard around the country.

JASON: *Pleeeeeeease?*
MOM: *Jason, no! You already have a Nintendo machine!*
JASON: *But this one's got 16-bit graphics . . . digital stereo sound . . . the latest Mario cartridge.*
MOM: *And a $200 price tag.*
JASON: *Hey, nobody said eternal bliss comes cheap.*

A headline in Scottsdale, Arizona, read: NINTENDO RISKS PARENTAL WRATH WITH NEW SYSTEM. In Chicago: NINTENDO COUNTS ON THE NAG FACTOR TO SELL SUPER SYSTEM. In San Rafael, California: NEW NINTENDO A COSTLY PAIN FOR PARENTS.

For the previous year (1990), the average amount of money spent on toys per child in America was about $225, according to Jodi Levin, a spokeswoman for the Toy Manufacturers of America. Since the year of the SNES launch, 1991, was a recession year, the figure wasn't likely to go up. The question was whether parents would spend almost all of their kid's Christmas budget on one toy. But by October 1991, despite predictions of an extremely bleak Christmas, Nintendo had decided it was going to make its goals for the SNES debut. It claimed it would sell all the SNES units it shipped in those first four months—2.2 million, according to the company. Peter Main was claiming he could sell twice as many if NCL were able to meet the demand.

These claims were disputed by Sega and some store owners, who said that they didn't sell out their stock, although others reported extremely good sales. Toys "R" Us's stock tentatively crept up on the strength of SNES sales. So did Babbages' stock. By the end of 1991, Nintendo claimed to have already sold more systems in four months than Sega had sold in two years. Sega's Al Nilsen argued that Nintendo had sold *through* only 1 million, 1.2 million at the most. Sega, he said, had sold 1.4 million machines during the same period, making the total number of Genesis systems in homes 2.3 million.

At the January 1992 Consumer Electronics Show, Nintendo boasted that it had already blown Sega out of the water, and, as Peter Main put it, "You ain't seen nothin' yet." He announced sales predictions for the year: 6 million SNES units in the United States. According to the company's calculations, this would bring total sales to 8 million, not a bad start considering that the original NES had sold only 1 million in that amount of time. Most industry watchers believed these figures were slightly optimistic, an attempt to jump-start sales by creating a bandwagon mentality. The fact was that SNES had little chance of approaching the NES's success since Sega had already assured that Nintendo wouldn't have the market to itself.

Sega lowered the price of the Genesis to $149 and the system was packaged with "Sonic the Hedgehog," which continued to propel sales. The game's immense popularity took Sega by surprise, but the company ran with it. It didn't quite stop the SNES in its tracks, but it gave both Nintendo and customers pause. Sega boasted that in a test conducted by a market-research firm, seven out of ten kids preferred "Sonic" to "Super Mario World."

In the Atari Corp. trial, which continued through the winter of

1991–1992 in San Francisco, Arakawa was asked, on the stand, if the inventor of "Sonic the Hedgehog" was a genius—like Miyamoto, the inventor of "Super Mario Bros." Arakawa cagily answered, "Yes," and then qualified it. "They looked at 'Super Mario.' They wanted to come up with something similar." Howard Lincoln would admit, however, "They came up with a darn good game. They're going to be a very strong competitor."

A Sega press release was headlined SEGA CONFRONTS NINTENDO IN $1 MILLION AD ROADBLOCK. Their television commercials, which vied for airtime with Nintendo's massive ad blitz, showed a kid resisting a pushy salesman's attempt to sell him the more expensive SNES. "Super Mario World" looked timeworn compared to "Sonic the Hedgehog," the game playing on a nearby monitor. The ads ran with *Beverly Hills 90210, Cosby,* and *The Simpsons.*

"This is war," boomed Sega's Al Nilsen, who steered and cheered the battle. It was the company's first and probably only chance against Nintendo. Sixty percent of his customers had defected from Nintendo, and Nilsen planned to keep them. His ads featured the "Sega advantage," 100 software titles for Genesis (made by Sega and twenty third-party developers) as compared to a mere handful for the SNES.

The close of 1992 through 1993 would have a great deal to say about the disputed figures released by both companies. Analysts and retailers predicted that a total of some 10 million Genesis and SNES systems would be sold by the end of 1992, with the two companies in a virtual dead heat. If this occurred, it would be a rousing victory for Sega, which could then continue a head-to-head battle. If Nintendo ended up way ahead, the status would remain quo. Nintendo would still control perhaps 75 to 80 percent of the market, and Sega would be a modest number two (though certainly stronger than before).

Choosing sides or attempting to play both against the middle, the licensees had a major stake in the competition. Nintendo's licensees continued to release NES titles—the market had weakened, but 35 million games for the NES and another 25 million games for Game Boy would sell in 1992, according to the company's predictions—and most signed up to make SNES games.

Nintendo planned to sign up 100 SNES licensees so that the SNES–Genesis game gap would cease to be an issue.

In the United States NOA sought to control the quality of its new licensees' product in a unique way. There was no longer an exclusivity clause (in part, at least, because of the FTC and the antitrust cases pending). Instead, Nintendo would allow licensees to make three games a year. But it also built in a strong motivation for companies to make excellent games: games that earned thirty or more points in the Nintendo rating system didn't count as one of the three games. Also, a more objective method for choosing what games would be covered in *Nintendo Power* was initiated: only those games with thirty or more points would be featured.

Some SNES licensees, as with NES licensees, would still buy cartridges from Nintendo, but some could manufacture their own games (they would, of course, still buy security chips from Nintendo). Acclaim contracted with a Spokane, Washington, company called Keytronics to manufacture games. Nintendo still earned its high share of sales, a 20 percent royalty based on the wholesale price, and this didn't include the security chips, which cost $1 apiece.

Although the new contract didn't restrict companies from releasing games for other systems—Electronic Arts released "John Madden Football" for the SNES and Genesis—there was speculation about whether Nintendo would ultimately exert the same kind of control, overt and covert, that it exerted over 8-bit licensees. Would Nintendo seek to hurt companies that aided and abetted Sega? Ratings of games would be one measure: Would it use its rating system, which presumably would continue to correlate with sales, to hinder companies that also released their titles with Sega? There were additional fears about the subtle ways in which NOA continued to "influence" distributors and retailers. The SNES was available in 16,000 retail outlets its first four months out, while the Genesis was only in 8,000 (although Sega later raised this number to 11,000).

Most Nintendo licensees liked the idea of competition between Nintendo and Sega, even though it meant they had to make choices about which system to support—or whether to support both. The attitude of some licensees was that anything that weakened Nintendo was good news. Part of this was resentment—it was always good to see a tyrant fall—but part was that it strengthened their positions. Whereas they had originally come to Nintendo hat in hand when they wanted to become NES licensees, now Nintendo needed them. Nintendo had to have a strong library of games to beat Sega, and the licensees were key.

In spite of their discreet rooting for Genesis, the licensees found neither openness nor benevolence at Sega, whose own licensing terms got tougher until they appeared to be nearly as strict as NOA's. The only thing some licensees had going for them was that they were being wooed by both Nintendo and Sega, which gave them some negotiating power. "As often happens, a revolutionary accomplishes a coup and becomes the next despot," a licensee said. "Sega was as bad as Nintendo because Sega wanted to *be* Nintendo."

Nevertheless, the competition between the two companies allowed software companies a kind of independence they hadn't had when NOA was the only game in town. It was analogous to the computer industry, where Microsoft and Lotus were cleaning up by selling software that ran on all the warring hardware makers' machines. The game software companies won whether Nintendo or Sega, or neither, ended up victorious. Electronic Arts was in the position to be the video-game industry's Microsoft.

"If Nintendo achieves its stated goals, this machine will almost double the market for 16-bit entertainment software in the U.S. in 1991," said Bing Gordon of Electronic Arts. "EA should be well positioned to take advantage of this new market, with our proven 16-bit properties from Genesis and PC formats." Larry Probst noted that he expected the Super system to add $4 million to his company's sales in 1991 and $10 million in 1992. Continued growth was anything but guaranteed, however. In 1991 short-sellers were betting that Electronic Arts would be a casualty of an Atari-like crash. They said that EA stock should trade at a single-digit toy-company multiple rather than at the "lofty double-digit multiples enjoyed by software companies like Microsoft," according to Eben Shapiro in *The New York Times*. Even stock analysts who were supporters were nervous.

EA's leaders, of course, claimed they had ever more tricks up their sleeve. The issue, Bing Gordon said, was whether EA could become "a $500 million company with a $50 stock in four years." Its position with the 16-bit platforms, plus CD systems of the future, made it a good bet. Gordon predicted that by 1995 companies like his would have strategic alliances with a Sony or a Matsushita, or be acquired outright by the consumer giants.

As the early months of the contest rolled by, SNES sales remained healthy, though nowhere near as strong as NES sales had been at their peak. The trade press remarked upon it as if they had found a chink in Nintendo's armor, and indeed, the first round of the 16-bit war proved that NOA was no longer infallible. The future was up for grabs, although there was one certainty: it would be a mistake to discount prematurely the Arakawa-Main marketing force and the resolve of Hiroshi Yamauchi.

In a lapse of his ability to foresee the long-term future, Yamauchi had realized too late the importance of sewing up the 16-bit market, even at the expense of some 8-bit sales. At stake was more than the $10 billion-plus that consumers would spend on video games in 1992 and the escalating amounts predicted for the following years. The video-game industry was changing. It had found itself at the center of a new industry, fast emerging as *the* consumer-electronics and home-computer industry of the future.

Multimedia was the buzzword in the consumer-electronics and computer industries as they entered the last decade of the twentieth century. Definitions of *multimedia* varied widely, but there was a shared certainty that the new technology would bring together such media as television, video games, stereo, and the VCR in combination with a high-volume storage device (a compact-disc player) and a central processor. Other components could be added, such as a digital photograph reader or a printer. A cable-television receiver, one that would manage and search through the thousand cable stations of the future, would probably be incorporated too. Key to all the elements that defined multimedia, however, besides the television screen, was the computer. It would be the clearinghouse of the huge amounts of audio and visual information and allow people to interact with it all.

Multimedia at first would offer higher-quality entertainment, whether stereo television or video games with CD-quality sound, ultrasharp graphics, and intricate animation. The biggest aid to sales of multimedia machines would probably be feature movies when they became widely available on CDs instead of VHS tape cassettes, since the quality of movies on CDs would be so superior (and because CDs, unlike VHS tape, would never degenerate). Advances in technology would have to occur—in the field of image processing and in data compression techniques—before movies would play on CDs, but they were coming (Sony showed off early versions in January 1992).

Movies, however, would be only the beginning. Television viewers had lots of options in the late twentieth century—dozens of cable stations, prerecorded videotapes, home movies, network TV, video games—but multimedia would bring an astounding array of choices. "There will continue to be more and more competition for percentages of time on the TV screen," predicts Trip Hawkins, an early multimedia visionary.

Multimedia would bring about interactive media, so that viewers could influence the outcome of films or television programs, for instance. Entertainment and education would intersect in novel ways, too, many of them still unimaginable. All media would have an essential new character: they would not have to be linear. "Viewers" of movies, television programs, and electronic books would no longer be passive. They could choose to sit back and watch a National Geographic special on the Andaman Islands, or choose to become involved, "paging," in any order, through sections of the program that were of interest. They could watch footage of the alcyonarians that dance below the giant sponge forest, or observe the remarkable spectacle of elephants swimming between islands. A boat trip between the islands could be experienced from the perspective of a passenger, who could choose where to stop and what to explore. It would be something like "playing" a movie as if it were a video game.

CDs would completely transform video games. Whereas an expensive, high-powered Nintendo or Sega game cartridge filled with RAM chips could have 8 or 10 megabytes of game instructions, an inexpensive compact disc could have more than half a *gigabyte*. All this mass storage could be filled with digitized full-motion video, so that games would feel more like movies, with incredible, lifelike sound, including symphonic music and real (digitized) human voices.

Multimedia video games could be either far more realistic, or else completely out of this world. Graphics and sound effects would be light-years ahead of those on 16-bit cartridge games. Just as soundtracks transform the experience of watching a movie, games could be made far more compelling with richly textured scenery, whether a jungle or an animated wonderland, and sound effects—jungle sounds or bone-chilling screams—would be realistic and intense.

CDs would also mean that an enormous spectrum of video games would be far less expensive to manufacture. Whereas a Nintendo game cartridge costs manufacturers $12 to $16, a CD would cost them $1 or $2. The difference would mean that software companies could proliferate and make games that appealed to small audiences. While no company would risk an investment in a Nintendo game about stamp collecting or an electronic version of canasta, many of them would if the cost of manufacturing were much lower. There would be CD games based on everything from self-help books to authentic archeological expeditions to lawn bowling to simulated sex. A new market would result: people who had never considered playing video games before.

Games could include layers of fantasy, intrigue, and challenge that were formerly inconceivable. The technology that would remove the wall between the game player and the game was in its infancy. Futuristic virtual-reality games would have *you*, the player, experience the "virtual" world inside a video game directly instead of vicariously. Perhaps wearing goggles, gloves, or body suits connected to the multimedia system, players would enter a fictional world and perceive it not as an outsider looking in but from within the game itself. They would not be controlling a tiny cartoon of Mario on a television screen; they would *be* Mario. They would look up, face a cherry-red Goomba the size of a VW Bug charging toward them through a surrealistic palm-tree forest, and they would have to react—quickly.

For all their uncertain futuristic promise, the multimedia- and interactive-entertainment businesses were coming; that much was unarguable. It was also generally accepted that multimedia was going to swallow up large segments of some of the world's largest industries, including much of the consumer-electronics and entertainment industries, as well as segments of the computer business. Analysts predicted that the multimedia business could eclipse that for the VCR by as early as the turn of the century. Estimates of the size of the industry varied widely. Some experts predicted that more than $3 *trillion* a year would be spent on multimedia by the year 2000. The largest multimedia corporations would own commanding shares of that stake. Needless to say, Nintendo wanted as much of it as possible.

If Nintendo could repeat the success of its 8-bit system with the 16-bit machine, it would be in an ideal position to emerge as a strong player in multimedia. At last it would have the opportunity to follow through on what Hiroshi Yamauchi had envisioned years before. His video-game system would transform into a multiuse, multipurpose home computer, the first truly pervasive home computer for the mass consumer market. While there were more than 50 or 60 million Nintendo home game systems in the world, the potential number of multimedia systems was larger still. How many? Trip Hawkins estimates that there are 300 million television sets in the world. At the extreme, there could be several hundred million Nintendo machines in homes throughout the world, all running Nintendo-made or -controlled software. Nintendo would emerge as a formidable global corporation.

Although companies such as Apple, IBM, Sony, Matsushita, Philips, Fujitsu, and Microsoft were scheming how to get shares of this market, Hiroshi Yamauchi had daringly announced early on that Nintendo would "define" the home-entertainment-systems industry of the future. The move, he said, was the company's "boldest departure yet from the antiquated perception of video-game technology." Howard Lincoln says, "Once the momentum begins, it becomes a self-fulfilling prophecy." A large customer base would attract licensees, who in turn would produce a wide variety of software. If the software was good enough and broad enough, the customer base would grow, bringing on more licensees, more customers, more licensees, more customers . . .

Other video-game companies, including Sega, were going to try to get a stake too, although they were unlikely candidates for the biggest leagues; Sega, primarily a software company, seemed too narrow-minded and lacked the resources to promulgate a standard on its own. Instead the competition would be among computer and consumer-electronics companies, or among newly formed joint ventures between them and entertainment companies. Hiroshi Yamauchi's appetite was large enough to imagine it. As *The New York Times* reported in early 1992, "Nintendo, sometimes underestimated, is not about to concede anything to larger electronics companies." Peter Main said, "We continue to be very aware of the Sonys, the Apples, and the Microsofts. We are not taking our past success for granted." Hiroshi Imanishi said, "The Super Family Computer will be the fastest-selling and, ultimately, the most widely spread computer of all."

To get there, Arakawa and Yamauchi kicked the pace into overdrive by mid 1992 (though their failure to blow Sega away with the SNES was a sobering cold shower). On one front, the Super system push was revved up. On another there was a heightened drive to do what Nintendo had done better than anyone in the past—create games that would keep its fans, and new generations of fans, intrigued. To that end, Yamauchi increased the research budget to explore the future of video games and multimedia. He also entered into secret alliances with technology companies and negotiated with entertainment companies for licenses to their characters and stories.

As has often been the case with mass-market consumer technology, multimedia was first developed by techies *for* techies to use on computers. A vision of multimedia on computers was outlined by Microsoft, a forerunner in the field, in May 1991: "A multimedia-equipped personal computer is still a personal computer. It can do the same things today's traditional personal computers can do. But because it adds the possibilities of sound, animation, and high-quality graphics, it provides richer building blocks for new, more

compelling, more engaging ways to use computers." Microsoft estimated that more than 15 million of the personal computers on the
market in 1991 were "multimedia-upgrade ready." To help get
them there, it updated its popular Windows 3.0 program to accept
extensions with the capability of driving multimedia. The
Microsoft multimedia team, sporting sweatshirts that read, I WOULD
TELL YOU WHAT I WAS WORKING ON, BUT THEN I WOULD HAVE TO KILL
YOU, planned to be at the center of the industry, whichever way it
went.

In general, the key addition to a computer for multimedia is a
compact disc player that plays CD-ROMs as opposed to traditional CDs. The difference is that CD-ROM discs can do more
than play great-sounding music. There are different formats of
CD-ROM discs, but they all have in common the ability to carry
music, moving and still pictures, voices, and computer programs as
well as the text and simple graphics normally associated with computers. Almost anything that can be digitized, encoded, and incorporated as data on a platter can be fodder for multimedia.

Sophisticated games are one selling point for CD-ROMs, but
the most compelling application of the first CD-ROM player
hooked to a computer was immediate access to enormous amounts
of information. For $700, Sony's Laser Library could be attached
to a personal computer and play normal CDs as well as CD-
ROMs, a sampler of which was included. On one disc was the
entire *Compton's Family Encyclopedia,* with text and pictures. On
the *National Geographic Mammals* disc was a reference book of
information about animals plus recordings of animal sounds and
video clips of animals in the wild.

Sony had pioneered CD and CD-ROM hardware technology
and planned on leading in software, founding Sony Electronic Publishing specifically to develop and license CD-ROM software for
multimedia. In addition to releasing the CDs that accompanied its
Laser Library, Sony Electronic Publishing planned to make games
and other entertainment and reference software on CD-ROM that
would play on many systems.

Hardware companies that are adapting computers for multimedia include Apple, IBM, Fujitsu, and Tandy. There are software companies as well. Microsoft's chairman, Bill Gates, has

founded a side venture called IHS to create multimedia software. Gates exploded the boundaries of the pre-multimedia definition of software and has attempted, for instance, to corner the market on the rights to digitize museums full of paintings. With those rights, Gates could offer, say, a CD-ROM tour of the Musée d'Orsay in Paris. A "visitor" could peruse the paintings in any order, at any pace, zooming in to study details, and calling up audiovisual biographies of the painters. The discs could contain audio commentary from critics and historians, and users could, perhaps, even "sample" paintings. Presumably one could place Whistler's mother's head on the body of the nude in *Déjeuner sur l'herbe,* decorating the picnic with bunches of Van Gogh's lilies.

Although the first multimedia systems have been centered around the computer, Commodore, Apple, and a steady stream of new entrants into the multimedia business have seen that more people would use systems for entertainment, education, and information if they were easier to use and had the trappings of a consumer product, *not* a computer. The personal-computer revolution has already proved that most people will continue to spend their leisure time around their televisions rather than around computer screens. It is a safe bet that multimedia will find its place in the living room rather than the office; multimedia machines will ultimately replace and supplement normal television, cable, VCRs, and the like. Computer companies will likely be left in the dust of the proliferation of multimedia that play on television sets unless they figure out ways to participate.

Apple, which has a growing R&D budget for multimedia, will probably launch a CD-ROM–playing machine—a unit containing a Macintosh computer processor and operating system that will, presumably, hook up to televisions, much like a VCR. Apple, however, has the disadvantage of not understanding the under-$500 consumer market. An earlier plan to develop a video-game system was reportedly abandoned. "They realized how important the software business was," Arakawa says. "Who is going to make it for them?"

Nonetheless, as it became clear that the consumer business is where the industry is going, Apple has set a course. To beat the drum for the company's first entry into the consumer business,

Apple chairman John Sculley showed up at the Consumer Electronics Show in January 1992. In a speech he observed that "the personal-computer industry and traditional consumer-electronics industry are converging on an inevitable and potentially wonderful collision course." Although he made no specific product announcement, Sculley indicated that an Apple multimedia machine was going to be at the vortex of the collision.

Commodore, makers of the Amiga computer, is already there. CDTV was released by Commodore in 1991, and Nolan Bushnell was hired to push it. He said that he had agreed because CDTV provided the kind of potential that his brainchild, the video game, never delivered: technology that would help dissolve the distinctions between art, entertainment, education, and productivity.

A comedown from the flashy corporate offices of his past—whether at Atari, Chuck E. Cheese Pizza Time Theater, or Catalyst during the glory days—in early 1991 Nolan Bushnell's office was in a dirty-beige, brown-trimmed complex on a side street not far from Hewlett Packard in Mountain View, California. His lobby could have been an insurance company's except for the fact that Bushnell's first commercial video game, "Pong," was set up along one wall. It was a classic, a black-and-white MGA television in a cabinet of fake wood veneer. Inside was a six-by-eighteen-inch circuit board and a coin box, on the front were two simple controllers and a yellow panel, and that was it.

In a purple, black-dotted shirt and gray tie, Bushnell sat behind a cluttered desk in his office. He leaned forward, resting his weight on his large fists. Near him was a Macintosh IIs computer with a large-format monitor. Piled on the desk and shelves were an Audubon Society book on whales, a stuffed Cupid Mouse from Chuck E. Cheese, a toy robot named Spybot, a Playskool Baby Monitor, "Scrabble Lexor," a Cherry Coke, and more books: *Android Design, World of Robots, Making Robots, Introduction to Artificial Intelligence,* and *Iaccoca.* There were also handsomely bound volumes commemorating Chuck E. Cheese Pizza Time Theater stock offerings—1,169,610 worthless shares.

From this office, years after Atari crashed, Catalyst folded, and Pizza Time went bankrupt, Bushnell ran numerous small companies. Vent made computer peripherals and software; Buffalo made laser-printer cartridges; Names and Faces made a coin-operated photo booth—like the ones in airports and arcades but employing no photographic chemicals; instead, the photographs, taken by a video camera, were digital pictures, spit out instantly from a laser printer.

In spite of his other projects, Bushnell had accepted a job as a consultant for Commodore in 1990. He said he was devoting "by far the biggest chunk" of his time to Commodore, "beating the drum and evangelizing for multimedia."

Commodore released CDTV in early 1991. The multimedia machine was actually an Amiga computer hidden inside the sleek box of a consumer-electronics product (much as the Apple multimedia player would probably be a Macintosh in a cabinet designed for living rooms instead of offices). Commodore's primary advantage was the Amiga; it had a readymade software library, and there was a software-development community that appreciated the machine.

The Amiga processor was attached to a built-in CD-ROM drive and controlled by a fancy remote controller, although a keyboard, mouse, and floppy-disk drive could be added. On its own, CDTV sold for $799. A couple of hundred dollars' worth of add-ons would transform a television set into a full-blown computer, although that was not what would drive CDTV. Instead, it would be CD-ROM entertainment and information programs, such as a multimedia atlas (with audio samples of languages, reference books of statistics, and picture slide shows, in addition to the expected maps), a multimedia version of *Grolier's Encyclopedia,* and multimedia children's books, such as the *"All Dogs Go to Heaven* Talking Electric Crayon" coloring book.

CDTV and the prospects of multimedia were revelations for Bushnell, who had founded the first wave of the video-game business with the belief that it would lead to new ways of learning. He said he had felt disappointed with his progeny—"not quite the guilt that Robert Oppenheimer felt, but guilt nonetheless. . . . Video games have not fulfilled the promise that I envisioned," he said. "The repetitive, mindless violence that you see on video games right now is not anything I want to be associated with. I don't know how to put it more kindly. I think it's shit." But CDTV, he added, "will finally interweave education and fun."

In spite of Bushnell's lofty vision and a major advertising campaign by Commodore, it was by no means certain that CDTV would be a player in multimedia. It had a head start on Apple and other contenders, but Commodore had a history of missing the boat when it came to large-scale marketing to consumers.

The Dutch electronics giant Philips N.V., the largest consumer-electronics conglomerate in Europe, had a better track record with consumer products, and it had backed its multimedia platform, the Compact Disc Interactive (or CD-I), with a huge amount of development money and a long-range view of the market. Philips released CD-I years behind schedule, in October 1991, months after CDTV, because of technical problems. The $1,000 "imagination machine" (the price later went down to $799) was designed strictly for multimedia, so there were no keyboards or floppy-disk-drive add-ons.

Owing to Philips's stature in the marketplace, it was supported by more software companies than CDTV. Released under its Magnavox label, the CD-I could boast a library of impressive programs. One was a better version of a multimedia children's storybook than the one released for CDTV. When kids popped in a disc licensed by the software division of *Sesame Street,* Bert and Ernie appeared on the television screen. Perfect hosts, they invited kids to explore their apartment, a treasure trove of fun. Learning was the subtext, of course; a child could read books along with Ernie and do fun math problems with Grover.

The library of CD-I was diverse, from "Caesars World of Gambling" to "ABC Sports Golf: Palm Springs Open" to "Treasures of the Smithsonian," a self-guided tour of 150 of the institution's attractions. A novel educational disc developed by Time-Life (called "Time-Life Photography") contained condensed versions of ten popular how-to photography books. In one lesson, a rower piloted a scull through the water and the "viewer" had to snap a photograph that preserved the background detail but conveyed motion.

In spite of the early software and the promise of great things to come, CD-I and CDTV posted unimpressive sales. The systems were too expensive and the market was too new. Still, Philips noted, in its first year CD-I sold more than either the VCR or CD player had in their first years. The executives at Sierra On-Line, which produced a CD-ROM game called "Mixed-Up Mother Goose," were happy to have sold 6,000 discs, a number that would have been considered an embarrassment in the video-game business.

The companies knew it would take time for consumers to appreciate multimedia, let alone invest in it. After televisions were introduced in the United States, it took a full decade before 1 million were sold. EA's Trip Hawkins summed up the problem: "If you're going to bring out a hardware system and it's going to cost $800 and the main reason why somebody's supposed to buy it is for interactive applications, which is something they really don't understand right now—I mean, that's just not going to work."

What might work, however, was a lower-cost machine with a popular application: games. "CD-ROM's popularity will increase dramatically when Nintendo rolls out a CD-ROM–based game machine," the influential computer-industry columnist John C. Dvorak wrote in September 1991. The thinking was simple: Nintendo had in its clutches a huge number of savvy consumers, unafraid of technology, who could be expected to do whatever it took to get their hands on the hottest new games. Kids clamoring for a new super-powered Nintendo CD game would do more for multimedia system sales than the most impressive multimedia encyclopedia.

Nintendo's entry would also, Dvorak imagined, bring down the prices of CD-ROM technology, essential for it to catch on. "Why couldn't manufacturing giants like Toshiba, Hitachi and Sony excite the market enough to drive down prices for drives?" he asked. "Even they seem to think that Nintendo can effortlessly (and overnight) accomplish what they failed to do. If that's the case, why doesn't someone just hire the Nintendo marketing genius."

In a sense, two of the big companies tried. Philips N.V. contacted Nintendo and entered into two-pronged negotiations. First, Nintendo was planning a CD-ROM drive that would hook on to the Super NES and Famicom; Philips wanted to help create this system. Second, and more significant, Philips wanted future Nintendo games to be playable on CD-I.

An agreement was painstakingly hammered out between Philips

executive Gaston Bastiaens, who headed the CD-I group, and Minoru Arakawa and Howard Lincoln. To finalize the deal, Lincoln and Arakawa headed for the Netherlands in May 1991 to meet Bastiaens at the Philips world headquarters in Eindhoven. The two companies agreed on a deal that allowed Philips to create a bridge so that Nintendo CDs could play on CD-I. Significantly, Nintendo would control the licensing rights to the compact-disc games in much the same way it controlled NES and Super NES licensing. For its part, Philips won a valuable affiliation with Nintendo that would help sell CD-Is. Philips also could win a lucrative contract with Nintendo to supply the CD-ROM drive that would be attached to the Super NES.

Arakawa and Lincoln then headed to Chicago for the June Consumer Electronics Show, where they planned to announce the deal. But sparks were about to fly; the deal Nintendo had entered into with Philips conflicted with another one it had made with Philips's competitor, Sony.

The two giants, Philips and Sony, competed in many consumer markets, including television sets, radios, videocassette recorders, cameras, and CD players. They had once been allies in the development of CD-I until reported personality conflicts and clashing visions caused an unfriendly parting of the ways. Philips released CD-I on its own, while Sony released its CD-ROM system for computers and planned to release a cheaper alternative to CD-I, a game machine called the Play Station.

When the Play Station was announced in summer 1991, it caused headlines. Acknowledging that the video-game business had grown too large to ignore any longer, Sony was entering the fray in partnership with the dominant force in the industry, Nintendo. Sony also had awakened to the fact that it could use a relationship with Nintendo to explore the CD-ROM business. The Sony Play Station seemed to be an ideal vehicle that would usher in the era of CDs since it played CD-ROMs, called Super Discs, as well as Super NES cartridges. It was an extraordinary alliance; two Japanese companies, giants in their respective industries, were joining forces.

But this was before Nintendo's executives realized the implications of the deal they had signed with Sony back in 1988. By 1991 it was seen as a disaster, one that contradicted Nintendo's cardinal tenet of giving nothing away. Nintendo, which had predicated its business on complete control of its game software, had granted Sony the right to control (and profit from) all CD-based software that played on the Play Station. When it had announced the Play Station, Sony emphasized that it was the "sole worldwide licenser of the Super Disc format," and Nintendo was left to twist in the wind.

Perhaps Yamauchi had been intimidated by Sony back in 1988, when the deal was signed, or perhaps he had underestimated the importance of CDs. But the situation was intolerable, regardless of his signature on the contract.

Sony announced the Play Station at the June 1991 CES in Chicago. The Play Station would have a port identical to the one on the Super Nintendo, as well as a CD-ROM drive that would play Sony Super Discs, CDs that contained up to 680 megabytes of data. The machine would play high-quality video games and eventually other forms of interactive entertainment. Olaf Olafsson, the chief of Sony Electronic Publishing, who was in charge of the Play Station, explained, "In order to promote the Super Disc format, Sony intends to broadly license it to the software industry." Olafsson, youthful, blond, and dapper, added, "We will draw on the creative and entertainment resources of the Sony family of companies, including Sony Music and Columbia Pictures, to develop and market an exciting lineup of software products."

Sony was going directly after Nintendo's customers. In *Fortune,* Olafsson was quoted as saying, "Primitive cartridge-based games brought in $4 billion to $5 billion in the U.S. alone last year. The video-game business won't get any smaller. It will get more sophisticated. With film and music, the games will be much more interesting. . . . By owning a studio, we can get involved right from the beginning, during the writing of the movie. We can get footage as it's being filmed. We can say, 'Can you film the backstage in a certain way, because we need it for a video game?' " *Fortune* reported that Olafsson was seen on the set of *Hook,* the Robin Williams and Dustin Hoffman movie, several times before the filming was completed. He was deciding what backgrounds to use in a "Hook" video game, which parts of the soundtrack to include, and how best to time the game's release.

Sony was in a position to cripple Nintendo, and this deal set the stage for it to use Nintendo to win customers, steal licensees, and then discard the company. But there was more. Sony was the sole supplier to Nintendo of a key chip—the audio chip—inside the Super NES. It had been designed by Sony so that its full capabilities could be taken advantage of only by programmers working on an expensive development tool created by Sony. As one industry consultant put it, "Sony had Nintendo by the balls. . . . It was not a tenable position as far as Hiroshi Yamauchi was concerned. Because of that, he instructed Arakawa to proceed with the Philips deal. It was meant to do two things at once: give Nintendo back its stranglehold on software and gracefully fuck Sony." *The New York Times* restated it more genteelly: "The move was widely seen as an attempt by Nintendo to enter the compact-disc market on more favorable terms."

Publicly Nintendo explained it had allied itself with Philips, makers of the original CD, because its CD-ROM technology was superior. Privately it was known that Yamauchi had allied himself with Philips because he had decided that Nintendo would not become dependent on any other company. No one, not even Sony, intimidated him.

Nintendo conspired with Philips to pull the rug out from under Sony, in front of the press and public, at the CES the day after the Play Station was announced. There was no doubt that the move was designed to embarrass Sony and make it clear to any doubters that Nintendo was fearless.

The Sony executives in Japan had learned about the pending press conference forty-eight hours earlier, and were, according to all reports, stunned. The company's chief executive, Norio Ohga, attempted to stop the deal by placing calls directly to Hiroshi Yamauchi at NCL and Jann Timmer, the new chairman of Philips, in Amsterdam. Bastiaens also heard from Sony brass. Nintendo and Philips presented a united front. They said a deal had been concluded and there was nothing to discuss. Though faced with the threat of lawsuits and "other repercussions," Nintendo insisted that the Philips deal did not interfere with the arrangement with Sony, and the protests fell on deaf ears.

Top Sony officials continued to attempt to head off the announcement. Howard Lincoln says, "There were tremendous efforts on a worldwide basis to keep that press conference from happening. They gave up on us, but they kept pressuring Philips." Timmer and Bastiaens fielded numerous calls. In the meanwhile, Sony's New York team considered canceling the announcement of the Play Station, but things had gone too far; also, it would have played right into Nintendo's hands.

The press conference began at 9:00 A.M. sharp. It was heavily attended by a huge gathering of press from the trades and major newspapers. It was expected that Nintendo planned to boost the Super NES and remark on the company's significant alliance with Sony. Instead, Howard Lincoln, at ease at the podium, announced Nintendo's plan to work with *Philips*. The buzz in the room was audible, but it was nothing compared to the frenzy at Sony. Olaf Olafsson was livid. "They stabbed us in the back," the young powerhouse from Iceland told a confidant. In statements to the press, he insisted that Sony had an exclusive deal with Nintendo that was being violated.

The shockwaves were felt throughout Japan, where the industry speculated about the implications of Hiroshi Yamauchi's stand, and the price Nintendo would pay for humiliating Sony. Also at issue was Nintendo's breach of the unwritten law of turning against a reigning Japanese company in favor of a foreign competitor. The repercussions could be severe.

In Japan there were meetings at both companies' corporate headquarters. In the United States lawsuits would probably have been filed under the same circumstances, but Japanese companies work differently. Nintendo had taken a strong public stand against Sony and Sony had to react. It had to be careful, however; it had a lot at stake. Nintendo was going to be instrumental in selling the Play Station because of the port for playing Super NES games, so Sony had an incentive to resolve the problem rather than exacerbate it.

Nintendo, meanwhile, could benefit from the relationship, too, if it could win better terms, primarily the control of CD games it made or licensed. Nintendo was tied to Sony because of the audio chip, and it was sure to suffer in one way or another if it alienated Sony.

While a contract between American companies is typically dozens of pages long, a contract between Japanese companies might be merely two or three pages, written, to Western eyes, with considerable ambiguity. These nebulous agreements are designed to allow room for continued negotiation over the life of the contract. Often a good-faith clause is included, indicating the parties' agreement to sit down and discuss problems as they arise.

Since the contract was vague on certain points—at least according to Howard Lincoln—NCL danced through the loopholes to extricate itself from the bad deal. Nintendo's Hiroshi Imanishi, one of the negotiators, would only say, "I did not make this decision. Mr. Yamauchi made this decision. I can't tell you all the reasons behind it. We made a business decision that we are not going to support Sony, although we will honor the contract we have with them. However, we decided that we would enter into this relationship with Philips, and it will prove to have been the correct decision."

In the weeks following the blow-up, Nintendo and Sony communicated with each other through the press. Although Olafsson criticized Nintendo in *The Wall Street Journal,* he also said, "The door is still open." "The goal now is for both companies to find a way to save face," an analyst surmised. "In some ways it appeared that Nintendo used Philips to either revise the Sony deal or to retaliate against Sony. It's not over, though."

Minoru Arakawa kept silent about the affair, citing a nondisclosure agreement in the contract. However, Hiroshi Imanishi said, "The only acceptable deal would allow Nintendo to make its own software with the same software control as with game cartridges." In the aftermath of the CES announcements, Howard Lincoln said, "It's safe to say that if we release 'Super Mario 6' on compact disc, it would *not* play on Sony's CD player unless Nintendo and Sony resolve their differences." Yet, he went on, "anything is conceivable." One thing, of course, was *not* conceivable: Lincoln insisted that there could be no deal that had Nintendo making games compatible with Sony's CD player exclusively or with Sony owning the rights to Nintendo's games.

In spite of all the CD machines that hooked on to computers and the expensive CD-ROM players that hooked on to televisions, the multimedia explosion would probably *not* begin until a company like Nintendo entered the business. Trip Hawkins, in 1989, said, "The VHS became successful not by being the best technology in the world, any more than a Nintendo machine is better than a personal computer. It was the cheapest technology that did the job. Multimedia would proliferate like the VHS business if there was a $100 video-game system that could be sold as an upgraded system to the 35 million households in America that are video-game literate. Of those 35 million households, at least several million are going to be interested in a product like that if it is released with good software. That is the way of getting the market going without having to beat people over the head to educate them about something completely new. They *already* understand interactive entertainment and it's in their price range." The product was the Super NES, Genesis or something similar.

"Then," Hawkins continued, "release a $200 consumer-grade CD-ROM player with a digital controller and some RAM that would also play audio CDs. If the price were low enough, a lot of consumers would buy it instead of a regular CD, figuring they would get the extra capability to play games for a little more money. The total becomes $300, which is an order of magnitude less than what the current PC companies are trying to sell multimedia for."

In June 1992, Nintendo announced that it would release a CD-ROM drive that would attach to the Super NES in 1993. Its press conference at the summer CES heralding this made it clear that Nintendo would be entering into the CD business and would come up with a licensing program very similar to the NES and Super NES arrangements. "We will exercise quality control over the CD games," Howard Lincoln said. "There will be a security system in the CD player, just like in the Super NES." In a press release, Nintendo announced that "[CD] licensees will need to submit their games to Nintendo for evaluation and approval and enter into license agreements with Nintendo." Some of the old controversies were set to rear their heads once more. "We'll definitely have the security system, and it is conceivable that we could have an exclusivity clause in the CD agreement," Howard Lincoln said. "See, I don't think there's anything wrong with Nintendo's exclusivity clause, and no court has yet determined that there is."

Nintendo worked to finalize a format for its CD player and to devise a security system. The machine had to be better and cheaper than competitors' offerings, of which there were many, almost all of them incompatible. NEC had arrived too soon with too little. Its hardware technology wasn't good enough, the CD player hurt the performance of the game system, and the player wasn't a true CD-ROM machine that could play real-time graphics and high-resolution video.

Once again, Nintendo's immediate and strongest competition seemed to be Sega. Its CD player was available by the end of 1992 for about $300. The initial price tag would keep most consumers away, but once again it was beating Nintendo to the punch. Sega made the additional announcement of a deal with Sony, which planned to create games for the Sega Multimedia Entertainment System based on its entertainment companies—the Columbia and TriStar studios and Sony (CBS) Records. Tom Kalinske, president and CEO of Sega of America, said, "We will be tying in with Sony artists. I just talked to [Columbia Pictures president] Peter Guber, and we talked about shooting footage for games at the same time as they're filming." The first example was footage being shot on the set of Steven Spielberg's *Jurassic Park,* based on the Michael Crichton book. "People talk about multimedia a lot in the personal-computer industry, but we're doing it," Kalinske told the *Los Angeles Times.* "The days of the dictatorship are over."

Although Sega and Sony would benefit most from the deal, the announcement also served as a slap at Nintendo by Sony. As one reporter put it, "Sony said, 'Okay, Nintendo, you chose to ignore our deal and team up with Philips, so we'll aid and abet your biggest competitor.'"

There were other CD players coming out that played both CDs and cartridges, including Sony's Play Station and ones planned by

NEC (combining the TurboGrafx and a CD drive) and JVC (which played Sega CD games). There were also stand-alone machines by Commodore, Philips, and probably Apple. All these companies were vying for a share of the high-stakes industry. Most entries would not survive, and there were rumors that CDTV might be the first victim.

Trip Hawkins decided that too much was at stake to sit on the sidelines. Back in 1989 he had predicted that there would be an everyone-for-himself attitude by the hardware companies, with dire consequences. "It will set the industry back a decade," he said. "Technology will be sacrificed and consumers will stay away." If a single standard were developed, he imagined tremendous leaps forward for multimedia; then software developers could focus their energies on what they do best by taking the technology and running with it to create unimaginable multimedia packages.

Since it seemed that the machine Hawkins envisioned wasn't likely to come from any one company, he was asked if *he* would attempt to create it. He answered, "We certainly will try to influence hardware companies to make it, but we're not a hardware company."

A year later a piece came out in *The New York Times* in which it seemed that Hawkins's attitude had shifted. He characterized the existing multimedia hardware as "low on performance and high on price," and complained that Sony and Philips were alienating consumers and forcing software companies to guess which format to jump on. As a result, the *Times* reported, "Mr. Hawkins is no longer just talking."

Hawkins had turned the day-to-day management of Electronic Arts over to president Larry Probst so he could concentrate on a new venture that was likely to focus on the issue of hardware design and compatibility. Although his new company, 3DO, was shrouded in secrecy, he told an interviewer, "It concerns me," he said, "that when I look at the hardware companies, there are the PC companies, who are clearly not going to do the right thing as far as I am concerned, and consumer-electronic companies, who don't know what to do. Then there are the Japanese video-game companies, who are too shortsighted to see where this is going. Obviously we would like to see a standardized system that a lot of hardware companies are supporting. I am trying to figure out if it makes sense to get a group together to create the next-generation system. We would then go around and attempt to license it to hardware companies."

Hawkins reportedly allied with Time Warner, Motorola, LucasArts, and Matsushita and its MCA subsidiary, in an attempt to come up with a powerful 32-bit CD-based game system that would also run multimedia encyclopedias with animation and interactive movies. The machine would probably connect to telephone and cable lines, which would pave the way for "pay-per-play" games. A Time Warner network might bring a video store's worth of movies into the home that could be selected at any time via the 3DO machine. It could potentially steal the VCR and video-rental businesses. Hawkins reportedly planned a 1993 shipping date and a $700 price tag. If he succeeded in creating the technology, he'd be going up against a formidable list of competitors, including Nintendo.

The preponderance of CD systems, with varying formats and no (or almost no) compatibility, threw the software community into a tizzy. The surfeit of warring companies baffled them; whom should they support? "I wish one of those guys would win, I don't care which," said Richard O'Keefe, now with Warner New Media. It was a sentiment shared by many.

Executives of several of the largest Nintendo licensees traveled to Kyoto to attempt to convince Hiroshi Yamauchi to reconsider his decision to oppose Sony. A Sony-Nintendo-Philips coalition might instantly create an industry-wide standard. They pleaded with Yamauchi to see the wisdom of promulgating a single standard that would be open, with complete access to all software companies, and further the industry. Consumers would jump in with excitement instead of waiting through the early elimination rounds. But "Mr. Yamauchi had already made up his mind," Henk Rogers says.

Yamauchi summed up his attitude in a meeting with another company president. "Nintendo believes in a standard," he said. "*Our* standard." The problem, Acclaim's Greg Fischbach says, is that "there's nothing we can do until it's sorted out. We'll sit in the cheering section and watch what happens."

Nintendo's negotiations with Sony continued through 1992. Because of the Philips deal, Nintendo succeeded in reopening negotiations over the right to control CD software that played on Sony's machine. An agreement was reached in October 1992. A source inside Sony says, "We concluded that we had to ally ourselves with Nintendo when we saw that it was going to be the 16-bit winner. We wanted access to all those Nintendo players."

After all the threats and announcements of deals with each other's competitors, Nintendo and Sony had worked out their differences, and their CD-ROM players would be compatible. (Philips still had the right to create a bridge from its CD-I system to Nintendo's format, but it was on the sidelines of the deal.) Nintendo's and Sony's machines would be more powerful than any other CD-ROM-based systems in the consumer market—like the rumored 3DO machine, built around a 32-bit processor. With that and the companies' marketing strength, Nintendo and Sony could very well be the ones to create a standard—a standard that they will control. A Nintendo-Sony standard could be almost unbeatable, and the companies could share a guaranteed position at the center of the multimedia industry.

The deal put Nintendo in the dominant position in the most lucrative arena: software licensing. Nintendo won the right to decide whether any software to be released on the CD-ROM systems is a game or not. If it is, even if it is one of Sony's own games, Nintendo will license it. If it is a non-game, anything from a multimedia reference "book" to a CD-based tax seminar, Sony has the right to license it but only *through* Nintendo. The upshot is that Nintendo receives royalties on all the software that will play on the Nintendo *and* Sony CD-ROM machines. Hiroshi Yamauchi had successfully stood up to Sony, and emerged with as much control as ever.

Software for some first-generation CD-ROM machines that worked with computers showed glimpses of why the disc would take over. For a game called "Sherlock Holmes, Consulting Detective," the ICOM Company filmed actors in the roles of Holmes and Dr. Watson as well as various Scotland Yard commissioners, witnesses, and suspects. This footage was digitized and included in the game. It was far more realistic than the animations in most video games. Syracuse Language Systems released "Introductory Games" in Spanish and French (Russian, Hebrew, English, Chinese, and Japanese were planned). A Syracuse University team consisting of an educator, a computer engineer, and a linguist researched how to use multimedia games to teach foreign languages to children. A reviewer of the Spanish game said it was "as thoughtful and imaginative a way to introduce English-speaking children to Spanish as any I've seen short of dropping them into a preschool with native Spanish speakers."

Still, in spite of the games by ICOM, Syracuse Language Systems, and a few others, most early CD games delivered nothing more than cartridge-quality games. Some had remarkably good sound and visuals—game music and sound effects, when piped through good speakers, added drama and realism—but the game play wasn't any better than a Nintendo game. What would ultimately lead to a rush for multimedia systems would be a quantum leap in game design. With revolutionary kinds of games, kids, and then a wider group of game players, would buy CD-ROM drives. If "Super Mario Bros. 5" or "The Legend of Zelda 4" came out on CD, and if the games delivered on the promise of the new technology, Nintendo could reaffirm its position as the dominant game company, if not as a dominant entertainment-software company. However, the revolutionary new games could just as well come from a third-party software developer. In that case, the benefactor would be whatever hardware company or companies licensed it.

Developers at all the software companies throughout the world had ideas about the possibilities of CD games. Essential to Yamauchi's strategy to remain the biggest game company, and therefore one of the biggest consumer-electronics companies, was Sigeru Miyamoto and his other designers. At the same time, of course, Yamauchi was investigating alliances with entertainment companies. He observed that software was what gave Nintendo its edge against the slew of larger, more established adversaries. "I don't see that most of the companies are capable of creating very good entertainment software," said Hiroshi Imanishi, echoing the chairman's view. "The company that can make great software is the one that has knowledge and experience in making games. That company is Nintendo."

Yamauchi felt that nothing short of Nintendo's future was at stake. "We learned our lesson from Atari," he said. "We are able to understand very clearly why Atari failed. No toy company ever became a truly big and great company by remaining a toy company. We have much more ambition than that. As the lines that limited video-game companies in the past disappear, Nintendo will play a larger role in the world."

16

BORDERS

Some of the multimedia capabilities of the future are based on the concept of computer networks. Multimedia systems would have a high-tech umbilical cord connecting them to mainframe computers, much as TV sets are already hooked up to cable. Early versions of information and entertainment networks were already available for computers, connected via modems to traditional telephone lines. Higher-quality phone lines of the future, called fiber-optic lines, will one day funnel in everything from tailor-made daily newspapers—you decide what's of interest to you—to a huge selection of games and movies. The user will make choices with a click of a remote control.

Nintendo plans a network for the United States, part of the global Nintendo Network Hiroshi Yamauchi envisions. The network in Japan will ultimately be part of a worldwide network that will connect families with a Nintendo system (NES or Super NES) everywhere. According to that plan, Nintendo will then take its place as a force in another field—communications.

Nintendo's huge customer base is again the doorway in. "Given the size of the installed base," says David Leibowitz, at American Securities, "Nintendo is the most appealing platform for a network." NOA could take advantage of all the NES households and also sell access to those homes to all kinds of businesses. "Open a video game and you find a computer, optimized for fun with special chips for zippy graphics and rich sound," wrote John Schwartz in *Newsweek*'s look at "The Next Revolution" of computer-based technology in April 1992. "That's a lot like the sort of machines that computer makers are hoping to make for the consumer-electronics market. Computer companies such as Apple will be hard-pressed to beat video-game companies at their game, says Mike Saenz, founder of Reactor, a Chicago-based entertainment-software company. 'If they think they've got a technology that can compete against the entertainment machines, they've got it wrong.' . . . Nintendo [in Japan] has already tapped the computer power within the game box: consumers use the consoles to bank at home, trade stocks and even to bet on lotteries and horse races. . . ."

The network is significant because it is the first time Nintendo demonstrated that it had plans to be more than a toy company. It is the first time Nintendo has actively stepped on the toes of computer companies since the NES co-opted the games market.

Yamauchi's plan was unveiled in 1989. "The network shows how the Famicom has outgrown its single purpose as an amusement system," he said. "It is our important business target this year." A Japanese business magazine proclaimed, "The Famicom is the first mass home computer. Soon you will be fighting for time on the Famicom with your children."

As with multimedia, it seems reasonable to conclude that a technology that was supposed to have been popularized on computers will win widespread use because of Nintendo. Computer networks, like multimedia, could enter the popular consciousness on the coattails of Nintendo. "The Famicom and NES will be the computer system that will popularize networks," Yamauchi says, and Howard Phillips noted, "The NES is the one piece of electronic gear that's in darn near every household, or at least the largest number of households around. It's kind of what the PCs wanted to be but didn't get to be." *Newsweek* added, "For all its billions in sales, the computer revolution has reached only 15 percent of American homes by some estimates." Meanwhile, Nintendo systems are in more than 33 percent. Arakawa said almost whimsically, "It seemed that we might as well take advantage of our position in the homes of America."

Powerful and easy-to-use networks, relying on a computer such as Nintendo's, will bring access to information, services, and other people worldwide into the family living room. The list of on-line services that would be available via the network is familiar: electronic banking (including bill paying), stock and bond monitoring and purchasing, shopping, airline reservations, and much more. Those services would be followed by a complete line of services—limitless possibilities—that would be available when fiber optics are in place.

When the Nintendo Network was initiated in Japan, it offered these basics and then some. Nintendo became embroiled in a battle between securities companies that wanted to use the network (Nintendo allied itself with Nomura, while the other three major companies used a competing, noncompatible system). Banks signed up too, beginning with Sumitomo and Kyowa. The Japanese telephone company, NTT, started a Nintendo banking service, as did Daiichi Kangyo Bank, which marketed a specially designed Famicom-and-modem combination (designed with Nintendo) called Convenient Boy. Over three hundred banks signed up.

There is a growing, eclectic list of other network services. By logging on to the Japanese postal system on the Nintendo Network, one can order stamps and peruse directories of postal codes and weight and price charts. Japan Airlines and JR Tokai, the railroad company, offer on-line schedules and reservations. With Mitsubishi Trading, one can buy and sell precious metals. Bridgestone offers the Famicom Fitness Center, and the Japan Racing Association provides at-home wagering on horse races.

But all the early hype and promise of the network (and Nintendo's lofty projections for it) remained unfulfilled by the early 1990s. A survey in Japan diagnosed that one of the biggest stumbling blocks was Nintendo's name. In spite of its size, it was still viewed as a toy company, and most adults cannot conceive of using a child's toy for business. In addition, as an article in a Japanese newspaper pointed out, kids refuse to relinquish the video-game controllers to their parents.

In the United States computer networks tie together several million people. Prodigy, pushed by Sears and IBM, is the largest, with 1.3 million subscribers as of January 1992. Prodigy users shop, bank, call up movie reviews, exchange information about their jobs and hobbies, sell cars, make airline reservations, and play games. NOA's plan for America is to launch a network that will dwarf Prodigy and its competitors, including Compuserve and General Electric's Genie.

It is a reasonable ambition. Hooked by the promise of access to the game players across town and across oceans, kids would be the pioneers, bringing modems into households. They would probably educate parents about the potentials of a network, but even if they didn't, the generation of kids who grew up with the Nintendo Network will one day be trading stocks and making airline reservations for themselves.

In a 1988 press conference, Minoru Arakawa said that a network in America had the potential to be far bigger than one in Japan. "Telephone communications play a bigger role in the United States than in Japan," he said. In addition, NOA doesn't have as much of a problem with Nintendo's image as a toy company. There is, in fact, a *growing* user base of adults in the United States who are at ease sitting on the floor in front of the TV screen, NES controllers in hand.

Jerry Ruttenbur, a senior vice-president in charge of prerecorded videocassettes at Home Box Office, was hired by Arakawa to create the Nintendo Network in America. In 1989 Nintendo announced that it planned to have an entertainment and information network on the market the following year, and that ten million homes would be linked by 1991.

Though Ruttenbur's background was in sales and marketing, he took over an R&D department. The reason was practical; before he could sell a network, it had to be invented. He was given little direction but a grand goal: to sell modems to the NES households in the United States and get them all hooked together for banking, stocks, shopping, and games.

Ruttenbur made inroads. While a team of engineers from Japan (borrowing from the experience there) and researchers from America developed the hardware and software, Ruttenbur set out to determine what companies would be interested in utilizing a network that connected Nintendo households. Fidelity Investments was the first to sign up; users could manage Fidelity portfolios from home. Peter Main announced that Dow Jones Professional Investors Report also had approached Nintendo. Banks were interested too.

AT&T entered into negotiations to become the carrier. In a conference of three dozen people at AT&T's New Jersey headquarters, all the presentations were given twice, in English and Japanese. (More than once the translator threw up his hands. "There are no Japanese words for these things," he said.) Still, before the meeting was over, AT&T was all but ready to sign up, although Nintendo made certain to announce that it had also been approached by other long-distance companies.

Nintendo designers came up with a cartridge that transformed the NES into a data terminal. The on-screen interface that appeared looked like a new version of "Super Mario Bros.," but in fact it was a menu of network choices. A click on an icon would have the modem connect to a bank, stock-brokerage firm, or whatever the user chose. The hardware package included a modem, a keyboard, and a low-priced printer. The NES was being transformed into a full-fledged low-end PC.

Market research with consumers revealed to Ruttenbur the biggest roadblock to his network. Not enough Nintendo users were interested in banking or trading stocks via modem. Seven percent of NES households had accounts with brokerages, and of these, most of the users interested in a network were already using one. Back to the drawing board.

Ruttenbur decided to attempt to launch the network in the spirit of Nintendo's past successes. At least initially, it would be a fun-and-games network, *not* a business one. The other uses would come later.

The designers came up with a menu of choices that would appear when a consumer signed on. One could get on-line game tips, and there would be a "chat line," where gamers could "talk" to one another. It was modeled after CB lines of other networks. Users would choose "handles" to identify themselves and jump into a dialogue with dozens or hundreds of people. On computer networks, the CB lines were dedicated to specific subjects such as cars, computer software, or sex. The Nintendo CB channels would be for game talk and socializing. Kids could have access to electronic pen pals throughout the country, and eventually the world. Ruttenbur's market research showed that Nintendo's prime audience would love to reach out and touch one another via Nintendo.

Beyond the CB service and game tips, the network would also have games that could be downloaded—that is, sent from a central computer into people's homes. To "capture" and save them, the NES needed a disk drive that could copy game programs onto floppy disks. When the drive, modem, keyboard, and printer were hooked to the NES box, the metamorphosis was complete.

The most exciting offering of a network for devoted game players is the possibility of real-time games—games that kids everywhere could simultaneously play against each other. The problem has been that Nintendo's engineers haven't been able to dream up challenging games with large numbers of players, and too difficult to program the large mainframes to manage huge numbers of smaller contests. Another problem is that initially on-line time is going to be far too expensive for kids to play time-consuming strategy and role-playing games.

Work on the network has gone slowly, but the projections Ruttenbur's team came up with show it will be worth the effort. One of Ruttenbur's plans, to form a club of network members who will pay a monthly fee, would generate "extreme profits," Ruttenbur says. "We would have made money in the second year of operation with 10 percent penetration of Nintendo users."

Still, in spite of the potential for all this money, there was resistance within the company. Ruttenbur couldn't get answers to his questions or okays for key decisions. Minoru Arakawa, preoccupied with everything but the network, particularly as Nintendo geared up to launch the Super NES, was rarely accessible.

Although the network is now on hold, Nintendo has done nothing to dampen public anticipation for it. At a press conference at the Consumer Electronics Show in January 1991, Peter Main, responding to a reporter's question, said, "Development of the network continues on a very active program. Two years ago there was much speculation about applications—financial systems, banking systems. They were floated around, and with that came the expectation that the real product was right around the corner. But our focus is the entertainment business, and the network applications we're planning first are primarily focused on entertainment. Development work is on a high priority, although we're not making any announcements yet."

However, in spite of a grandiose vision and sizable investment, there was no sign of a Nintendo network through 1991. In part, the reason was technology, but Ruttenbur felt the technical problems could have been solved. Frustrated, he left Nintendo for MCA's videocassette division, and subsequently a start-up software company.

The network quietly died, or so it seemed, until September 1991, when it showed new signs of life. Nintendo had been asked to join a venture with Control Data Corporation, the company that runs lotteries in many states. Control Data had proposed a system that would boost lagging lottery sales in Minnesota by making it possible to play without leaving home, via Nintendo. The NES would be transformed into a lottery-playing machine with a special "game" cartridge. A modem, similar to the one designed by Ruttenbur's team, would connect to the NES via the port in the bottom of the machine. Through the phone lines, homes could have a direct link to the Minnesota lottery.

Nintendo couldn't have invented a better scam. Its dormant network would be on the map overnight. The large number of people who play lotteries meant that the potential user base was enormous; NOA would quickly become the largest network in Minnesota and any other states that followed suit. Once it was established, it would be able to expand to other on-line services, including games, shopping, and more. Otha Brown, a vice-president of Control Data, noted that the new system could increase lottery volume by 10 to 15 percent. It would also raise Nintendo's prominence in the network industry by 100 percent, and it would be sanctioned (and partly funded) by the state of Minnesota.

A limited test was announced. Ten thousand homes would be hooked up. Participants would pay a service charge of $10 a month for the software and modems that would allow them to play all the state's lottery games. After players set up accounts with the lottery, they would use the NES to select numbers, which would be stored electronically in the central lottery computer and in a file in the system at home. One analyst, Gary Arlen, of Arlen Communications, said, "I've been looking for the killer [on-line] service. This is the kind of thing that makes sense."

It did, but the lottery plan was too good to be true for Nintendo. The state's attorney general criticized the program. In spite of passwords and other safeguards built into the system, he feared it offered too much opportunity for children, some of them quite adept with computers, to gamble. In addition, state-promoted at-home gambling would set a bad example for kids. Although the Minnesota announcement had brought Control Data a flood of calls from other states interested in a Nintendo Network, the lottery director bowed to pressure from the state legislature and canceled the test. After all the excitement over Minnesota, the Nintendo Network was back where it had been: nowhere.

Probably *some* on-line network will be almost as prevalent as telephone lines early in the twenty-first century. Modems that can cheaply send larger amounts of information are forthcoming, and fiber-optic lines will carry more information reliably. The hardware and software behind the future networks will be powerful, easy to use, and inexpensive. Bill paying will be simple, less time-consuming, and cheaper than paying bills by hand, and volumes of paper will be saved. Shopping, travel planning, and countless other tasks will be done easily via networks. Entertainment possibilities, including phenomenal games, will be routinely available on networks. The question is not whether mass networks will arrive but whether there will be a *Nintendo* network.

By the end of 1992, it appeared that Nintendo would miss out on the enormous opportunity. If it did, the reason would be a lack of vision and commitment to it. Jerry Ruttenbur said the network never flew because Minoru Arakawa didn't support it in spite of a sound business plan and projections of huge profits. "It was Yamauchi's dream, not Arakawa's," Ruttenbur said. "Arakawa never bought into it."

In 1992, when reporters quizzed Arakawa about his vision of NOA's future, he was emphatic about only one thing: "No one is better at entertainment software than we are." But he seemed unsure about Nintendo's role in the larger consumer-electronics, multimedia, computer, and communications industries.

Hiroshi Yamauchi had built the Famicom with the potential to expand. He launched a network in Japan and formed technological alliances with Sony, Philips, and other companies. Because of this foresight, Nintendo was poised to be more than a video-game company, but Arakawa appeared to be uneasy in these leagues. The deals with Sony and Philips were pulling him into multimedia even though he didn't seem to understand exactly what it meant. With vague ideas about a network, Nintendo had the potential to become a communications company, but Arakawa was unable to set a clear direction. Uncomfortable in the world of keyboards and CD-ROM drives, he was in a position to transform Nintendo into a leader in the converging computer and consumer-electronics industries, but he seemed to have fallen into this position haphazardly. This was very different from his invasion of the U.S. video-game industry, which had been marked by a vision and tenacity.

Nintendo isn't the only company that doesn't know how to prepare for the future. Many companies—Sony is the best example—are going in as many directions as they can at once, in the hope that one of them may be the right one. Others, such as Philips, are betting on a single product. By the time there is a shakeout, Arakawa may have set NOA on a clearer course. Throughout 1992 it seemed that he was trying to find this new direction. In the meantime, Nintendo has been carried along by the momentum of the NES.

Arakawa's apparent lack of resolve was deceptive, however. "We'll let the others fight it out," he said in an interview. "Consumers will decide. We'll just continue to do what we do the best." By following this limited approach, Nintendo may be in the best possible position when it comes time to commit to a specific new direction. In the meantime, Nintendo was working in secret with many companies on the cutting edge of technology.

NOA had commissioned a study by Market Data Corporation, and its results sobered Arakawa and company. Top-level staffers met in the Zelda conference room, settling into cozy purple chairs around a massive table made up of smaller tables that had been pushed together. Representatives from Leo Burnett, the ad agency, and Nintendo's top brass, including Arakawa, Main, White, Shigeru Ota, and Gail Tilden, were all present. Explaining the results of study, the man who addressed the group uttered the words "slow growth" and "erosion" so many times that the mood turned dark. Nintendo had been used to meteoric growth and increasing excitement over its products. Now, the study said, many of its players were abandoning Nintendo for Sega, while others had lost their interest in video games altogether. The study of eight hundred people revealed some good news too. Of those families with systems in their homes, 90 percent actually play them. Also, more girls between the ages of six and fourteen were becoming primary players, and their level of satisfaction was intensifying.

But the fact was that the primary players Nintendo had counted on were beginning to age, and their satisfaction with the product was lessening. Kids still played a lot—in the sample, they spent an average of 2.3 hours a day playing Nintendo five days a week—but they were "more apathetic, less involved" with it. The worst news was that "it isn't as cool, not as much the thing to do."

The survey was by no means conclusive, but it did indicate that Nintendo needed new strategies in order to retain its dominant position. Segmentation was a key—the company would have to market specific products to specific groups. One group that would remain a relatively easy sell was younger kids, for whom Nintendo was still the coolest.

The respondents rated Nintendo and the competition in a number of areas. It scored highest in terms of fun, choice of games, and excitement, but the numbers were less dramatic than in earlier years, when it had an absolute lock on its target group. There were glimmers of hope, insofar as seven of ten people in the survey wanted to buy the Super NES: 30 percent "probably" would, and 42 percent "definitely" would. (Peter Main cracked from the sidelines, "Good! So no advertising.") A message also came in about Mario: 96 percent of the people surveyed knew Mario and 83 percent liked him ("Good! We'll exploit him more," Gail Tilden chimed in). On the other hand, there was great concern that a third of Nintendo's heavy users were becoming bored.

The Nintendo chiefs brainstormed and set a course designed to win back the ground they had lost. Younger kids were bombarded with Nintendo advertising for the SNES, and a new product was released in early 1992, the Super Scope 6, an awesome wireless infrared bazooka that came with six shooting games. Next, in fall, came "Mario Paint," a remarkable program that allowed kids to make their own animated cartoons set to music they created themselves (using preprogrammed "instruments," including a pig snort). Kids who wanted the Super Scope 6 or "Mario Paint," and there were many, would also need an SNES. It was important that Nintendo win the younger kids, since they were the ones who would determine much of the industry's immediate future. A changing of the guard was coming. A new group of kids was coming of age (that is, of *Nintendo* age) while older Nintendo users— the first wave of the Nintendo generation—were discovering Guns N' Roses, sports, books, and even girls.

Since Nintendo had been around before their time, young kids did not view the company as the cutting edge of culture, but it was still a significant part of their world. The frenzy over video-game playing that had arisen back when Nintendomania first possessed America's youth had calmed, and video games seemed to have taken a place among kids' (and many adults') lives as just another form of entertainment. But the market for video games continued to expand, and Nintendo still had the largest chunk. Sales were living up to NOA's expectations, and it had every reason to believe that they would continue to grow—and not only because of the ways the industry would expand, such as CD-ROM or multimedia networks. U.S. census figures showed that in 1989, the most recent figures available, there were already 2 million more children under age five than there had been in 1984, a total of 18.4 million. Many of them were the children of older parents, who traditionally spent more on their kids than did younger ones. These kids were entering Nintendo's clutches in the early 1990s.

On the other hand, competition and a sagging economy had shaken Nintendo, and its hold on the industry had been loosened. Nintendo now had to give mark-down money to retailers, for instance. Older games were put on sale, sometimes for as little as $15 or even $7 for games that had originally been priced at $40 to $50. There were no more opportunities for creating shortages because retailers weren't complaining that they couldn't get enough product, and Nintendo had eased its restrictions on licensees.

Industry analysts predicted that each year was going to be Nintendo's last good year. "Expect Atari all over again," one said in 1989. "This was their last year as a force," another said in 1990. In 1991: "A crash is imminent." But each year NOA did better and made more money. If the rate of growth wasn't as high it had been, the slowdown was viewed as a disaster. But the fact was that in spite of a maturing industry, Nintendo raked in higher profits than most companies in any industry in any country.

NOA's new 360,000-square-foot, robot-controlled distribution center, which became operational in 1991, was shipping 600,000 Nintendo systems and games directly to stores every day. When *Playthings* magazine announced the top-selling toys for 1991, Nintendo or Nintendo-related products held eight of the top twenty spots. The SNES was the top seller, the original NES was number three (behind the Genesis), and Game Boy was fifth (after Teenage Mutant Ninja Turtle action figures). Three Nintendo games— "Super Mario Bros. 3," "Teenage Mutant Ninja Turtles 2," and "Tetris"—were on the list, as was Galoob's Game Genie, which worked in conjunction with the NES.

Sales in 1992 of at least 2 million NES units (below Nintendo's projection of 4 million) was remarkable, given that the SNES was out and many analysts expected the 8-bit machine to disappear. Whether Nintendo would be a $10 billion-a-year business by the turn of the century was still a question, but in mid 1992 it remained "one of the strongest Japanese corporations," according to *Nihon Keisai Shimbun*. "Its performance continues to astound as the Japanese index of stocks weakens and other Japanese companies are affected by the worldwide recession."

Minoru Arakawa worked to make up for Nintendo's late entry into the 16-bit race with the sheer force of his marketing machine. At the January 1992 CES, Peter Main announced an advertising campaign that Sega couldn't hope to match: $60 million worth. He planned the launches of the Super Scope and such promising new SNES games as "The Legend of Zelda: A Link to the Past" (also known as "Zelda 3"), and "Mario Paint." Sales would also be sparked by the release of some great games by licensees, including Sunsoft's "Lemmings," Capcom's "Super Ghouls 'n Ghosts," and their incredibly violent but popular "Street Fighter 2," a home version of the most popular arcade game since "Pac-Man." Spectrum Holobyte was one of the newest Nintendo licensees, releasing games for the SNES, including a "Tetris" spin-off called "Wordtris." LucasArts released "Super Star Wars," with a plot that mirrored the movie and with the phenomenal John Williams score. At the summer CES in Chicago, Nintendo announced that seventy-five new Super NES games would be out by the end of the year.

In the weeks leading up to summer break, neighborhood playgrounds, perhaps the most reliable oracle for the industry, began humming with talk about the Super NES. It continued in schoolyards in September, and the buzz continued to build for the Christmas season. It seemed that Arakawa might well be on his way to taking back his commanding share of the video-game business, which in 1990 had been at least 85 percent. It had dropped in 1991 to 79 percent, but Nintendo projected it would be up to 82 percent (of $5.5 billion) in 1992. This 82 percent was a composite figure that represented 95 percent of the 8-bit market, 85 to 87 percent of the hand-held market, and 65 percent of the 16-bit market. (Sega disputes all these figures, but analysts feel they are only slightly exaggerated.)

Along with the new-product releases designed to jump-start sales, Peter Main and his team plotted a barrage of merchandising and promotions (including a massive tie-in with Pepsi). There was price slashing, a doubling of retailers' ability to earn credit in John Sakaley's merchandise-accrual fund, and new billing options, including September dating, which meant that some 1992 orders didn't have to be paid until the third quarter of the year. A groundswell was under way that was reminiscent of the first Nintendo invasion, quiet at first, but building: Son of Nintendomania. By the end of the year, just under 20 million SNES units had been sold throughout the world.

In spite of what happened in the marketplace, Nintendo was still in a precarious position due to the Atari Corp. lawsuit, which hung ominously over the company through the early months of 1992. John Kirby was staging an impressive defense, but juries were unpredictable, particularly in the midst of an American recession that some politicians, businessmen, and economists blamed on Japan. The implications of the suit were so enormous that neither Arakawa nor Lincoln could conceive of losing, yet there was speculation in the courtroom that the outcome was leaning in Atari's direction. For Nintendo it would be painful to part with all that money, but even more agonizing to have to give it to the Tramiels at Atari Corp. Worst of all, though, would be the precedent. Other lawsuits, including the Atari Games suit, and continuing investigations by the Federal Trade Commission, would be heavily influenced by the outcome of the trial. If Nintendo lost, it would confirm the view of the company that saw it as ruthless, immoral, and illegally profiting at the expense of American companies, and the vultures would descend.

The jury began deliberations during the last week in April, at the same moment that Nintendo was being scrutinized by the ownership committee of major league baseball.

Lincoln and Mino Arakawa had been invited to a Mariners baseball game as guests of Trip Hawkins, who had reserved the box of the team's owner, Jeff Smulyan.

A friend of Hawkins's father, Slade Gorton, Washington's U.S. senator, happened to poke his head into the box. Lincoln knew Gorton from Nintendo's lobbying efforts. The meeting was friendly as the men distractedly watched the Mariners lose. The team's owner, Smulyan, also stopped by to say hello.

Months later Smulyan announced that he was selling the Mariners; the most likely buyer planned to move the team to Florida. Slade Gorton and a group of local politicians and businessmen searched for ways to keep the Mariners in Seattle, and the major corporations in the area were approached. Microsoft's Bill Gates was one wealthy local businessman who declined.

Gorton, who had been dealing with Nintendo over the years, met with Arakawa and Lincoln because he felt they might be able to find investors in Japan who would take a majority interest in the club and agree to keep the Mariners in Seattle.

Arakawa asked the senator if he felt that Americans might react badly to the idea of Japanese ownership in America's favorite sport, but Gorton said no, not as long as people understood that the investment was passive and that the motivation was to keep the team in Seattle. Arakawa said he would see what he could do.

Arakawa told his father-in-law that he had been asked to help find investors in the baseball team. Yamauchi had no hobbies other than *go*. He had never played baseball nor had he ever seen a game. Nonetheless, he said that he would buy the team.

In December 1991, Lincoln called Slade Gorton to tell him this news. Yamauchi would put up the cash, and Arakawa would oversee the investment. Gorton arranged a meeting with other potential investors who were interested in the team. Between them an offer was put together, based on Yamauchi's majority share in the partnership. The group offered $100 million. In addition, they agreed to invest at least $25 million in the club, an amount of money that had the potential to drastically improve the perpetually losing team.

Yamauchi's partners would include a Microsoft executive named Chris Larson and John MaCaw, a businessman who ran MaCaw Cellular Communications. Smaller investors would be John Ellis, the chairman of Puget Power, and Frank Shrontz, the head of Boeing. Shrontz gave weight and credibility to anyone who was unimpressed by the other investors, and Ellis would actually run the operation. Larson, the second largest investor, would have sizable voting power, but the centerpiece of the offer was Hiroshi Yamauchi's 60 percent interest, for which he was putting up $75 million in cash.

This offer was conveyed to Smulyan, who was ecstatic. He would need certain approvals before he could accept it, but he knew he had to act quickly. According to his timetable, a deal had to go through by May if the team was to remain in town. Smulyan accepted the offer before the baseball commissioner and ownership committee approved it.

On January 25, 1992, commissioner Fay Vincent announced that the deal would *not* be approved. Baseball, he said, could not allow foreign ownership. The opinion made no sense—Canadians owned baseball teams—so he qualified his stance: baseball could not allow *non–North American* ownership. The deal would not be allowed even though Yamauchi had agreed to give an irrevocable proxy of his voting interest to Minoru Arakawa, a fifteen-year resident of the Seattle area.

Seattle's citizens and politicos were furious. The ownership requirement was ludicrous, they said. Yamauchi would keep the team in Seattle, but he couldn't have the team because he was Japanese, while another new owner, an American, planned to move the team to Florida.

This was just the beginning of the storm that raged in the United States and Japan. The baseball commissioner's attitude was decried as racist, or at least reactionary. Only over some people's dead bodies would a baseball team fall into the hands of the Japanese, who were buying up all that was dear to Americans—real estate, movie and record companies, golf courses.

Moreover, the offer came when the American economy was reeling in a worsening election-year recession and the Japanese were viewed as monsters who had lost the war but won the peace. The trade deficit with Japan was more than $40 billion when President Bush, flanked by the heads of the Big Three American car companies, headed to Tokyo to wrestle it down. Wittingly or not, the baseball commissioner placed Yamauchi's offer smack in the center of the trade issue, making Nintendo a lightning rod for America's hostility toward the Japanese. In Japan, banner headlines portrayed NOA as a victim of the latest round of Japan-bashing, and in the United States it was either a perpetrator or victim of the trade wars.

Frenetic lobbying commenced. Washington State lawmakers and businessmen called on the baseball commissioner to relent. Anti-Nintendo forces charged that NOA was itself a racist company, as evidenced by its hiring practices, and that it was involved in gambling, thus an inappropriate bedfellow for professional baseball. Bill Giles, who owned the Philadelphia Phillies, even worried aloud that the sale would be akin to selling baseball to "Middle East oil sheiks."

"Never in my life have I ever been discriminated against," Howard Lincoln volunteered. "The only thing that stopped this cold was that Yamauchi was Japanese. I realized what it must feel like to be Arakawa or any member of a minority, where you get ready to grab the ball and somebody pulls it out of your hands because you're black or Hispanic or Japanese."

Baseball's ten-member ownership committee would be the ones to recommend the sale's approval or disapproval to the group of all the owners. The committee was made up of a number of team owners, including Giles and George W. Bush, the Texas Rangers' owner and son of President Bush. Slade Gorton told a *Seattle Times* reporter that if Mr. Bush voted against the Yamauchi-led offer, "he would not be helping his father's reelection campaign at all." Certainly not in the state of Washington, nor in a growing number of places throughout the country where the Nintendo offer was in favor.

In Kyoto, Yamauchi was surprised when he heard that the baseball commissioner had rejected his offer. He didn't care about the team one way or the other—it was an investment and good PR—but he had never expected to be embroiled in such a public controversy. Yamauchi, who had always tried to keep a low profile, now appeared on the front page of *The New York Times* on February 7, and people across the United States were suddenly asking questions about him.

Allegations of racism made by some African-Americans and other minorities were quickly dismissed by Howard Lincoln, who sent the baseball commissioner a copy of Nintendo's affirmative-action policies and their results. But the charge that Nintendo was involved in gambling was more difficult to defuse. The Minnesota lottery deal that had fallen through was cited as proof of Nintendo's intention to become involved in gambling. Critics also noted that the company had its origins in gambling cards and that the Nintendo Network in Japan offered a horse-race betting service.

Responding to these charges, Howard Lincoln released a statement: "We don't have any interest in racehorses, casinos, sports-betting parlors, card rooms, racetracks, or other gambling activities," he said. "The implication that Nintendo or its executives are somehow mixed up with gambling is absurd."

The ownership committee continued to meet in secret, although there were sporadic leaks. Rumors that the deal was dead were followed by one that said that the committee had suggested to Nintendo that Yamauchi's investment would be approved if he lowered his interest to under 50 percent. Yamauchi reportedly refused to become a minority owner, causing some to wonder how sincere he had been in the first place. Yamauchi said he was stepping in to help Seattle—"to pay back" some of what Seattle and the United States had done for him. Then why wouldn't he be more flexible? The fact was that Yamauchi was unlikely to agree to a minority interest in anything.

As the deadline neared, popular support in the press and public for the Yamauchi-led offer seemed to influence the baseball owners, and it looked likely that the Nintendo chairman was going to get his team. He would have majority ownership, although he agreed to have less than a 50 percent vote. There was an additional reason that the deal was likely to go through, according to an associate of Senator Gorton: he and other politicians had intimated that baseball's tax exemptions might be reviewed by Congress if the commission blocked the deal.

The acquisition was formally approved by the club owners on July 1, 1992. In Seattle, if not in the rest of the country, Nintendo was viewed as a savior in a town that had been frustrated by the large company's lack of involvement in local society and philanthropy. Nintendo gained favorable attention not only on the business pages of newspapers around the country, but on the sports pages as well.

The Consumer Electronics Show in January 1991 was significant for Nintendo. The party NOA threw was more than a reward for employees, distributors, buyers, and licensees. It was a statement of might. It was not the first time that the company had taken over an enormous section of the CES with its own tent, choreographed production numbers (with dancing characters from popular games), and heavily attended press conferences. But headlines suggesting that Nintendo was in trouble despite record profits had persisted through the preceding Christmas. It was crucial for Nintendo to do something to counter the growing sentiment that the company might be about to stumble.

On the first day of the CES, Nintendo held an 8:00 A.M. press conference. Besides heralding new numbers, projections, and sales figures, Peter Main announced that NOA cartoon shows were reaching 40 million viewers a week, and discussed the Kool-Aid tie-in and several other promotions. He also announced the *Super Mario Bros.* movie. It was significant that Nintendo was moving into films, according to a journalist who asked if Nintendo would be going after more of Disney's turf—theme parks, in particular. Peter Main had no firm answer. He said the company had been approached and that he had left the possibility open. "The value of characters such as Mario are very strong. As we go down the road, you're going to see many applications."

At the Nintendo party that evening, after singer Kenny Loggins left the stage, Peter Main and Howard Lincoln stood in front of the microphone. Main had glitter in his hair, Lincoln wore a slicked-up punk spike surrounded by a glow-in-the-dark halo, and they both wore black silk Nintendo jackets. With Bruce Donaldson, head of sales, they gave away a Chevrolet Geo and several other prizes to winners of the Campus Challenge. Main then called Arakawa to the stage, and people who knew him almost choked on their beers and canapés. Arakawa's thick nest of hair was teased high on his head and it was shimmering with glitter. Wearing huge glasses made out of bent fluorescent tubes, he grinned from ear to ear. A DJ played "Allie Oop" and Arakawa, Main, Donaldson, and Lincoln sang and danced, swaying and singing, karaoke-style. It was a memorable night, an unabashed display that Nintendo's leaders felt they could do no wrong.

At the next CES, in June 1991, Nintendo was once again supposed to be about to topple, according to industry insiders. Nevertheless, Nintendo executives made higher sales predictions for the year than anyone expected. The company also announced its controversial alliance with Philips Electronics, and at the same time took on Sony.

This was the first CES at which Nintendo's living mascot wasn't present. Howard Phillips had left the company, a fact that NOA tried to minimize. Although Arakawa had continued to be amused by Phillips and to realize his value as Game Master, there were reports that Phillips was ruffling feathers. "He'd grown up," a colleague said. "That was the problem. He was getting big-headed and people didn't like it."

It was not the amicable parting that both sides pretended it was. Phillips knew things were changing; he admitted to his colleagues that he was growing bored. Still, it was a shock when he announced that he was leaving for a job at LucasArts. He wanted to pursue educational aspects of video games, he said, though he added, "It's hard to leave. It's like giving up a good pair of tennis shoes."

Tony Harman defended Phillips after he was gone. "A lot of people say they won't miss him, but I will. He gets engrossed in things and forgets about all the details. He annoys a lot of people and they get ticked off. But he's one of the few people who believed in the company with that childlike passion. We could argue about whatever happened and come up with a better solution. He made me hate life sometimes but I miss him."

Nintendo Power magazine bid Phillips farewell in its own way by having the cartoon character of Howard ride off into the sunset, leaving his partner, Nester, on his own.

Phillips, who had grown up at Nintendo, didn't last long at LucasArts. He had moved his family to a San Francisco Bay Area suburb, but he missed the seclusion of his retreat in Seattle. When he left LucasArts, he returned to his farm, set on a brook amid a forest of pine trees, and found a job at a Nintendo licensee.

At both 1992 Consumer Electronics shows Nintendo put up an invincible front in the face of renewed speculation about the company's looming demise. One sign of trouble was the announcement that the once sancrosanct policy of no paid advertising in *Nintendo Power* was being reversed. The SNES push seemed to be working, but the positive signs were a bit shaky, and Nintendo's position in the CD business partly hung in the balance. Arakawa admitted that Nintendo's success contributed to the problem. "We are still having growing pains," he said. "There are quite a few people who have grown with the company, but we also have had to hire people from outside. Then it becomes a different company. We don't have time to teach them how to run the business. Now we have fifteen hundred employees, and we have to catch up with them."

Arakawa admitted that he had stopped enjoying the company's annual picnic because it was too big, too impersonal. Gail Tilden felt the same way. "We've gotten so big we don't know everybody anymore," she said. (Arakawa's favorite picnic was still the one back in 1986, when he was the target in an egg toss.)

As part of his vision of Nintendo's future, Arakawa bought land where he planned to build a Nintendo "university," a center at which game designers from all over the world would come together to work. "It is where we will find and cultivate the geniuses of the future," he said. Another idea was being pursued with Disney: for fantastic Disneyland and Walt Disney World rides tied to the summer 1993 *Super Mario Bros.* movie. But these grandiose plans were on hold until the jury returned its decision on the Atari lawsuit.

Yoko Arakawa worried that her husband was taking his job too much to heart, working longer and longer days, juggling more and more people, attempting to balance an overwhelming amount. He kept a routine that he rarely varied. He was up at five for a bath, steam, and sauna and then a brisk walk with his dog, a black Lab named Pippin. From his study, he placed morning calls to Europe —the day was ending for Shigeru Ota and Ron Judy there—and prepared for the day's meetings. He headed to Redmond at eight or nine and returned anywhere between nine at night and three in the morning. The last few hours were often spent communicating with NCL in Japan (he talked to Yamauchi almost every night). Yoko waited up for him and kept dinner warm.

Yet no matter how hard he worked, Arakawa was never as compulsive and humorless as Hiroshi Yamauchi, and he always managed to find time for his children. In spite of the pressures at Nintendo, he took vacations with his family. They skied once a year and spent time in Hawaii at the houses he had built on the Big Island. Arakawa reserved one for his family each Christmas, and Howard Lincoln took another. They were there again at Easter 1992, just before the Atari case ended.

They returned with trepidation, as the mood in the courtroom had them expecting the worst.

Meanwhile, Yamauchi sat back and watched from Kyoto, planning for a variety of outcomes.

The live-in maids at the Yamauchi family home were gone. A day maid and a cook came each morning and left after supper. Another maid came a few times a week to clean. Michiko Yamauchi had run the household since her husband's grandmother, Tei, passed away in 1979, a year after Hiroshi's mother, Kimi, died. Michiko ran the home informally. Some modern furniture was brought in, and the teahouse was used as a storage closet. Sada had never invited people in, but Michiko enjoyed entertaining. The household lit up with parties and visits from friends and relatives.

Hiroshi Yamauchi avoided his wife's parties and usually kept to himself. On rare occasions he might reluctantly accompany Michiko to a family wedding or a relative's dinner party. He no longer caroused at the Gion. His only relaxation came from a tumbler of Scotch and a game of *go*. His most frequent partners were older masters, men like Yoshio Komeda, who owned the Yasaka Taxi Company and a foreign-car dealership. Otherwise, Yamauchi was preoccupied with Nintendo, which had been a wedge between him and his family for more than forty years. His three children had always found him an elusive, angry father given to unpredictable fits of anger. Age had mellowed him, but only slightly. He was a better grandparent than he had been a father, but although he enjoyed having the Arakawa children around, he never played with them. When his rare good moods were followed, unexpectedly and unprovoked, by outbursts, Michiko would sigh and tell her grandchildren not to worry. "Grandfather is tired," she would say. "He means nothing. It is his style."

Yamauchi's obsession with Nintendo had been rewarded as the company became one of the most successful in Japan's history, yet he observed year after year of record-breaking sales and profits without celebration. He said he didn't take seriously the findings of a book indicating that Nintendo was better run than Sony, Mitsubishi, and Toyota (and had far better productivity); he took this for granted. His family came to realize that none of the affirmation Yamauchi received over the years meant anything to him.

With his company facing potentially devastating threats from lawsuits and government investigations that could cripple his American operation, Yamauchi moved in another direction. His drive in Europe was explicitly designed to cushion the blow of the potential weakening of NOA. The plan was to make Nintendo, like Sony, so large throughout the world that the political "irregularities," as he called them, in other countries would have less impact than when Nintendo was dependent on NOA.

Although Yamauchi was attempting to do many things with his company, one analyst in Japan noted that his primary goal remained "Take the money and run." "Yamauchi has no ambition to be a lord of a video-game industry," said the analyst. "He wants to make an enormous amount of money and wield influence in Japanese society. No one should forget that."

Shinichi Todori, stern and rugged, seemed disdainful of Westerners, yet Yamauchi put him in charge of Nintendo's growing international business. For years Todori had overseen a number of distributors who sold Nintendo's products in Europe, Latin America, Australia, and select countries in Asia (although much of that continent was lost to counterfeiters).

Todori was in charge of Europe, the biggest push for expansion, where Nintendo was only one of many Japanese companies to view the changing face of the continent—the maturation of the Common Market and the opening of Eastern Europe—as an invitation. Although Nintendo was no stranger to European markets, the effort there lacked direction. In some countries, Nintendo had fallen behind Sega; in others, video games were less popular than computer games. European computer-gamers were younger than in Japan and America; they tended to be under sixteen, the same kids who were America's and Japan's biggest video-game customers. Sales of floppy-disk games far outpaced cartridge sales. In fiscal 1991, Electronic Arts earned 35 percent of its profits from Europe, mostly from computer games.

The NES had been available in Europe since the early 1980s. The company had had a distributor in the Scandinavian countries and one in Germany since the introduction of the Game & Watch. In 1987, Mattel agreed to distribute the NES in Italy and the United Kingdom. Mattel successfully distributed Nintendo products in Australia and New Zealand, but its European operation treated the NES as if it were a toy and expected quick, short-term profits. A Nintendo launch required a significant investment, which Mattel never made in those countries. As a result, by 1991 Nintendo had almost no presence in Italy, the only territory that Mattel retained.

The Scandinavian distributor released the NES at the end of 1986, but 1988 was the first year that any European companies had significant distribution. Since then, Nintendo had done best in Scandinavia and France, where 10 to 12 percent of the homes had the NES system.

France and the Benelux countries were Ron Judy's domain. He had left NOA when Peter Main took over as vice-president for marketing. "I'm an entrepreneur," he says. "By the end of 1986 my job was policing, overseeing, reading reports, writing memos." In 1983, Judy had set up distribution for Nintendo's coin-operated games in Europe and had fallen in love with Paris. He decided there was no reason Nintendo should not be successful there, and Arakawa allowed him an exclusive distributorship.

In April 1987, Judy took an apartment in Paris and opened up a small office above the Champs-Élysées. His profits from selling Nintendo products in France, Belgium, the Netherlands, and Luxembourg were huge. There was every indication that sales would expand through the later 1990s; he believed the NES could eventually end up in 25 percent of the homes in those countries. He also saw serious potential in the United Kingdom, where he took over the distribution after Nintendo's obligation to Mattel ended.

Judy's Nintendo International was a private company that maintained an independent relationship with NCL in Japan. Todori was his contact, although he also consulted with Minoru Arakawa about many decisions. Judy was a big customer of NCL, buying NES systems and games and paying in advance. He had the games translated, then commissioned NCL, which took its usual large profits, to manufacture them.

While the NES had modest sales in most European countries, Game Boy went through the roof. Almost three times more were sold in France its first year than were expected (1.4 million instead of 500,000). Game Boy took France by storm. Some schools forbade kids to bring it to school because children were playing, not working. *Club Nintendo* magazine, produced in France and distributed (in several languages) to Nintendo users in the countries under Judy's domain, had a circulation of 800,000, and was shooting for 1.5 million in 1992. The circulation through Europe would likely exceed *Nintendo Power*'s, since *Club Nintendo* was free. It was already Europe's most popular magazine for children.

As Europe opened up in the early 1990s, Yamauchi realized he needed to have a more organized international organization. From Todori he learned about the peculiarities of the markets in each country. He was warned that it was a mistake to attempt to market to Europe as a whole, since each country was so different, though a concerted strategy was certainly possible.

The discussions led to the decision to establish a Nintendo of Europe to direct the operation. It would allow autonomy for Nintendo's exclusive distributors in various countries, but would direct and assist them. Distribution was unique in Europe by country, as were the kinds and sizes of stores that carried Nintendo's product. In France, Judy sold crateloads of machines to "hypermarkets," single stores as large as five American Kmarts, as well as smaller orders to toy and appliance shops. Merchandising was developed specifically for the various sales locations, and Nintendo of Europe provided displays and other merchandising paraphernalia. Advertising sensibilities were also remarkably different, so distributors hired local agencies to come up with plans.

Yamauchi decided he could steal players from computers because the NES made it so much easier to play most kinds of video games. He was confident that Mario could be used to convert game players to the NES and ordered a push to establish him in Europe.

NOE headquarters were to be located in Germany, a vastly underexploited market. Shigeru Ota, the former bookkeeper who had been monitoring the sell-through of Nintendo products in America, was tapped to open the office, and was given ambitious goals. First, he was assigned to get Nintendo systems into 20 to 35 percent of German homes. Second, he would help coordinate operations in Europe, working with Ron Judy and other distributors. NCL bankrolled the German subsidiary and planned a $10 million introduction. It was slow going at first, although Game Boy sold as strongly in Germany as it had in France.

Although the 16-bit war had spread to Europe, Nintendo decided to sell as many NES units as possible before confusing the market with the Super system. Waiting meant that the Sega Genesis was gaining a large foothold, but Ron Judy was unconcerned. "We still need to develop the game players here. The 8-bit system is still selling very well; it's nowhere near its peak. The bigger the 8-bit base, the more potential customers for the 16-bit machines," he pointed out. Still, it was a risky posture. Nintendo could fight the 16-bit systems with marketing dollars, but it was unlikely that it could devastate its competitors with old technology.

In Kyoto, Todori oversaw the international operations and supplied goods to Judy's and Ota's companies as well as the other distributors in Scandinavia, Italy, and Spain. There is speculation that Todori might move to Europe in spite of his reluctance to do so because Yamauchi wants more control over the operations there and Todori's presence would accomplish this.

Ron Judy's ambitions for Europe exemplified the direction of the Nintendo push. "When we started NOA, I had the goal of being bigger than Nintendo of Japan. We did that. When I came to Europe, my goal was to be bigger than NOA, and we will be." Arakawa and Ron Judy raced to sell more Game Boys. Arakawa had had a head start, but by 1992 Judy was catching up, as he turned the Nintendo business in France, the U.K., and Benelux countries over to Bandai, and he returned to Seattle. Bandai, a Japanese toy company that had NOA and NCL licenses, planned a push in its countries equal to NOE's in Germany.

For the parent company, this was all good news, and Hiroshi Imanishi said that NCL's dependence on NOA was shifting. In

1992, NCL's shipments to Europe exceeded the shipments to the United States for the first time. Units worth $300 million went to Germany in NOE's first year, and in 1992 a total of 6 million Game Boys and 3.5 million NES units sold in Europe. The Super NES was introduced in some European countries, and Nintendo sold about 3.5 million in 1992. Floods of software, cautiously parceled out in Europe, would make up the difference in dollars. "Once the whole of Europe becomes excited, we will grow four times," Imanishi boasted. Then, he said, Nintendo would make a push in South America and more of Asia. "We don't find any difference in kids' feelings nationwide or worldwide," Imanishi says. "Our R&D is thinking about the world as a target for each of their products." It was not, he said, a question of if but of *when.* Marketing and production capabilities were all that held Nintendo back. "We would like to see Europe for now as the main target," he said. "When we have taken our rightful place there, the rest of the world is waiting."

Eastern Europe was a low priority, but it was not being written off. The Austrian Nintendo distributor introduced the NES and Game Boy into Hungary in 1991. Game Boy took off there, although the numbers were necessarily modest compared to France and Germany. Australia and New Zealand had rapidly expanding markets; in those countries Mattel, the local distributor, saw the potential of household penetration as high as in the U.S. Korea, Mexico, and Latin America were also all targeted for expanded operations.

Whether Todori headed to Europe or not, there was one strong indication that much of the operation would still be managed from Redmond. Although all contractual arrangements in Europe were between NCL and the different entities on the Continent, NOA's legal department reviewed the deals and advised NCL on them. With Howard Lincoln's assistance, Arakawa was overseeing more of the major decisions regarding NCL's policies, from promotions to marketing plans. "The United States is not the only important market I'm looking at," he said.

"We do not see borders in this business," Hiroshi Yamauchi said several years ago. "Some countries may be too poor or have heavy tariffs on imports, but with those exceptions we will go anywhere in the world. There are no borders."

If an individual week can foretell the future, the one that ended May 1, 1992, said volumes about the fate of Nintendo and many of America's and the world's electronics industries. On that Friday, after four days of deliberation, the jury in the Atari Corp. suit reached its verdict. The San Francisco courtroom was packed. Sam Tramiel seemed cocky. He was sure he had won.

There was an audible gasp in the courtroom when, against all expectations, the jury found Nintendo not guilty.

Howard Lincoln seemed more shocked than the many representatives of the other companies present in the room. When it sank in, he hugged John Kirby and dispatched an associate to get Arakawa on the phone. "Mino," Lincoln teased, in a sober voice, playing it out, letting his gloomy silence imply the bad news until he couldn't stand it any longer, "we won."

In his Redmond office, Arakawa was too shocked to say anything at first. Finally he whispered, "We did?"

They planned a celebration before hanging up. Then Lincoln composed himself before calling Japan, where it was 3:00 A.M. He woke up Hiroshi Imanishi's colleague Yasuhiro Minagawa, who spoke English, and Minagawa, as instructed, called Yamauchi "We won," he reported.

Yamauchi said only, "That is good."

The jury unanimously concluded that Nintendo's licensing program had not hurt Atari. They deadlocked on two other issues relating to monopolization and restraint of trade, but John Kirby was confident. "I think [it] will be disposed of very rapidly," he said. "We're not only happy, we're delighted."

This was an understatement. At NOA, thirty or so top staffers, from Peter Main and Bill White to game counselors, were dancing in the hallways. A week later Kirby's predictions came true: Judge Smith dropped the remaining charges against Nintendo. *The Wall Street Journal* reported, "The case may prove to be a landmark for Nintendo along the lines of IBM surviving a series of antitrust cases. . . . No other company may be able to summon the resources for another broad attack like Atari's. . . ." Sources reveal that Atari agreed not to appeal so that it would not have to pay a costs bill from Nintendo of between $500,000 and $1 million.

The decision came in a week that had begun with a two-part front-page article, also in *The Wall Street Journal*, on the troubles inside the major Japanese industries. The first installment, written by Jacob M. Schlesinger, was a look at the decline of the Japanese electronics industry, which had been considered indestructible. "It was only a few years ago that a group of Japanese companies with deep pockets and superior factories drove most American competitors out of products like television sets, radios and critical memory chips," Schlesinger wrote. "An American company invented the videocassette recorder, but it was Victor Co. of Japan Ltd., known as JVC, that mastered its mass-production and helped the Japanese dominate the field. Japanese companies threatened to do the same with computers, high-definition television and a range of other advanced products. Their triumphs created panic in the U.S.: Japan, people said, would soon control the world's technological fate."

However, Schlesinger reported, those Japanese companies were stumbling, "and not just because Japan and its overseas markets are in an economic slump." JVC was expecting its first annual loss since 1951. Fujitsu, Hitachi, Matsushita, Mitsubishi Electric, NEC, Sony, and Toshiba were all expecting to report that their combined profits for the year were down to half of what they had been the previous year. One of the reasons for the decline was a shift in the importance of hardware in favor of intellectual property, or software. "Notably, one of the few Japanese electronics companies currently doing well is Nintendo Co.," Schlesinger wrote.

Also on May 1, Minoru Arakawa announced a reduction in the price of the Super NES. Sega then lowered its prices, and Nintendo followed suit, until both systems were available for under $100. At this price neither company was making much on the hardware, but the idea was to blow each other out of the water. The Christmas season was coming, and analysts predicted that the

Super NES would outsell any other product. Arakawa's goal of dominance in the 16-bit video-game market seemed reachable, although Sega was putting up a good fight. If Nintendo did succeed, it would continue to be the biggest threat to the array of American, European, and Japanese companies that were fighting for the electronics industry of the future.

At Nintendo's annual shareholders' meeting in Japan in June 1992, Hiroshi Yamauchi made his report. In 1989, Nintendo had celebrated its hundredth birthday. Now, three years later, its second hundred years had begun with promise. Today, both Sony and Matsushita were reporting sharply weaker annual results whereas Nintendo's pretax profits were up 14 percent from fiscal 1991, to almost $1.25 billion, on sales of $4.3 billion. Nintendo had sold a total of 114.2 million hardware systems to 40 percent of all the homes in Japan, 33 percent in America, and a growing percentage in Europe. This included 64.2 million NES units, 32.2 million Game Boys, and 17.8 million Super systems. The households with all those systems were buying unprecedented numbers of software from Nintendo and its licensees.

The coming year, Yamauchi promised the shareholders, would show even greater sales and profits. Within the decade, Nintendo would double in size.

In late 1991, a reporter asked Yamauchi if he had chosen a successor. "I have not decided yet," he answered.

Yamauchi's son, Katsuhito, had entered the family business at Nintendo in Canada, but lasted only a year there. Language was a problem, and the business was not his forte. When the job didn't work out, Arakawa helped him form a company in Vancouver that would sell Nintendo products in World of Nintendo kiosks in shopping centers and malls.

Yamauchi's daughter Fujiko had married a doctor and lived in Japan, quite content to be separate from Nintendo. Other possible contenders for Yamauchi's position were his top officers, Shinichi Todori and Hiroshi Imanishi, though Imanishi himself admitted, "I am better to serve the president."

In the final analysis, there were few doubts that Yamauchi's successor would be Minoru Arakawa, although there could be no certainty. Even Howard Lincoln, Arakawa's biggest supporter, felt more comfortable with Yamauchi in charge. "As long as he's the boss, the better I'll sleep at night," he said. Arakawa acknowledged modestly, "Mr. Yamauchi is the best president for Nintendo."

If Arakawa were to take over, there would be changes. The company would become less of an autocracy. Arakawa relied on others for the decisions that Yamauchi controlled by intuition and fiat. One of Arakawa's assets was that he knew his limitations. "I do not have the product sense of Mr. Yamauchi," he said. But his critics acknowledged that he had an exceptional ability to hire talented people. Also, while Yamauchi was inaccessible and often terrifying, Arakawa was available and responsive. The difference worried some people, who felt that Arakawa was too nice, that he lacked the dynamism required to lead the company. Nintendo had not succeeded by being nice. Arakawa had proved that he was determined and effective, however, even if his style was unassuming. He had, after all, led the drive that brought in up to 60 percent of NCL's profits before sales in Europe took off.

Speculation about Yamauchi's retirement continued, but there was no reason to believe that he would do so before the late nineties. Although he was transferring some responsibility to Arakawa in Redmond, he wasn't ready to give up complete control to anyone, and he was as confident as he ever had been. After a slowdown in 1991, 1992 turned out to be the video-game industry's biggest year yet, with Nintendo regaining its considerable share. Nintendo was pulling ahead of Sega in the 16-bit race and, allied with Sony, it was ready to do battle for the multimedia market. Yamauchi had secured the power to control CD-ROM software with a licensing agreement that could be as restrictive as the one he had used to build Nintendo, and the most significant attempts to stop him had thus far failed. The threats to his dominance in the massive American market were evaporating one by one (in December 1992 the FTC announced that "no further action is warranted" in its Nintendo investigation and the case was closed), and the European invasion had begun. "No one can stop us," he said. In the U.S. Arakawa said, "I don't think the other companies understand that they do not have what Nintendo has. It is why we will grow. Maybe the growth will not always be as fast as it has been, but it will continue." Howard Lincoln added, "We are internationalizing. The momentum has begun. Anyone who chooses to underestimate us will lose." The tiger was poised, with renewed confidence, ready to take on any new threats.

EPILOGUE: FOREST
OF ILLUSIONS

Henk Rogers had founded a joint-venture company in Moscow in 1989 as a way to fund new projects and help his friend, Alexey Pajitnov. Rogers's joint venture worked with Russian companies such as Doka, the trading company that handled the rights to "Welltris," Pajitnov's quasi-three-dimensional version of "Tetris." A *New York Times* reviewer of "Welltris" noted the circuitous route by which the game reached America: " 'Welltris' was . . . created by a Soviet mathematician, licensed by Doka, a Soviet trading company, to Bullet-Proof Software, a Japanese company, which licensed it to Spectrum Holobyte, an American company . . . 'Welltris' is not a game, it is an obsession."

This joint venture allowed Pajitnov and his colleague, Vladimir Pokhilko, to retain rights to games they built and, potentially, to earn money for them. The two had other ideas for games, and for "human software," psychology-based programs that would be advanced implementations of ideas Pajitnov had explored in his Biographer program. Pokhilko, who wore a longer version of Lenin's beard over a flushed, round face, had an idea for a program that explored genetics in an on-screen, computer-generated aquarium. They called the project "Elfish," for electronic fish. When Rogers agreed to fund it, development commenced in a lab in Moscow.

The two Russians worked jointly on new games. One was a video game called "Hatris," another relative of "Tetris," with puzzle pieces that dropped from the sky. This time the pieces were hats (bowlers, top hats, fezzes, cowboy hats). They had to be moved quickly around and stacked in piles. Programmed into the game were two animated "helpers," who could dash over to pull hats off piles that were in trouble. The helpers were digitized pictures of Alexey and Vladimir.

Another game, for computers, called "Faces," was sold through a joint-venture company called ParaGraph to Spectrum Holobyte. "Faces" had scrambled puzzle pieces made of slices of the faces of famous scientists, painters, and politicians. Players unscrambled the falling pieces to create complete portraits. One could end up with a face that combined Mikhail Gorbachev's bald head, Margaret Thatcher's eyes, and Ronald Reagan's chin. "Faces" also had digitized images of cartoon characters and paintings (a winking Mona Lisa, for instance, and a Van Gogh self-portrait). Players could also scan pictures of themselves and insert them into the game. The *Los Angeles Times* reviewer lauded it: "[It doesn't] encourage you to destroy worlds, and errors do not result in gory deaths. . . . While playing, you do your part to improve U.S.-Soviet trade relations."

To promote "Hatris," Henk Rogers arranged U.S. visas and flights for Pajitnov and Pokhilko for the January 1991 Consumer Electronics Show in Las Vegas. The two men had mixed emotions about leaving Moscow just then. War seemed imminent in the Persian Gulf, and President Gorbachev's swing from Nobel Peace Prize–winning divinity to local despot was unsettling Muscovites. Still, they said good-bye to their families, though they wondered whether they should be taking them along.

Although by now Pajitnov was a veteran of international travel, Pokhilko had never been outside the U.S.S.R. The closest he had ever come to flying was in a flight simulator he once tried in Moscow, so he was on the edge of his seat as the Aeroflot jet winged over the Atlantic Ocean.

At the Excaliber in Las Vegas, which looked like the castle at Disneyland, the receptionist welcomed the Russians to "the largest hotel in the world, covering over half a million square feet." After they checked in, the hotel staff offered greetings of "Have a royal day."

Amid moats, hourly jousting matches, and restaurants with names like Lancelota Pasta, the two men had a day free to get over their jet lag. In his black corduroy pants, white polo shirt emblazoned with the BPS logo, and woven sandals, Pokhilko tried his hand at blackjack. He won fifty dollars and then lost a hundred. Pajitnov enjoyed the spectacle as he sipped a fresh cocktail.

The next day the two men met Henk Rogers at the BPS booth, which was inside Nintendo's grand tent, where the American was showing off the NES and Game Boy versions of "Hatris." The three of them sat on director's chairs in the booth, wearing hats— Pokhilko, a white top hat, Rogers, a black cowboy hat, and Pajitnov, a smart ruby-red beret.

By now Pajitnov had quit smoking, and he showed some heft as a result. He said he had quit for his health—America's antismoking sentiment had influenced him—and in any case, it was impossible to get cigarettes at home. When a friend asked him if the scarcity was part of a campaign to improve the health of the Soviet people, Pajitnov laughed heartily. "Yes," he said, "and the reason we don't get food is to keep us from getting fat."

When Pajitnov laughed, his eyes shone with a gentle bemusement, but there was also a trace of disquiet there. He seemed frail despite his bearlike physique. Perhaps it was all overwhelming for him.

Nintendo threw a party on the second night of the convention. LucasArts's party the night before had been impressive—the bar from *Star Wars* had been re-created, and Darth Vader posed for pictures—but NOA took over the entire convention center at Caesar's Palace for an unparalleled bash. Near the entrance were Michael Jackson and Cher lookalikes. Off to the side, hairdressers stood poised to spike, punk, and glitter up the hair of willing industry guests. Games soon followed, with prizes that included a

Chevy Geo. Arcade video games were set up throughout the ball-room, and a blow-up Mario that could have wafted in from the Macy's Thanksgiving Day parade floated overhead.

Vladimir Pokhilko stood at a table eating shrimp and musing with Alexey about America. (There were also tables full of tacos, pasta, pizza, barbecue, and desserts, not to mention half a dozen bars). The news from home was distressing. The Soviet Union was retrenching after the first years of glasnost. Gorbachev was tight-ening controls on the press as his troops marched into Lithuania in the first major crackdown since the Soviet experiment with "open-ness." Wolfing down succulent shrimp, Pokhilko was thinking of his family in Moscow. He said, "I think it would be very good if our families came here and we stayed one, maybe two years. We could see what happens from here, where we would be safe."

"Tetris" fans came by the Bullet-Proof booth throughout the next few days of the convention, and Pajitnov gave interviews and autographed copies of his games. Through it all, he admitted that he had a difficult time dealing with such attention. "At home I know none of this," he said. "This is like what a rock star might get."

Thereafter he and Pokhilko left Las Vegas for Seattle, where, determined to lose the weight he gained, he fasted for ten days, drinking only a little fruit juice. From there he flew to the Bay Area looking trim and fit, and he and Pokhilko worked at the offices of Spectrum Holobyte in Alameda. They stayed nearby, at the Spec-trum Holobyte company apartment, but ventured into San Fran-cisco on a few occasions. Pajitnov had memorized much of the city's layout from a jigsaw puzzle he had made with one of his sons in Moscow. The two Russians took an interest in popular Ameri-can pastimes, such as tennis and basketball. They accompanied a gang from Spectrum Holobyte to a Warriors basketball game at the Oakland Coliseum. Pajitnov delighted in learning new expressions in English; a favorite was "oops." The behavior at the basketball game stunned him. "Such exuberance," he marveled.

When word got out that Pajitnov was holed up in Alameda, he was courted by representatives from companies everywhere. He remained mistrustful; he had come to believe that the video-game business was full of dishonest people. He continued to work with

Gilman Louie and Phil Adam, but counted Henk Rogers as his best friend outside Russia. Most people flattered him when they saw his new creations, but Rogers was brutally frank. "This is shit," he would say if he disliked a game. Pajitnov would be taken aback, but he understood he could trust Rogers. He and Pokhilko agreed to sign a contract to work full time with BPS while maintaining working relationships with Spectrum Holobyte.

Together the two Russians made the decision to leave the Soviet Union forever. Henk Rogers arranged for their families to come to America and helped set them up with salary advances. Alexey's wife, Nina, had visited the United States once before. After she returned to Moscow, her friends asked her what she had thought of America and she couldn't answer; she could only cry. Pajitnov knew that she would work with him to begin a new life here, but he worried about his aged father, who he knew would never agree to leave Moscow.

Pajitnov found an apartment in a suburb of Seattle, near Bullet-Proof's office, which was also not far from Nintendo's headquarters. Pokhilko settled in Palo Alto, where he worked on Elfish and pursued his primary interest, psychology, utilizing the resources at Stanford University.

The Pajitnovs moved into their comfortable two-bedroom apartment, on the second floor of a condominium complex called The Lakes. The "lakes," Alexey said, laughing, were actually yellow and murky, with noisy ducks and geese. But his children were in heaven as they chased the birds and squirrels. Nina took her time in acclimating to life in the United States. Once the children were settled in school, she planned to look for work in academia; if nothing else, she could teach Russian. Alexey, meanwhile, worked on new games. His life, though bittersweet with memories and the thought that he had had to leave his father behind, was very good.

Ideas for games still came to Pajitnov at unexpected moments. One day he was driving through a thick coastal fog on a narrow road that wound through redwood trees as it ascended a mountain. At one point the road shot up so steeply that he had to shift down into first gear, hold on tight to the steering wheel, and gun it after a deep breath. At the approach to the steepest part of the hill, where the last of the fog vanished, the road appeared to swoop up in front of the windshield at a near perpendicular angle. Directly at this spot was a triangular road sign the color of French's mustard that warned, in the blackest of lettering: HILL. Under this word, in messy script, someone had added the only appropriate commentary: NO SHIT.

A short drive beyond the sign was the crest of the hill and a turn-off to the right. Farther on, varieties of conifers, with trunks more ashen and needles more indigo, seemed to huddle closer together. The air rushing in through the window was of a pine scent no air freshener would ever re-create. Pajitnov breathed deeply. This woodsy Eden was an open-and-shut case for his new home, the Pacific Northwest, where resplendent nature existed right in town, a mile away from a mall and a wonderful sushi bar.

Pajitnov slowed the car to a crawl and gazed out his window to see the pointed tops of the trees that arced slightly in an eastward breeze. Behind him, the fog looked like thick frosting on a big white birthday cake. He pulled the car onto the shoulder, switched off the engine, and got out. Craning his neck back, he observed how the evening sun, filtered through the clusters of needles, illuminated columns of dust in the air.

Burying his hands in the pockets of his gray parka, Pajitnov walked into the forest. It grew darker as the sun steadily sank and the trees there grew progressively taller and more closely grouped. It was quiet except for his footsteps on the hard-packed dirt and the swelling and receding of the evening forest noises.

Soon he noticed something extraordinary. From the molasses-colored bark of the trees were escaping, like bubbles from a bubble pipe, perfect tiny icosahedra, spinning on their axes. They drilled in wherever they landed, into the bark or the mulchy ground, creating small gorges the color and consistency of liquid mercury. The rivulets merged into an oval hollow that had the shimmer, smoothness, and luminosity of a mirror, though it produced no reflection. Pajitnov stared at the astonishing sight in a reverie, and it was well past dusk, and impending darkness had painted a maroon wash over the forest, when he briskly turned back toward his car.

Driving down the hill, Pajitnov was lost in thought, musing about the difference between the imagination's view of the forest at night and what was really there. "To re-create the emotions that accompany that experience," he said, "would be quite wonderful, I think." As he drove, he conjured up an image of a video game that could simulate not only the process of discovery but the emotions accompanying the experience of groping along in the darkness, which call one back to one's childhood. "Yes," he said. "It would be quite an interesting and wonderful game."

AFTERWORD
TO THE VINTAGE EDITION
THE REAL GAME
IS JUST BEGINNING

In an industry that was known for excess and hyperbole, the launch of one 1993 game nonetheless stood out. Children were breathless, and parents were worried, as Mortal Monday approached, the day the home version of "Mortal Kombat," an arcade game, would hit the stores. "Mortal Kombat" was the most inane, repetitive, and violent game yet—and it was wildly popular. It consisted of relentless button thumping by the player with the aim of pummelling one of the beefy, repulsive fighters. If it taught any lesson, it was that mercy was wrong; you battered your adversary into unconsciousness and then were instructed to "finish him off." This could be accomplished by pulling out his still-beating heart or tearing his skull off his body and then holding it up as a trophy, the spinal cord dangling.

Acclaim Entertainment, makers of the game, announced that four versions of "Mortal Kombat" would be released: for the Super Nintendo system and Game Boy but also for Sega's Genesis and Game Gear systems. Here was incontrovertible proof that Nintendo no longer wielded the clout it once had. Acclaim had been one of Nintendo's most prolific, successful, and loyal licensees but, like almost all of the other licensees, it had signed up with Sega, and there wasn't anything Nintendo could do about it.

Apparently, Nintendo had blown it. There were now about as many Sega players as Nintendo players, and Sega was considered the cooler brand. NCL was still making more money, much of it because of Game Boy, but it was struggling in the 16-bit race. If there was any doubt, the two months following Mortal Monday showed how much: three million "Mortal Kombat" games sold, breaking sales records for 16-bit games, and the more violent Sega version outsold the Nintendo version two to one.

Hiroshi Yamauchi seemed to be watching helplessly while his company lost more and more ground. First Sega beat him into the 16-bit market, which put him in the position of playing catch-up when the Super NES was launched eighteen months later. The Super NES was a good machine, with better graphics than the Genesis, but there was no software that compared to the games that had spirited the NES revolution. The biggest-selling game for the Super NES wasn't even made by Nintendo; it was Capcom's "Streetfighter 2," another violent fighting game, in which one could bite an opponent's skull and draw blood. There seemed little room in the high-stakes industry for imagination and risk, which had once been Nintendo's strongest suits. Aside from "Super Mario Paint," little truly new or inventive had been announced by the company in several years.

While Nintendo fought to retain its market share, Sega upped the ante again when it released its CD-ROM player. Just as it had when Sega launched the Genesis, Nintendo did . . . nothing. Its public posture was to dismiss the first wave of CD games as no big deal, but Sega sold a million of the units within the year.

Licensees' games sold a lot of Nintendo's and Sega's systems, with sports games selling steadily, games based on movies selling sporadically, and fighting games selling enormously. When Acclaim was ready to launch "Mortal Kombat," Nintendo, apprehensive about mounting concern over the violence in video games, insisted that the licensee release a watered-down version of the game for the Super NES. In spite of—or perhaps because of—less gruesome graphics, this was soundly trounced by Sega's bloodier version.

There were more defeats, such as the failure, after all the cautious planning, of the "Super Mario Bros." movie at a time when Sonic the Hedgehog cartoon shows took off. Further, Nintendo of America's marketing and p.r. wunderkind Bill White left the company over a disagreement about strategies—and wound up a couple of months later at Sega. But the worst news for Nintendo came with the 1993 "Q" survey. It showed that Super Mario was still more popular than Mickey Mouse, but that Sega's hedgehog was the most popular of all.

Although Nintendo was still raking in a fortune, Sega was enjoying the kind of growth NCL had experienced during its ascent. Sega Enterprises still earned a third less than NCL, but it had momentum that, analysts predicted, would increase its 43 percent share of the total U.S. home-video-game market to 54 percent in 1994. It had been unimaginable a year earlier that Nintendo would be in the number-two position.

Sega also seemed to be more ready for the future, as evidenced by a string of announcements about exciting new technologies, including virtual-reality goggles that hooked up to the Genesis; a full-body controller; a gamers' network called The Edge, developed with AT&T; and a television cable channel developed with Time Warner. After all of Yamauchi's boasts about a global communications network, the Edge, for on-line game playing, and the Sega Channel, over which "Sonic" and other Genesis games could be downloaded, was a bigger step forward than any Nintendo had made. In addition, Sega seemed poised to do battle against the next-generation machines that were coming from companies such as 3DO: with Hitachi and Microsoft, it was developing a 32-bit system that would be ready in Japan by 1994.

But the game was nowhere near over. First of all, both Sega and Nintendo were faced with smaller profit margins because the yen was so strong in the last half of 1993 (Nintendo's earnings forecast was slashed to $1.14 billion). Moreover, the promised wave of competition had begun to hit.

3DO's machine, manufactured and distributed by Matsushita

under the Panasonic label, was released in the middle of 1993. The next year AT&T and Sanyo released 3DO machines as well. Trip Hawkins had not only created an elegant 32-bit system, but had managed to sign up 350 top developers, so software would be plentiful. In spite of a prohibitive price tag, 3DO had a good chance of succeeding because its licensing requirements were less restrictive than Nintendo's and Sega's, and would allow small companies with innovative software ideas but little money a chance to get into the video-game business. The breadth of software that could result would be 3DO's strongest asset. The video games that had sold the most hardware in the past were not invented by engineers in the R&D labs of the big companies. "Tetris" was created by a mathematician, and "Super Mario Bros." by an artist. Now the Alexey Pajitnovs and Sigeru Miyamotos of the future were more likely to be able to do business with 3DO rather than Nintendo or Sega. For example, a young designer in San Francisco came up with a fantastic new kind of video game—an interactive comic book—and easily obtained a 3DO license but found it impossibly expensive to work with Nintendo or Sega.

Other well-backed competitors were on the way. New systems were coming from Commodore and Pioneer. A revised Sony Play Station, dubbed PS-X, would be out soon, and Atari Corp. was back again with a contender dubbed the Jaguar, an impressive 64-bit machine.

As with any new industry, a shakeout would come, but in the meantime, there was a lot of money to go around. Worldwide, the industry's sales for games alone in 1993 were more than $10 billion in spite of the global economic recession. In the U.S., video games once again beat out the movie business (by $400 million). But as video games were being redefined by the new technology, these figures were nothing compared to what would come. The multimedia and interactive television revolutions being speculated about were closer than ever, and they were growing out of the video-game industry.

Now, more quietly than ever, Nintendo prepared to make its move. As Sega made inroads into Nintendo's business and new companies were proliferating, Hiroshi Yamauchi changed directions. The results of a new series of negotiations were finally revealed in the fall of 1993. First came the Nintendo Gateway System, a multimedia system for captive audiences in planes, hotels, and ships. Travel information, shopping, and games would be offered through a Nintendo-controlled network. Then, following this announcement, a press conference was called at the Mark Hopkins Hotel in San Francisco. Representing Nintendo, Howard Lincoln was joined at the podium by representatives of a company that had never been in the video-game business, nor in any consumer business whatsoever.

At a time when the video-game companies were attempting to make their games look and feel more like movies, Silicon Graphics, a $1 billion California-based computer company, was *making* movies. The company's state-of-the-art computers had been used to create the special effects in *Jurassic Park, Terminator 2,* and many other feature films. Because of their chips, Silicon Graphics computers create more sophisticated special effects than any other computers in the world.

When Nintendo announced its press conference with Silicon Graphics, an executive of an electronics company that was entering the video-game business shook his head. "Just when we thought it was safe to get into the water," he sighed. The marriage of Silicon Graphics' highly advanced technology and a Nintendo home machine was made in video-game heaven. The companies were also a good match; Nintendo instantly had the best possible game technology in the world, and Silicon Graphics had a first-class entry into the consumer business.

At the press conference, Lincoln and Silicon Graphics' chief executives revealed that a machine was coming that would surpass any innovations the competition was contemplating. It would incorporate Silicon Graphics chips, including a 64-bit processor, that would allow "players to step into a real-time three-dimensional world." There would be unbelievably realistic graphics, high-fidelity sounds, and phenomenal processing speed. Priced at $250, it would be ready for 1995. It was reasonable to assume that this could be the best video-game machine ever seen, leap-frogging the 16-bit, 32-bit, and even other 64-bit systems. Conceivably it could trump everyone else.

If—the biggest if of all, of course—Nintendo and its licensees could come up with games and multimedia applications that would take advantage of such power, Hiroshi Yamauchi could not only return Nintendo to the position of undisputed leader of the video-game industry, but he could take it beyond the limited video-game business to be one of the strongest multimedia-entertainment companies of all. The industry had grown too big and complex for this to be a certainty, but the announcement proved that one thing had not changed. Howard Lincoln was correct: It would be a fatal mistake to underestimate Nintendo.

—David Sheff
January 1994

History of Nintendo

With a history that spans more than 100 years, Nintendo's electronic-game success has evolved from its roots as the manufacturer of stylish Japanese playing cards or hanafuda.

Nintendo expanded its line of playing cards in 1959 to include a set for children that featured Walt Disney characters.

Nintendo's Laser Clay Ranges, light-gun shooting galleries first opened in 1973, were hip spots but nearly led the company to collapse when Japan's economy tightened.

Driven by the budding calculator market, Nintendo launched a series of pocket games called "Game & Watch," which featured a game and digital clock in one diminutive unit.

New Game & Watch units became a great way to introduce new hand-held games as well as keeping Nintendo-created characters

As its characters became more popular, Nintendo builds on its worlds by releasing new products featuring derivative characters and story-lines, such as those inspired by Mario and Donkey Kong.

With the Japanese launch of the Family Computer (or "Famicom," as it was commonly called), Nintendo started a wave in home-entertainment that would make it a household name worldwide.

The Nintendo Entertainment System (shown here with some accessories, including "Robotic Operating Buddy" or "R.O.B.") revived the video-game market in North America after its release in 1985.

Game Boy, a portable gaming system, was introduced in 1989. While it wasn't a technological marvel, offering fairly crude monochromatic graphics, it has gone on to spark many spin-off products and has sold around 70 million units worldwide through 1998.

Nintendo followed up the Famicom/NES success in 1990 with a more powerful game system called Super Famicom. Shown is its North American counterpart, Super Nintendo Entertainment System (better known as Super NES or SNES), released in 1991.

In 1993, ten years after the original system came out in Japan, Nintendo released a sleeker, less-expensive version of the Family Computer. A similar updating of the NES was unveiled in North America.

The Super Game Boy cartridge, released in 1994, enables gamers to play Game Boy cartridges on their Super NES system.

ACKNOWLEDGMENTS

This is my opportunity to ackowledge the invaluable assistance I received in the preparation of this project. Thanks go to many people for making the new content in this updated edition shine.

First and foremost, to David Sheff, for creating a comprehensive and thoroughly entertaining trip through a wild era in the development of the interactive entertainment industry. I hope my new chapters are a good fit and will carry the torch a little further.

I wish to thank all of the people at Nintendo of America, who opened their company's doors in many ways and enabled us to gather the material for this book. Their names are too numerous to mention, but this list would include everyone from the company's head executives to the receptionists that greet visitors and callers to the Redmond complex in Washington.

The opportunity to undertake this project was an exciting one, and I have to thank President/Publisher Hal Halpin and Editorial Director Mike Davila for trusting me with it. The hard-working staff of CyberACTIVE Media Group, Inc. should be commended for their tireless efforts in getting this project from raw word-processing documents and photographs to what you are holding in your hands now.

A deep bow and a tip of the cap go to Chris Bieniek, Zach Meston, Dan Amrich, and Dave Karraker for their contribution to the factual accuracy of the new chapters. Similarly, much appreciation goes out to the many nameless and faceless creators of websites on the Internet—professional and amateur alike—whose various archival and historical accounts enable quick and valuable research.

Last but not least, my debts are emotional as well as intellectual. I'd be remiss not to recognize my family, who, year after year, share with me many new project—and the gripes, groans, and grins that go along with them. I couldn't possibly thank them enough for the equal doses of patience, interest, encouragement, and proud support they provide.

—Andy Eddy
January 1999

PREFACE

I became very good at predicting the Next Big Thing, not because of any supernatural or even analytical abilities, but because of an oracle in our house: my eldest son, Nicolas. Nicolas has uncanny instincts to glom onto many fads that eventually become sensations. His track record is impressive. Nick never cared much about many products that were promoted as the NBT but came and went after their fifteen minutes of fame, he became obsessed—months if not years before they caught on in the mainstream— with a succession of winners: Pound Puppies, My Little Ponys, Teenage Mutant Ninja Turtles, Transformers, and myriad revivals of toys based on X-Men, Star Wars, Batman, and Star Trek. He was quoting the Simpsons, Beavis and Butt-head, and X-Files at least a year before they were on the cover of national magazines, and he played (loudly!) CDs of bands, like Metallica, Beck, and Nirvana, before they became bestsellers. It's true that he briefly adopted the Kris Kross backwards-clothes thing, but no one's perfect.

Nick, when he was six or so, introduced me to Nintendo—the video-game system had barely arrived on American shores. I first noticed the word "Nintendo" in some of his odder conversations with his friends. Typically, he would say something like, "May I go to Alex's to play Nintendo? We're at the fourth level where there's a mini-boss who can't die unless you hit him with four bombs from the Princess." I'd shrug. "Fine," I'd say. "Be home for dinner."

I didn't pay much attention to Nintendo until that Christmas season, when Nick included a Nintendo Entertainment System in a painstakingly considered letter to Santa Claus. Since he had been "good" that year, or so Santa concluded, it was my job to shop for the items on the list. I dutifully headed to the toy store and found the bright box containing the NES and the original "Super Mario Bros." game along with other games on his list called "The Legend of Zelda" and "Metroid."

He was very happy on Christmas morning.

I fussed with the connectors and hooked the system up to the TV set. It was only then that I began to comprehend the importance of Nintendo. First, I watched the compulsive way that Nick and his friends played the games. If I presumed to interrupt them, only to ask if they wanted a snack, they'd grunt. Their conversations, when not playing, were about the games (in their new language, Nintendo-ese), and they increasingly seemed to live in an alternative universe, populated by the characters and facing challenges from the games.

But I didn't truly fathom the power of the NES until I myself sat down to play with him. Nick was an ace, whereas I was pathetic, instantly "dying" as I manipulated the tiny on-screen Mario or Link off a cliff. I did improve. Soon I learned to maneuver around cliffs and dodge fireballs, and I could hold my own against Nick and his friends, though I could never beat them. Then there were even nights when I found myself playing the NES by myself after Nick was sound asleep, dreaming (presumably) about Goombas and Princess Toadstool.

This wasn't just another toy. It was an obsession.

We spent more time playing the games, discussing strategies, inventing stories based on the characters, calling the Nintendo game counselors for tips, and postponing dinner until we beat a particularly challenging "villain." When, after countless hours, Nick and I actually won a game, we were elated and celebrated the victory with a trip to the toy store to buy more games, including the most addictive of all, "Tetris."

Were we, in our small corner of the globe, the only ones obsessed? The question was answered one morning when I picked up a newspaper. This was in 1990—George Bush was in the White House, Iraq had invaded Kuwait, *Goodfellas* was released, and Nick had turned eight—when the current "Q" poll results were reported in USA Today. The "Q" ratings, based on surveys, rank the most popular personalities and icons in the nation. That year's poll indicated that American children recognized Nintendo's star, Mario, more than Mickey Mouse. More than Mickey Mouse! Something big was under foot. Most adults were oblivious, but Nintendo had infiltrated the consciousness of our nation's children.

The insight caused me to do some reporting, and I came upon facts that would later "stun" *The Wall Street Journal*—readers of my book. Nintendo, the company that had appeared from nowhere (as far as the American computer and video-game industries were concerned), controlled 90 or so percent of America's $8 billion video-game industry. According to an article in a Japanese business journal, Nintendo had become Japan's number-one competitive company, surpassing Toyota, which had held the record for years. Nintendo was also Japan's most profitable company, with nets that were higher than IBM's, Microsoft's, Apple's, and all the movie studios combined. Other stunning facts were that Nintendo was using more than three percent of all the semiconductors made in Japan and its magazine, *Nintendo Power*, had become the largest subscription magazine in the U.S. for children and teenagers.

I saw a story here and convinced the editor of a new magazine, *Men's Life*, who sent me out on my first visits to Nintendo of America in Seattle and Nintendo Company, Ltd. in Kyoto, Japan. These were my first encounters with the masterminds of Nintendo's invasion into the U.S., European, and Japanese markets. In the course of researching the article,

I also interviewed numerous other people in the video-game industry, including its founders and representatives of its relatively new competitors such as Sega and Electronic Arts. I spent time with game designers, ad men, toy-store managers, teachers, and kids.

The article appeared in *Men's Life*, and I heard from a wide array of people. Some were impressed by the business acumen of a company that they had barely heard of. Some were horrified that another American industry was being lost to foreign competition. Some parents and educators worried about the impact of games on kids' health. Many insiders in the game industry had stories about Nintendo; the company was both praised and vilified. No one was neutral about Nintendo.

I realized that the article examined what was the tip of a business and social iceberg. Based on the responses and the voluminous material I'd gathered that was beyond the article's scope, I proposed a book about Nintendo and the burgeoning video-game industry. My editor at Random House was Joe Fox, who was widely revered in the publishing industry. But Fox, who had worked closely with Truman Capote and many other great writers, was no techie. In fact, he used a typewriter, not a computer, and played one game—chess. Nonetheless, Fox agreed that there was a book in the phenomena I'd discovered through Nick.

The following year and a half was devoted to researching and writing *Game Over*. When it was ready to go to press, Nintendo, after originally agreeing to allow Random House to use an image of Mario on the jacket, withdrew its permission. Phone calls between the game company and publishing house were many and furious, but Nintendo refused to reconsider. So the Random House art director hastily constructed an effective cover around a new image: a child staring hypnotically at a glowing TV monitor.

When the original edition of *Game Over* was published in 1993, the video-game industry was having a banner year of worldwide sales: $8 billion in games alone. Nintendo was still the industry's leader, though Sega was coming on strong. In addition, Sony and a new company called 3DO were about to enter the industry. Congressmen were beginning to reflect the latest threat to children: Nintendo's system.

The book received a great deal of positive attention. Besides gratifying reviews, I heard that it had become an assigned reading in a growing number of university business courses around the country. In addition, it was viewed as required reading for anyone entering the video-game industry. Parents and teachers, concerned about the impact of video games on children, packed my promotional appearances around the country and flooded the caller lines on radio talk shows. (Their concerns lead to a spin-off book, *A Parents' Guide to Video Games*, in which I addressed the social, psychological, and political issues raised by Nintendo and its competitors.)

In 1994, Vintage released the paperback edition of *Game Over* just as Nintendo launched the best game system the industry had seen up to that point (better than the newest Sega and Sony systems), the N64, with a phenomenal new "Super Mario" game. Nintendo no longer owned the industry, but it was pushing it forward, setting a new standard. Nicolas, who was 12 now, had grown tired of the earlier NES and its successor, the Super Nintendo Entertainment System, but was excited by the N64; he and I were once again having competitions, some lasting until late into the night. One thing hadn't changed: Nick still beat me mercilessly.

Parents know that a few years make a big difference in the lives of children. Nick is now 16. He's interested in music, movies, books, his school newspaper, water polo, and swimming. He's once again predicted a new sensation when he became obsessed with surfing, one of the country's fastest growing sports, at least on America's coastlines. But Nick hasn't played a video game in a couple years.

However, Nick has a much younger brother, Jasper, and sister, Daisy. I haven't yet determined whether they have their older sibling's ability to be an earlier adopter of the biggest trends of the future, though two of Daisy's first words—she's only two—were "Beanie Babies." Meanwhile, the other day, Jasper, who is four and a half, climbed up onto my lap and told me he's been thinking about starting a letter to Santa Claus in anticipation of next Christmas. I asked what he would include and he recited the list. "I want a pirate hat, sword, and hook," he said. "And Dad, what I really want is "Banjo-Kazooie." It's an awesome game." "Banjo-Kazooie," for the uninitiated, is a recent hit Nintendo game.

The publication of this special third edition of *Game Over* by

CyberACTIVE Media Group, Inc. arrives at an exciting time in the still-thriving video-game industry, which now has sales of $8 billion even as the Internet and new multimedia systems continue to push the limits of gaming. It also arrives at an interesting time in our home. At the end of '98, I burrowed into our storage closet and pulled out our dusty N64.

—*David Sheff*
January 1999

Chapter 1

RIDING THE HARDWARE SEE-SAW

In its short history, the video-game business has seen many hardware platforms hit the market. As technology has taken leaps and bounds, it has driven the release of more powerful game consoles, each generation providing better sound quality, more colorful graphics, and increasingly immersive gameplay.

Cartridge systems, like Atari's 2600 and Mattel's Intellivision, which started the home game craze, were cutting edge in the early '80s, but the evolution of gaming hardware in just a few years made them look like antiques compared to Nintendo's Nintendo Entertainment System and Sega's Master System of the late '80s.

As a result, the realism of game software has taken directly proportional strides forward. The stark black-and-white of "Pong" advanced to better animation and color in "Pac-Man," which in turn gave way to more vivid imagery in recent games, like Sega's "Sonic the Hedgehog" and Nintendo's "Super Mario" series.

However, as the software continues to improve, there has to be a platform

444 PRESS START TO CONTINUE

that the software plays on—one that is continually improved so as to enable the game designers to add more hues to their electronic paint brushes; to boost the quality of the animation away from jerky flip book-like scenes; to smooth vision that hopefully will make the players forget they're watching a game; to create lush soundscapes with realistic sound effects and chatter from the onscreen characters.

While it's the software that seems to impress—indeed, the software is generally what people *remember* and talk about—there needs to be a solid hardware platform. The problem in the gaming industry is that each generation of hardware lasts roughly five years or so. Indeed, as computer technology has made serious strides in the last 20 years, those leaps can be equally tracked in the game consoles that have hit the market, as the number of "bits" indicated of the system's brain power.

Nintendo's NES and Sega's Master System filled living rooms in the mid-to-late '80s. Those were followed by improved systems, like Sega's Genesis, NEC's TurboGrafx-16, and Nintendo's Super NES (SNES), in the early '90s. Subsequently, Sony's PlayStation, Nintendo's N64, and Sega's Saturn were the entries for the mid-to-late '90s. Indeed, in the fall of 1998, Sega has already introduced its "next-generation" Dreamcast platform in Japan, with the plan of bringing it out in North America in late 1999. Both Sony and Nintendo—being challenged by other companies—expect to have their new game console at the turn of the century.

With the rapid leaps made by the system hardware manufacturers, it's a real battle to keep the player interested in one brand and develop a unit that will adequately compete with similar products for the next half decade. Perhaps Nintendo's near monopoly with the NES during its run gave the company the feeling it was invincible, or maybe it was just a blend of cockiness and confidence that caused it to falter initially in its next battle with Sega: The "Genesis vs. Super NES" match started with Sega's having a two-year head start on the SNES' 1991 launch, creating a see-saw ride that would see both companies jockey for the lead spot in hardware sales.

Nintendo's Peter Main, executive vice president in charge of sales and marketing, saw it as a planned pace that ended up working out in Nintendo's favor. That the battles along the way didn't give Nintendo a market share to the extent that it had with the NES—a situation that is likely to never happen again in the gaming industry—doesn't matter as much to Main as the end result of the entire war.

"We hate being last on any count, share-wise or any-wise," Main states. "But our history speaks for itself in that we have been very reluctant to sacrifice what we saw as a necessary amount of time to get the product right—to get the software right or the hardware right. We were coming off the NES, we were enjoying a very strong share position there. The other guys decided to go to market early in an effort to try and jump-start the [next generation]. We really had to look at how much business was still to be done on the 8-bit [NES], but more importantly how much work was still required to produce the initial software that we needed for the 16-bit [SNES]."

"So would we have liked to come out of the starting gate together? That would have been nice, but I don't think at the end of the whole 60-month process where they got a jump—[TurboGrafx-16] fell right out of bed, and Sega fell out of bed in the back half of that program. We ended up in a dogfight that had two strong guys splitting the market 55/45 or something like that, when the dust had settled. We met our objectives in terms of how much we wanted to sell and financial rewards that went with that."

But the battles were hard-fought, and if the hardware is equated to a razor, the software is the "blade" that the gamer needs—with fresh blades required on a frequent basis. Sega had lined up some enticing blades for the Genesis, which really took off when "Sonic the Hedgehog" was introduced and Capcom made a version of its popular "Street Fighter" arcade game for Sega's system.

Sega also went on a marketing tear, hammering consumers with catchy phrases and the "Sega Scream" ending each commercial. An underlying message to gamers was Sega was the more grown-up system, while Nintendo's offerings remained more like a toy for youngsters.

"In 1992, 1993 and on into 1994, we had it taken to us," Peter Main recalls. "Sega said, 'Sega does what Nintendon't, and we've got more aggressive product: martial arts, sports and racing games. They don't...they're just relying on that kid's stuff. We're for older people, and they're for children'."

"I think the industry, first of all, has certainly gone a long way from where it was in the '80s, when it was seen as a bit of an outgrown of the arcades. We were trying to balance between the public's perception of what kind of person hung around the arcades with a very strong image of something that was close to motherhood: 'This is great. Your kids will love it. You can rest assured that this is solid, solid entertainment for them.' We still believe that that's a very important, key attribute of our mainline product. I think, however, as the industry did grow, as game players went from a median age of eight or nine to a median age of 16 or 17, that started to say things about the types of products and types of appeals that would be made to attract those players to it."

The market was also reshaped by the increased scrutiny that Nintendo faced from the government and the ensuing media coverage over its business practices. In the NES generation, Nintendo managed to keep a lot of the exciting software titles and developers locked up for itself through exclusive contracts, leaving Sega to rely mostly on its own games; unfortunately, it wasn't enough to compete with any real success. However, Sega's headstart with the Genesis coupled with Nintendo's backing off on some of the shrewd—some would say illegal—practices it used to create the NES' market ownership made for a whole new ballgame. It put Nintendo in the unfamiliar catch-up position.

Nintendo did catch up, however, at least to a mostly neck-and-neck position with Sega by the mid-'90s, on the strength of its software offerings. This came courtesy of a drive to create solid software that would bring players of all games back to the Nintendo camp; for example, the Ken Griffey baseball games, brought a slice of the older sports market. The company also instituted an image change that slid the company from its toy foundation into a grungier, more raw edge—dare it be said, "cool."

Sega tried to bridge into the next generation in mid-1994 with 32X. This system, instead of being a stand-alone console, piggybacked onto the Genesis with the promise of boosting the 16-bit system into a 32-bit powerhouse. While that may have been what was stated in the business plan, little software was released for it, and there was no vivid demonstration of the technology boost that the hardware was supposed to deliver. As a result, software developers never jumped on the 32X, and it disappeared from the scene not long after its introduction.

Sega immediately turned its efforts to concentrate on its next console, to which 32X was supposed to pave a smooth path. Though 32X didn't jazz the game community, getting a jump on the competition seemed to be Sega's formula for success, and the company was confident that the mid-1995 release of its 32-bit Sega Saturn would turn heads—and open wallets—as Genesis had after its 1989 introduction.

However, hot on Sega's heels was a company that was destined to draw attention. Though new to the game-hardware market—and not having made a killing in its previous software efforts for Sega and Nintendo systems—its name did bring some cachet in the consumer-electronics arena.

Remember Sony, and how its PlayStation was originally designed to have a cartridge port that would accept Super NES cartridges?

Sony followed through on its promise to bring out the PlayStation, but the Fall 1995 model turned out to be an entirely new game system from what had been initially planned. The previously described falling out between Sony and Nintendo did keep SNES compatibility out of the plans, prompting Sony to go with a proprietary system—and undoubtedly a mission to make Nintendo regret its decision not to follow through on the originally promised partnership.

Nintendo's Howard Lincoln looks back on the decision with mixed emotions crossing his face and tingeing his words. Though no one can tell what would have happened if Nintendo went with Sony instead of going with Philips. "At the time, we were doing what we did at Mr. Yamauchi's direction," Lincoln remembers, "and it all made a great deal of sense, and we had a heck of a lot of fun doing it. But we certainly knew we had deeply offended Sony. I don't think there was any question about that—it didn't take years to know that we'd upset them. At the time it seemed like the correct business decision."

In fact, Lincoln continues to espouse his belief that after the dust settled on this messy chapter in the Nintendo's long history, the end result served to fuel the industry's overall growth, which plays a positive role in the revenues of all involved.

"Quite frankly, I don't regret that they got into the video-game business, because I think that's actually been positive for the industry overall, and I think it's actually made Nintendo a better company by having that level of competition. We are doing a much better job in the marketplace. We're making better products. We're executing across the board better, in large part because we have a very tough and able competitor."

Peter Main concurs on that point, emphasizing the company's philosophy not to over-analyze what's been done, instead, working on upping the quality of the current and future product line. "Having a strong entrant like Sony has helped grow business to a bigger level," Main points out, "and as we chase our own fair share of it, we're realizing some good rewards out of that. We don't spend a lot of time thinking about deals from '92 or '93."

A head-to-head-to-head game-system competition between Sega, Sony, and Nintendo was shaping up, but it would have to wait until the Fall of 1996 in North America, as Nintendo again paced itself by continuing to sell its Super NES and prepping its Nintendo 64 for release. Prior to that, Nintendo took a mid-1995 spin with an innovative yet odd 3D concept system called Virtual Boy.

Virtual Boy was another brainchild of Gunpei Yokoi, who had invented the Game & Watch and Game Boy while at Nintendo. It was not, however, a natural evolution from other game systems before it, either in look or in gameplay. Sitting on a tabletop atop a couple of stylish legs, Virtual Boy looked like some futuristic microscope, high-tech binoculars or a mechanical creature out of *Star Wars*. To the player who looked through the visor, Virtual Boy provided an added depth with its simulated 3D in games that included boxing, tennis, and even a variation on "Tetris."

While making itself a unique niche in Nintendo's arsenal of products, Virtual Boy never caught on with consumers. Most would point to the red-on-black graphics, which were vivid yet simplistic, as a major negative to the system's success. Additionally, the fact that you had to press your face up against the visor to see what was going on meant it wasn't an easily shared activity.

When it came out, Virtual Boy was priced at about $180, hardly in the same affordable realm as the Game Boy. And, again, the lack of a user base contributed to light development commitments, which kept the software library very low, particularly in North America. With all that against it, Virtual Boy was soon selling at major clearance prices, exciting few others than collectors.

Nintendo certainly was disappointed by the failure of the system, but in retrospect Peter Main sees the bright side of it, not only for what the company learned from Virtual Boy's failure, but also as an opportunity to relish the competitive aspects that surfaced around it. "The good news is, in its launch in the Fall, it sold more product than Saturn did, which says something," Main mentions with a laugh. "I was sorry that certain key executives at Sega left before I was able to deliver the 50-pound bag of dog food to them, they having described this as the 'dog of the year.' We outsold them and we got out."

"Interesting idea, and somewhat of a focus on breakthrough technology consistent with our ongoing beliefs that you gotta do it differently, and the difference has to be a real perceived difference by the consumer. This one had the potential for doing that, but was never adequately supported in terms of full development effort. One thing that I guess we learned out of it, as Sega did before us, is that working on two brand new platforms simultaneously with a finite number of creative and competent game developers is a very, very tough proposition."

Having a product fail in the market isn't necessarily a life-or-death situation—particularly in Nintendo's case, with plenty of finances and other resources that enable testing the waters with a new product and not succeeding—but the Virtual Boy debacle was seen as the impetus for Gunpei Yokoi to leave his accomplishments at NCL behind. After departing NCL, Yokoi started the Kyoto, Japan-based Koto Co., which was about to launch his own toy company. Unfortunately, he was involved in a minor car accident in October 1997, and, upon leaving the vehicle to check the damage, was struck and killed by another car passing by.

Howard Lincoln details a chain of events that fell into place after the Virtual Boy bust, ending in a situation that left the company and many of the veterans in the industry heavyhearted. "It was a very sad situation," Lincoln says. "Mr. Yokoi spent a tremendous amount of time on that product, and I think it's fair to say that many of us here were very close to him—he was a wonderful guy. We thought that the product was strong enough to launch, so did NCL. And it just failed... it just didn't do well. We got out of it as quickly as we could, taking the minimum hits. So did NCL, and we moved on. I think Mr. Yokoi was terribly bothered by that experience, and I think it had a lot to do with him leaving NCL and going out and setting up his own business—and probably he'd still be alive today because he wouldn't have been on that highway if he had still been working [at NCL]."

Particularly as a result of what happened in the shadow of Virtual Boy, the system has become a collector's item for serious gamers. Peter Main reveals how the company relished what it accomplished with this unique product— then moved on with full attention to the next challenge.

"It was a little quirky in retrospect, but it was representative of a very genuine and honest attempt at...how you present real three-dimensional, real-time game play," Main states. "We got it out of our system and got everybody's focus on N64. I think in any business, the sooner you leave it behind and get on with the next one, the better you are. There are no lingering scars or callousness about it."

With Virtual Boy out of the way, Nintendo didn't have any distractions to keep it from concentrating on the upcoming release of Nintendo 64, its "next-generation" system originally titled Project Reality. Ahead the console's release in Japan in mid-June 1995 (it came out in the Fall of that year in North America), Nintendo boasted of the most amazing graphics, thanks to a partnership it had established with Silicon Graphics Inc. (SGI). A stalwart of the burgeoning Silicon Valley, SGI was best known for its powerful workstations, which had been used prominently in creating the groundbreaking special effects for such movies as *Jurassic Park* and *Terminator 2*.

Certainly with the support of the graphic power SGI would provide to N64, the system would be a blockbuster and carry Nintendo up in the market again. However, a lot of people felt that Nintendo's hype machine was at work to serve up a heavy dose of "FUD." A phenomenon that is quite common in the fast-paced and competitive computer market, FUD—which stands for "fear, uncertainty and doubt"—is employed by a company to get consumers to hold off purchasing a competitor's product in favor of your own. In this case, Nintendo hoped it could get gamers to wait for its N64 console rather than buying a PlayStation or Saturn. Even planting such a seed in a consumer's mind would be a plus.

By the time N64 hit store shelves, the battle had already defined itself pretty well in the North American market. Sega didn't follow up its Genesis' success with a hit in Saturn, rapidly seeing weak sales versus PlayStation. Sony had a good name brand, a strong product, and a very solid wave of gamers behind PlayStation—as well as the support of a lot of the established game-development community. The razor blades were in great supply.

To the confusion of the industry, Nintendo was going in somewhat the opposite direction. Instead of striving for as many licensees as it could get to create product for N64, it said it was establishing a "dream team" of developers, gradually announcing carefully selected development houses to build a software library. In reality, many within the development community expressed concern over Nintendo's choice of delivery: It was sticking with cartridges instead of the compact disc that Sony and Sega had both moved to with their respective systems. According to Minoru Arakawa, the choice to go with cartridges was a difficult one to make. "There are good and bad about the cartridge. The bad things are that production takes a long time and production cost is high. Good things are less counterfeiting and the memory size is smaller, so programming cost is cheaper, and response time is much quicker than CD."

While the N64 struggled a bit with the cartridge model, as the system had become more popular with consumers, game publishers have fewer fears that they can recoup the investment costs of creating an N64 game. Peter Main contends that the downside of cartridge use contributes to this somewhat, thanks in large part to beneficial drops in the cost of elements essential to cartridge construction, such as the memory chips that store the game's program code.

"[N64 suffered from the decision to go with a cartridge] a little bit, although I think that we've eliminated several of those negatives. Delivery time from a production standpoint was a little longer, and as we got into this generation we probably endured more out-of-stocks than we should have.

When you add to that a limited library, it undermined our ability to grow the system as quickly as we should have. We've shortened that dramatically with local assembly and we're now [using air freight to bring] the stuff in. [Another benefit] was that the continuing decline in the price of silicon has made it much more price-competitive. Today, we're still $10 apart [from the retail price of PlayStation games]. We think we can live with that given the unique nature of our product."

Early on in the product cycle, Nintendo gave indications that it would add a peripheral to the N64 that might improve the memory concerns. Named the 64DD, this removable disk drive would give users the ability to store files on a cartridge, much like a Zip disk drive found on many computers. However, more notably it would lend itself to supplementary elements, such as added levels released at a later date, as well as larger game software.

Plans for the 64DD, at least in North America, seemed to have shifted as cartridge prices have dropped and fears over software piracy come to the forefront. Lincoln is very firm in his concern for delivering a peripheral that may not be seen as a necessity, especially as games that were announced as slated for 64DD are appearing on cartridge. Outside of alternative and replacement controllers, peripherals for game systems have never been a big hit anyway, as Lincoln points out with Sega's experience not long ago. "I don't think [64DD will] see the light of day outside of Japan," Lincoln states. "A lot of it has to do with the cost of manufacturing. The need is not as great, and we haven't found compelling software. Putting it out there would just be foisting an accessory on the consumer, and the memory of 32X is still vivid [to Sega]."

The 64DD situation brings to the surface the realization that there are vast differences between the markets in the various locales that the game industry serves. These contrarieties have resulted in some products hitting big in one area while striking out in another. Lincoln notes that Nintendo's recognition of the market differences enables the company to avoid bad business decisions, and investing in outside development houses provides an opportunity to shape software as it's being created to fit a particular mold.

"In the beginning years of the video-game business," Lincoln says, "we were entirely dependent upon the software that was developed internally at

NCL. There really were no outside game developers that we could look to. That's been completely changed in the last few years with our investment in Rare, Left Field and others, where we have gone out and had games developed that are specific for the American and European market as opposed to the Japanese market. What we're now seeing is that there are games that are good for the Japanese market and may not be good for this market we can avoid—and vice versa—and we have sources for games for North America and Europe that we didn't have before."

Taking control of its own destiny was something that Nintendo of America has had an increasing ability to do, even from just a few years ago. With most Japanese subsidiaries, decisions are made in the main headquarters and passed along to the auxiliary firms to be carried out. Certainly with the North American market providing such a large source of revenue to Nintendo and its recognizing market differences, an evolution has taken place that enables NOA to hold the reins for its own benefit.

"[Nintendo of America has] always been more autonomous than most Japanese subsidiaries that I know about," Lincoln asserts. "Certainly, if I compare NOA in its degree of autonomy as compared to Japanese companies that are in the video-game business, there's always been a greater degree of autonomy even in the 8-bit and 16-bit days. If anything, that's grown even more—although what we've found interesting is while the autonomy is either the same or more, the relationship between our people and the NCL people has changed a lot. In the early years, the NCL people looked at us as a marketing subsidiary, but over the years, the relationship of the people all the way up and down the line at NOA is so strong and our work force so stable."

Undoubtedly, Nintendo has ample competition from the PlayStation, but one more platform that has to be considered in the mix is the personal computer. While more expensive than the home consoles, the PC also offers great flexibility and power. Augmenting the system hardware is easier with each new generation of computer, giving the ability to "plug-and-play" such add-ons as an enhanced sound card, 3D graphic accelerator boards, and other devices. Increased hard-drive capacity enables game developers to store large and lush graphics for quick recall by the software application, providing the player with elaborate worlds in which to journey.

With the advent of the Internet, PC users also have more opportunities for remote gaming. The person you're competing against—whether it's in checkers, chess, a first-person shooter, like "Quake," or an online adventure, like "Ultima Online"—could be across the street, across the continent, or on the other side of the world. And network games also enable larger counts for multiplayer action, with some accommodating hundreds or thousands of people at the same time.

While multiplayer gaming has been catching on with the home consoles, with a split-screen display giving a few individuals to keep track of their onscreen characters, there have been few attempts at console gaming over a network. Whether it's an issue of gamers racking up large phone bills or concern over the effect of lag—delays in the data flow over the phone line that results in poor synchronization and gameplay—remote gaming has been low on the priority list for system manufacturers.

Still, it's an aspect of gameplay that more people are interested in, if personal computers are any indication, and it's likely that it'll have to be at least an option if the consoles are going to stay competitive with the PC in the long run. Yes, game consoles have had the advantage of low price and ease-of-use, but players are going to want to break through the walls that currently keep them from battling with others as computer games enable them to do.

The lure of network gaming doesn't escape Peter Main's thinking, but again it requires an investment that few are willing to take, though it's understood that it's the next step. "Few have taken the jump, and they've found it to be very deep water," Main states. "We're all looking at that aggressively. For anyone to deny the reality of the web is to defy gravity. The web is real and it's going to be an ever more important part of everybody's life. Discovering the appropriate business model for this sector of the business world, I think, is on everyone's lips, and we're all working at it very aggressively."

By tackling remote gaming, Nintendo and other manufacturers also open the door to non-game services, such as online banking. It's a rocky road to travel, though, as Nintendo found out when it announced a lottery-at-home plan in the state of Minnesota that was designed to turn the NES into a gaming terminal. Not long after the agreement was announced, all parties seemed to back off from the plan, likely because of concerns that a youngster—already familiar with and unintimidated by the NES system—would be able to place a bet.

The first step, according to Howard Lincoln, is to pin down the game aspects and consider "value-added services" later. And Nintendo is most definitely keeping its eye on the PC market to see what is attracting players. "You have to assume that network gaming of some kind is one of the things that we recognize as part of the future of the video-game business," Lincoln says. "So I think it would be fair to say that we're spending a lot of time and money in areas like that. We haven't made any announcements and we don't have anything planned here in the near future, but we're very interested in what a company like [Electronic Arts] is doing with Origin. The fact of the matter is that, on N64, one of the things we've discovered is that this four-player capability is a huge benefit, whether it's sports games or "Goldeneye." We know that. Everything is pointing us in that direction."

It's interesting to note that, while Apple is said to have feared Nintendo as a threat early on—a conflict that hardly seemed likely with Apple's apparent unwillingness to pursue a gaming market with the Macintosh, something on which it capitulated in early 1999—Nintendo now looks over its shoulder in a similar fashion.

"I think we certainly have, I wouldn't call it fear, but certainly we regard as a serious competitor not only Sony but Microsoft," Lincoln notes. "Sega to a lesser degree."

Peter Main also notes that the "PC is a stronger competitor than it ever was," but believes that the console-game manufacturers are delivering better entertainment. However, he does yield to a strategy that takes into account a serious look at remote gaming.

As 1998 came to a close, Nintendo was self-assured that it was closing the gap between itself and Sony in the console wars, just as it had in the Genesis/SNES battle before. It released yet another "Zelda" game, an N64 adventure called "The Legend of Zelda: Ocarina of Time," as well as another installment in the *Star Wars* series called "Rogue Squadron" in collaboration with LucasArts.

No time to rest, however, because Sony is working within its own camp to release the most compelling software it can get for PlayStation; and Sega is just around the corner with the North American release of the Dreamcast. While no one discounts that anyone has a fair shot at being king of the hill in the next generation, Main is confident in thinking that Nintendo has the ability to continue producing the most desired razor blades. "We have to continue to be pragmatic in understanding both what has worked for us and what hasn't," Main states. "Nobody has a bigger set of history books than Nintendo in terms of the last 12 years in this industry. We know what has happened to those players that we nurtured when they were in their pre-teen years, and they're now young adults. We have a pretty good idea of the kind of products we can take back to that group, and continue to excite them. We can't lose sight of all that great learning while applying that to an environment that has also changed. I think our key will be, we know a lot about this business. We know what it takes to make great games. We're investing against expanded capabilities to do that, and that's going to be the key."

Regardless of what happens, one person who's sure to enjoy the fight is Howard Lincoln. Veteran of many boardroom battles as a lawyer, he seems to be entertained by the give and take the ongoing competition creates—especially when Nintendo comes out on top. As exacting and demanding as the business dealings might become, it's hard to lose sight of the fact that the focal point of all this is games—they may be *serious* games if you're sitting at a desk in the trenches, but they're games nonetheless.

"The amazing thing about it is, we go through all of these years, we're still making a hell of a lot of money, and we're still having a lot of fun," Lincoln offers with a big grin. "The beauty of this thing is that it is unlike any other business, it moves with the speed of light, it's fast-moving and the amounts of money that are involved are huge. Think about a business where you can sell two and a half million cartridges at $50 a crack in the space of 39 days or so. That's a fun business. That hasn't changed—if anything the numbers have just gotten bigger. And it's still better than selling cat food."

Chapter 2

CH-CH-CH-CHANGES

The early years of electronic gaming were equated—many would say it was a "stigma"—with being "child's play." It was a situation certainly perpetuated by Nintendo at its most popular, as its game characters looked more like Saturday-morning cartoons than anything else. However, the sophistication of today's software content and graphics indicates that a game console can hardly be considered just a toy relegated to the kids' playroom. In fact, closet gamers have come out in increasing numbers, spanning a wide cross-section of ages and gender.

As a result of this broadened appeal, there's a definite pursuit by the game companies to expand the user base by grabbing gamers from outside what's still a core teenage market. And the game market has found its way into diverse groups, ranging from college-aged players right up to various adult segments, both as a result of carefully crafted marketing pitches and inspiring new games.

Over the last decade, the interactive entertainment industry has gone through a rapid maturation, with hardware and software making amazing strides in graphics and realism. Nintendo has found itself caught up in the wake of that wave, which at first thought doesn't seem likely for a company "at the top of its game" for so long. A series of humbling experiences, however, prompted Nintendo to redefine and refine the way it appears to the "outside world," the products it offers and even how it does business.

Though mostly a "behind the curtain" change, Nintendo made a staggering announcement in early 1994 that Howard Lincoln, then Nintendo's senior vice president, would be promoted to the position of chairman, while Minoru Arakawa remained Nintendo's president. Many saw this as more than a shot across the bow by Hiroshi Yamauchi.

Lincoln notes that the move would balance the company. "It was agreed that Mr. Arakawa and I would be jointly responsible for the overall administration of Nintendo of America," he said. "That's essentially what has happened, but the operating VPs—marketing, administration and so on—report to Arakawa."

"I think some people probably think that this wasn't going to work, but Arakawa and I have been doing stuff together for so many years, it really wasn't that significant a change in terms of the personal relationship we have. There are a lot of things that he's good at; some things that I'm good at. It's been a very collegial experience."

After Nintendo lost so much ground to Sega in the 16-bit battle—a situation that probably did more than hit the company's bottom line, but was also a source of embarrassment to some of its executives—staying the course didn't seem to be an option. Things needed to be done that would shake up the company, and this change did more than hit the higher echelons of Nintendo's organizational chart. The impact went down through the ranks and was just one in a series of moves that Nintendo made to tilt the playing field back in its favor.

Some moves brought the company more in phase with an older audience through various marketing campaigns. For instance, to address the "professional" demographic, Nintendo's Game Boy has been advertised in in-flight magazines as an ideal diversion for a long airplane trip; indeed, it's not uncommon to see a business person take the portable game system out of an *attaché* case or purse to pass the miles. Game-related advertisements also appeared in close proximity to alcohol and car ads in "grown-up" magazines, such as *Wired* and *Playboy*.

Indeed, while Nintendo still has cartoony icons, like Mario, Luigi, and Donkey Kong, it has strived to show a different face in recent years. A good example is the change in Nintendo's internal code for its games, which it established during its run with the Nintendo Entertainment System. These consisted of a set of rules that its licensees had to adhere to when creating a new product; the rules determined what couldn't appear in the game. Specifically, the mandate dictated a prohibition on such taboos as swearing, bloodshed and religious symbols.

The code tamed many titles that crossed over from the arcade, personal computer and even the offerings of Nintendo's parent company in Japan— and at times caused confusion to players and frustration to game developers. For example, when id Software's "Wolfenstein 3-D" was ported from the PC market to the Super NES, Nintendo demanded changes from id, alterations that the Texas-based developer publicly vented over. This included taking the vicious guard dogs from the original computer rendition and turning them into giant mutant rats, as Nintendo didn't want a gamer to be put in a situation that required harming a creature resembling a family pet.

A far more prominent alteration came when Midway's "Mortal Kombat" was brought to Nintendo and Sega game systems in 1993. The Genesis version, for instance, featured all the arcade game's gore and bloodletting— including the ability to put a "fatality" or special finishing move onto the opposing player's vanquished character. While it included such gruesome displays as ripping the spine out of the foe's body, it also maintained the "integrity" of the original game for home play.

However, Nintendo made Acclaim Entertainment—the licensee that published "Mortal Kombat" in home formats—take the frequent red blood spray that accompanied blows and fatalities in the original arcade game and convert it into gray "sweat" to soften it in the version for Nintendo's systems. Though seemingly a minor point, many players were outraged with the differences between the game versions, and accusations of "censorship" and "sanitization" flew through messages posted on the Internet.

It's said that hindsight is 20/20, and Howard Lincoln admits to not only the scope of the backlash, but acknowledges the cumulative effect that it brought to Nintendo's future business practices. "We definitely used the 'Mortal Kombat' experience as a learning experience," he says. "Instead of getting a lot of letters back from parents praising our position, we got a huge amount of criticism—not only by gamers, but even by parents saying that we had set ourselves up to be censors. [They claimed] that wasn't our job, and that they were offended by that."

However, while Lincoln admits that "[Nintendo was] hurt in the market-place by 'Mortal Kombat', [it] continues to have internal guidelines." He further explains, "There are some things that we just have a hard time with, but in the main I don't think we'll put ourselves in another 'Mortal Kombat' situation, and we'll be much more selective on what battles we do fight."

It was yet one more time that Nintendo felt the heat of Sega's success with Genesis, which revealed the stark differences between the two companies with regards to how consumers in the marketplace perceived them. It also set in motion Nintendo's renewed push to close the gap between the compa-nies—a situation that Nintendo wasn't accustomed to being in, that of trail-ing in market share.

Among the more publicly apparent changes in Nintendo was that to its image, possibly as a counter to Sega's edgy ads for its Genesis system in the mid '90s, which caught on partially because of the "Sega Scream" that marked the end of each commercial. School kids started punctuating their playground banter with a sharply delivered "SE-GA!," a sign that Sega's marketing staff had created as a pop-culture catch phrase akin to "I want my MTV" and "Be a Pepper." It also was another element in the wave that Sega's Genesis system rode to successful sales, which has sold more than fifteen million units in North America.

The response from Nintendo, in the form of a drastically altered image, was hardly subtle, perhaps taking the company like a slingshot past where it needed to go to appeal to a "cooler" audience. For instance, one TV ad cam-paign—which featured a continuing theme, "Play it Loud"—included images of grungy skateboarders, one guy getting a tattoo, and featured the music of a popular "alternative" band called "Butthole Surfers," complete with bleeped-out lyrics.

This change seemed particularly strange for a company that had been so firmly rooted in more or less wholesome youth-oriented fare, but it certainly proved to be a move that started a catch-up phase for Nintendo. "We're going to do whatever we have to from a competitive standpoint to protect our brand and our image," Lincoln says, "and if we have to get a little bit more hip and a little bit different in our marketing approach, we'll do that."

The strategic redefinition of the company image certainly caught people's attention, but also brought about an ethical dichotomy with which Lincoln confesses Nintendo executives have been struggling to this day.

"I think it is true that we are concerned about the kind of products we're putting out there," he says, "and certainly we've got to live with ourselves. One of the nice things when we had 90% market share was we really could say very easily, 'We're not going to allow that stuff,' and we could look at ourselves and say we're really saints to do this. The reality was, since we had the high market share, we could dictate that."

"The flip side of it is that it's just like the movies: Do I think that people in Hollywood deliberately set out to make every single movie as crude as they can, as sexually explicit as they can, as violent as they can? I don't think that's true. But what causes that is the competition. I do believe that, as a more mature industry, we have a responsibility to the public to maybe pull our punches a little bit and not compete with each other to see who can make the crudest, most violent and most sexually explicit game. This has been a very difficult problem—I know for me. I've been scratching my head about it, and I don't have a solution."

While the games were getting wilder, and the companies were making efforts to one-up each other, it was also a time that the industry was thrown into the spotlight with pressure from two congressmen: Senator Joseph Lieberman (Connecticut) and Senator Herb Kohl (Wisconsin). The outrage expressed by the senators over such games as Digital Pictures' "Night Trap"—what many labeled as an interactive vampire movie parody—resulted in Congressional hearings at the end of 1993. The focus of the hearings put violent game content in the spotlight. The interactive industry was threatened with potential government intervention and regulation if it couldn't pull things together by itself.

The hearings swirled with accusations by the gaming industry of game violence being presented out of context, enabling the senators to illustrate their points better by highlighting the "problem." Among them, Tom Zito, president of Digital Pictures, wrote an editorial for the *Washington Post*—where he had been a reporter years before—in which he detailed how pulling "only 30 seconds of nightgown in 100 minutes of video" made "Night Trap" look like a menace. Tom Zito argued in a 1996 interview in *Digital Video* magazine, "We never received a single letter from anybody who'd bought ['Night Trap'] saying they thought there was anything awful about the product. But, having been a journalist, I can certainly see why and how media-savvy politicians in Washington would have an easy time taking 30 seconds out of context in 'Night Trap.' These guys actually said in the hearing that the object of the game was to stalk and kill women. That is clearly not the object of the game." He further argued, "You could take 20 seconds out of *Bambi* and make it seem like the most horrific product ever developed. And you could similarly say, 'How could the Walt Disney Company sell this horrible *Bambi* to children?'"

Zito states that he was passed over to testify because, in one senator's words, Zito wasn't in attendance. However, he claims that he was present and ready to give his statement. "I spoke up and announced my presence," Zito says in his *Washington Post* editorial, "but was ignored. Maybe there was a communication problem—some staffer failing to pass the word that I was present—but in any event, it certainly made for a better show if I didn't complicate the script by giving my side of the story."

Though it always had been assumed that Nintendo and Sega hated each other, the hearings brought their conflict right out into the open for everyone to plainly see. Nintendo's Howard Lincoln showed the differences in the versions of "Mortal Kombat" for each company's systems. Sega's Bill White—formerly a driving force behind Nintendo's NES marketing—showed that, while Nintendo may have pulled the blood and gore out of its "Mortal Kombat," it left in the violent fighting and put it out for all ages. Sega had already instituted a rating system for its own games, and "Mortal Kombat" carried a rating suggesting play for those over age 13.

Lincoln also blasted Sega's Justifier light gun as a violence-enabling device. White countered by bringing out Nintendo's Super Scope 6 bazooka-like light gun. Although Nintendo avoided using the term "gun" whenever it referred to SuperScope in promotional materials, its purpose couldn't be denied.

It wasn't the best showing for the game industry—as much as it tried to fight the political machine, it seemed apparent that something had to give. In fact, rather than face an unknown fate at the hands of government regulation, the game industry decided it had to band together, a move that seemed to pacify Congress, at least for the time being. In April of 1994, the Interactive Digital Software Association (IDSA) was created, a trade organization that promised self-regulation through a rating system. More importantly, it also brought about the cooperation of the very companies that had been at each other's throats, for the good of the whole industry.

It took a while for the key companies to find enough civility to work together on industry issues, because they were so used to working as single entities out for their respective benefits. Though it was difficult to vault those hurdles, starkly reminiscing about the experiences brings a grin to Lincoln's face. "I can remember the first time that all of the industry leaders got together at Consumer Electronics Show to talk about forming this trade association," Lincoln recalls. "That was a pretty competitive group, and there were a lot of knives on the table that day—we certainly had ours. This industry was not at peace in that people had some side agendas and all that."

Things have changed slightly. Lincoln observes, "Now, the industry leaders can all sit down in the same room. They're all just as competitive, but there's no personal animosity—[we can] talk on the telephone or have a board meeting at the IDSA in which we can actually decide things that five or ten years ago we never would have decided, so it's changed a lot. But, it's still fun to figure out some way of acing out Sony and Sega—it's just absolute joy to do that, and I'm sure they feel the same."

An independent group of judges, under the name of the Entertainment Software Ratings Board (ESRB), was established in the Fall of 1994 to scrutinize the products and administer ratings for prominent placement on game boxes and in advertising. Though it was hoped a rating system matching that

of the Motion Picture Association of America (MPAA) could be adopted—
it was rebuffed due to claims that the cinema ratings were protected by trademark law—the ESRB system featured a similar range from "Early
Childhood" to "Everyone" (formerly "Kids to Adults") up to "Adults Only."

The same senators, who previously slammed the industry, applauded the
creation of the rating system. Of course, it also opened a Pandora's Box of
sorts, enabling game publishers to bring out games that skirted the boundaries of violence, sexual innuendo and language—much as the movie industry releases a large percentage of R-rated movies each year. As an example,
a game developer, Gillian Bonner, starred in her company's game as scant-
ily clad heroine "Riana Rouge," and also put her "software" on display in a
Playboy pictorial.

While the end result of the Congressional hearings was a sense of unity
within the industry, one more step showed that the interactive entertainment
industry had grown enough to be taken seriously. The industry never had a
trade show it could call its own, instead being a large segment of the then
twice-a-year CES. In fact, while video- and computer-gaming companies
consistently occupied large chunks of square footage at the shows in the late
'80s and early '90s, they were still collectively treated like a redheaded
stepchild. At one show in Las Vegas, where the traditional facilities at the
Hilton Convention Center couldn't contain all the booths, the game industry
was put in temporary "pavilions." The "pavilions" ended up being a large
tent with little protection from the weather, leaving attendees scrambling to
protect their booths from a rainstorm.

"CES asked for it in spades. They were being heavily supported by the
video-game category, and we were at the back of the bus—we were falling
off the bus in some of the booths. I remember one in Las Vegas, I thought
the damn thing was going to blow over. They had it in a tent in a windstorm,
and nothing was level," Lincoln describes.

The rumblings about the shabby treatment of the gaming industry had
been a post-CES norm, and a movement was started to break away from
CES. In its place, a trade show would be created, which would be devoted
solely to interactive entertainment. Thus, the concept for the Electronic
Entertainment Expo or E3 was born, an annual gathering that launched in

1995, at the Los Angeles Convention Center.

However, before the E3 came to be the industry's official show, there was the issue of rallying the entire industry around it—indeed, embrace it—a move that required confidence in the E3 concept and a commitment to bail on CES.

Pat Ferrell, then president of Infotainment World—best known for publishing *GamePro* magazine for video-game enthusiasts—worked toward getting the E3 show rolling by pitching industry leaders. "Sega was one of the first to say, 'We'll back you on this'," he says.

Not surprisingly, the road to creating E3 was fraught with politics and fighting. According to Ferrell, Nintendo wanted to stay with CES, and a vote on the shift to E3 came up 7-2 among the companies on the IDSA board, with Nintendo and Microsoft being in the minority.

Though E3 could have gone on without all of the major players committing to attend, it wouldn't have had the impact of a unified show. While the thought of an E3 vs. CES fight each year wasn't too thrilling to Ferrell, he got a lot of support that kept him pushing. Ferrell states that Sega told him, "Let's go ahead with the show—fuck those guys."

In fact, Ferrell took that approach and ended up booking a number of possible dates and venues for the show. In the end, he saw the straw that broke the camel's back with the scheduling for the Los Angeles site, which fell on the same days that CES was slated to run in Philadelphia.

Ferrell fondly recalls the date—October 10, 1994—when E3 became a "real" show. He received a call from Gary Shapiro, president of the Consumer Electronics Manufacturers Association (CEMA)—CEMA produces and manages CES. Curiously, he picked up the phone to take the call from the foe with whom he'd been engaged in a high-stakes staring contest. "I said, 'Hi, Gary,' and [Shapiro] simply said, 'Hi, Pat. You win'," Ferrell remembers. It appeared that the fragmented support for CES and an upcoming payment due to the Philadelphia venue for space was too much, and CEMA blinked. Ferrell adds that within hours, he'd received calls from Nintendo and Microsoft, summoning him to meetings in Redmond, Washington, where coincidentally both companies are headquartered.

Ferrell recalls that the meeting with Nintendo wasn't all that happy.

Howard Lincoln and Peter Main, Nintendo Executive Vice President, didn't like the company's placement in the show halls—generally a hard-fought chess game of power and position among larger companies in an industry. "Howard and Peter roll out the map and ask, 'Where are we?' I told them [South Hall] was sold out, and they said, 'But we're Nintendo.'"

Ferrell didn't waver, while Lincoln and Main remained incredulous that the industry's show wouldn't bend to accommodate Nintendo's placement in the event's main South Hall. His unwillingness to move attendees around provided a form of scolding for the game giant's protracted holdout. While he indicates that the meeting continued to be tense, when the show finally took place, it satisfied Nintendo's execs. "When they saw the traffic," Ferrell says with a chuckle, "they loved it."

Lincoln concurs, not at all shy about Nintendo's reasons for originally preferring CES to E3. "That had to do with the fact that Nintendo had an experience having a very large market share, but it didn't have any experience in dealing as one-of-many participants in an industry. It took a little bit of time to get our mind around that kind of change, but once we did it—and we were so pleased with that first E3 show—we said, 'Why did we ever fight this?'"

E3 show was an instant success, as everyone in the game industry wanted to participate. In fact, at its first show, E3 organizers couldn't even accommodate all of the companies that wanted to exhibit their products, relegating some to meeting rooms and others unable to exhibit at all. Meanwhile, CES changed its schedule to a once-a-year Winter show.

The gaming industry rallied around E3, and the media took to this promised merger of Hollywood and Silicon Valley. There was extensive coverage of the glitzy presentations that comprised each show, filled with pounding sound systems, elaborate light shows, attention-getting song-and-dance performances and high-profile celebrity appearances.

The industry had cause to celebrate, not only because it provided unity and a professional appearance, but it also was seeing amazing growth in its revenues. In recent years, the industry has staked some stunning business figures—for instance, entertainment product sales in 1997 were around $5.1 billion, putting it in the same arena as the movie industry (approximately $7 billion in 1998 and nearly $6.4 billion in 1997).

In fact, it's just such statistics that the game industry uses to remind people how much interactive entertainment has become part of today's lifestyle. After Nintendo's release of "The Legend of Zelda: Ocarina of Time"—one of the most sought-after products for Christmas 1998—the company issued a press release stating that sales of the game in the last six weeks of 1998 had surpassed the box-office receipts for some of the season's top movies.

Outside of all of its business changes, Nintendo's evolution through the mid-'90s has also altered its mood and approach. It doesn't seem to be the same company that, in the late '80s and early '90s, was willing and able to throw its weight around without regard for who or what was knocked aside. While it still carries the business and financial clout from the days when it held a market share in unfathomable 90% range, it's far from playing the part of the "nine-hundred-pound gorilla" it had been. To Lincoln, the differences are rooted in the industry's attitude change toward Nintendo—and, particularly, fewer lawsuits. "The difference is, for a period between '85 and '93, certainly that period we had a market share that was very high and we had a lot of people who were very envious of that market share," Lincoln notes. "I think we had a policy of protecting ourselves in litigation, and we were exposed to a lot of litigation not of our making. It wasn't that we were suing everybody. We were getting sued by everybody, and we were very successful in defending ourselves, whether it was Alpex or the Atari litigation.... We have a good record when it comes to litigation. I don't think anything's changed in terms of our outlook or our approach to litigation. I think all that's changed is that the business has become more competitive so we don't stand out as the 900-pound gorilla anymore."

In a nutshell, Nintendo doesn't appear as the playground bully because it's not in an overpowering bullish position, given the somewhat humbling competition it's faced from Sega and Sony in the '90s. Instead of debilitating the company, the various battles seemed to drive Nintendo's executives into introspection, a renewed strategy for success and a lower—though still high—profile.

"If you look at what Nintendo has done, say between '93 and '98, we made a lot of money," Lincoln states. "It was very profitable, and we've had a lot of success. It's just that we haven't had people taking shots at us all the time, and because there are other players who have large market shares and there's more competition in the industry and because there's a trade association, like the IDSA—that has a lot to do with it. Now the whole perception is, instead of a Nintendo out there, there's a video-game industry and Nintendo's a part of it and Sega and Sony are part of it. It's just a different perception. We haven't changed."

No matter what Lincoln and the other higher-ups at Nintendo may say, things have changed in Redmond. Perhaps it's better to label this change as an evolution—a process that has kept Nintendo from becoming a dinosaur or a has-been. There's no doubt that Nintendo realizes that it's still a primary mover in video gaming, which in and of itself represents an amazing change from its roots as a playing card company.

Furthermore, while its success has subdued and buried some companies, it has also brought riches to countless others. Undoubtedly, those companies on the plus side were willing to swallow Nintendo's stiff business terms and approach so that they would be able to benefit from the association.

"We have a well deserved reputation, not only as being a tough competitor, but we're tough on ourselves," Lincoln states. "We are regarded as fierce competitors. We're regarded as tough people to deal with by our retail customers. All of that's still the same—I don't think there's been any change there. But the industry has grown and we're not standing out there like we used to. Some of the stuff that came out, this idea that it was really tough to deal with us and, God, we held these guys with our hands at their throat. All of these guys ended up driving out of our parking lot in Mercedes."

It could be said that Nintendo has been like the mythical goose that had laid the golden egg. Yet, Nintendo's modern portrayal as the goose has the creature far more cognizant of its worth, careful in how and when it doles out its bounty and deliberate in maintaining a major share of the profits in dealings with others.

That scenario frequently brought resentment from some corners, especially when the pressure to bring out a product came to bear. There have been lots of rumors and innuendo that Nintendo gave priority to its games over its licensees' wares, playing both sides against the middle. "You have all of these companies that are competing to get their products produced for Christmas…and you've got this guy that's got the 90% market share—and he can make you or break you," Lincoln says. "We never tried to screw [any third-party publisher] in terms of where their product was versus ours. There's always been a Chinese wall between our licensing department—the third-party publishing people—and the rest of the company, and that's always been the case."

"Still, we got a bad rap, but we actually haven't made any significant changes that I'm aware of. It's still the same people running the program, the same attitude, we still meet with these people, we still have a Chinese wall, prices have come down, costs have come down. I don't think we've really changed. I know I haven't changed. I don't think I was regarded as a bad guy—or Arakawa, for that matter—in the '80s."

A deeper excuse is given, one that Lincoln thinks might stand to rub people the wrong way—or at least not go toward creating a touchy-feely relationship with its licensees. "One of the things that's true [is] we don't socialize with these third-party publishers…we don't do a lot of going out and seeing them and all of that. We don't schmooze with them. That has something to do with it, but we're just not the schmoozing type."

However, another argument is that along the way the goose forgot how much it needed others to help produce its valuable eggs. It seemed to momentarily lose sight of the fact that such dealings come from a symbiotic relationship. Even though one may wield more power and leverage, leaning the partnership heavily in one direction, the fact of the matter is that there's a benefit to both parties through such a collaboration.

It's this situation that brings a pensive reaction from Lincoln. After having a ton of licensees during the NES era, which coat-tailed pretty well into the SNES days, there were far fewer who jumped onboard the N64 program. It may have come from a Nintendo initiative or the disdain over Nintendo's choice to stay with the more costly cartridge over the increasingly popular and less-expensive compact-disk media. Still others would say it's an industry reaction to years of what many would call abuse, but Lincoln downplays it to a large extent.

"I don't think that that was a mistake to go with a lesser number of third-party publishers," Lincoln surmises, "but I do believe we left some major third-party publishers...they weren't there when we needed them—I'm thinking of companies like Capcom, companies like that—as compared with the prior hardware platforms. Those that did get in, like Midway and Acclaim, have done extremely well and made some great games."

"I think that there are third-party companies that I wish we had been able...to get to support N64. I think that some of them were worried about a financial model that they didn't need to be really worried about if they could make a really good game, because the financial model didn't hurt Acclaim or Midway or anybody that's really gotten behind N64. I think it's fair to say that I was disappointed that some key third-party publishers didn't support us on N64, and I think part of it had to do with the cartridge model. You can quarrel with it and say, 'Well, you're nuts. The model is fine.' The reality is that they didn't support it, and so you have to be disappointed and say, 'Well, we got to make sure that that doesn't happen again'."

How to prevent that from happening again is a question that seems hard to answer. To come up with a solution, you have to understand what is the cause of the initial problem. Lincoln indicates that it's not obvious, especially in comparison to its competition. "I don't think it's a relationship thing," he says. "It's more because I don't really think that Sony has done that great a job in terms of relationships with third-party publishers. But they sure as heck have gotten a lot of third-party publishers to support PlayStation, and there's no question in my mind that that's been a positive factor in their success."

In addition to Capcom, a major loss to the Nintendo software library came from the defection of Square Soft from the Nintendo camp to Sony's side of the battlefield. Square's commitment to PlayStation culminated in the highly touted release of "Final Fantasy VII" in 1998. The game is the latest entry in a series that always draws large throngs of game buyers, particularly in Japan where new "Final Fantasy" releases would be accompanied by lines of buyers waiting for stores to open.

"Final Fantasy" games went hand-in-hand with Nintendo platforms, so the announcement of Square's departure acted like a see-saw in effecting a rise in interest for the PlayStation system, while bringing about a dip in N64.

Lincoln sheepishly reveals his disappointment that Nintendo let Square slip through the cracks—but tempers it with optimism and a steely determination not to allow this kind of thing to happen again. "Yeah, you know...(chuckle) I'm sure it's bothered Mr. Yamauchi and Mr. Arakawa as well. But those things happen in business. You just have to move on. I think that had more of an impact in Japan in terms of 'Final Fantasy' than it did in here. There's always next year. The door is open across the board. I think it's really important that we maintain relations with all of these people."

Lincoln chalks up much of the difference between the video-game industry five years ago and today to parity as more companies have come in and taken a chunk of market share. Though in 1993, you wouldn't likely have Nintendo willing to share pieces of the pie, the present-day Nintendo concedes that the pie is being split with others, and Nintendo is simply looking for its "fair share." As a result, fewer people are setting their crosshairs on Nintendo, at least not as much as they had in the past.

"People don't just like to have one company in a dominant position—even if you're really nice about it, you still tend to piss people off, and they tend to say nasty things about you," Lincoln says. "And then, when they say nasty things, the Federal Government starts to look. And everybody in the media gets very critical. Whereas, in a more competitive environment, you just don't see it."

Though its success has been tempered by the competition from Sony and Sega over the past few years, Nintendo is still a force that few will turn their backs on. In 1994, Howard Lincoln was asked during a visit by group of gaming and mainstream press how Nintendo would compete against Sony's and Sega's respective hardware. His answer belied the company's confident position: "We have three billion dollars in the bank and no debt."

Reminded of that in 1998, Lincoln grinned and noted that the statement has pretty much remained true—though Arakawa puts the money figure at "about five or six billion today." The confidence continues, and he's philosophical about the sorties his company has found itself engaged in.

"The interesting thing is, a lot of people still think, 'Nintendo really got clobbered by Sega, and now they're getting clobbered by Sony'," Lincoln stated. "First of all, we didn't get clobbered by Sega. We got clobbered by Sega for a short period of time, but we came back and won the war. In the case of Sony, they were in the market for a year before we were. N64 did tremendous in '96 and at the end of '97, the market shares for hardware—PlayStation vs. N64—was about 50/50. But the reality was we had been leading all of 1997, while they got ahead of us at Christmas. They got ahead of us because they had good games, and we had some games, like 'Conker,' 'Banjo-Kazooie,' and 'Ken Griffey,' that were delayed. Now we go into '98 and we're behind, while they're ahead something like 60/40 in their favor."

"I do think that there is a difference—a big difference—between competing with Sony and competing with Sega. While I think we were able to beat Sega with good 16-bit games, like 'Donkey Kong Country,' it also did a lot of things that didn't have anything to do with us. They simply did a lot of things wrong, and they paid for it. Financially, they had some significant losses in the United States and Europe. Sony, on the other hand, is not going to bankrupt themselves and shoot themselves in both feet like Sega—they are a much more sophisticated competition and a multinational company with 50 years of experience dealing in markets of the world. That's a different kind of competition."

Though seen as trailing Sony in the race of PlayStation vs. Nintendo 64, Nintendo is hardly complacent—the hardware market in the game industry is one that will rush by if it's not carefully planned out ahead of time. And it's a fickle consumer base. Most would point to Sega Saturn as a prime example of wasted momentum and opportunities, openings that Sony and Nintendo took advantage of to blow by Sega.

The question remains, what is the future growth path for the industry, and what will Nintendo's role be in it? Though understandably closed-mouthed at a time before a new system has been officially announced, Nintendo executives would only acknowledge that it was working behind closed doors to develop its next-generation machine, the system that will challenge Sega's Dreamcast and the console that Sony is working on to replace its PlayStation. Arakawa mirrors the company's self-assurance that it won't rush a product out just because Dreamcast will beat it to market (slated for Fall 1999)—though in the backs of their minds, Nintendo's leaders surely remember the head start Genesis received by preceding Super NES.

"We are working on the next-generation hardware, as well as many other projects, and our target is 2000," Arakawa reveals. "But until we have the hardware and until we have the software and until we feel this is much better than the existing system, we will not introduce it."

It's hard to believe that Nintendo can still claim that it's the same after comparing visits to the company's headquarters four years apart. Though little has changed in its corporate structure, there seems to be a distinct difference in the mood among the staff members. Everyone from the company's executives down to the "worker bees" had a more casual demeanor, which translated to their dealings with others. It's certainly a reflection of a "new and improved" Nintendo that's neither a company with a bull's-eye on its back nor thumbing its nose at others from the top of the hill.

Perhaps Peter Main's commentary about the selling philosophy at the company best reveals Nintendo's mindset as it heads into the millenium. It's devoid of the cutthroat, take-no-prisoners tone that Nintendo was seen as having, yet it's calmly confident in discussing competition that seems to continue to be robust.

"We're obviously not selling a necessity of life," Main explains. "It's a product that we hope appeals to all ages, both sexes and is adequately compelling to have people choose a Nintendo hardware system and a piece of software over one of those other discretionary things they do: attending a concert, going to a movie, buying a new pair of high-tops. We're competing with a lot of people, but I do believe that, as we enter this new millennium, we're dealing with an audience that is more conditioned to interactive experiences. We know a lot about that, and we know a lot about it in the sense of what constitutes great game-based interactive entertainment. Anyone who was hoping to steal dollars from that should be fearful about Nintendo's role because we don't plan to get out of this business."

The company shows no sign of getting out of the business after more than a century of providing entertainment, but Howard Lincoln does hint of further changes in the organizational chart—this time less due to improving the business structure and more as a passing of the torch. His reminiscing over

16 years at Nintendo unveils sentimentality about the industry that few would have expected from the lawyer who took the company into nasty legal tangles with such entities as MCA and the U.S. Government and brought it out smelling rosy.

"I've been very, very lucky, and I've had a lot of fun in the process," Lincoln says. "I've been lucky in the sense that I've met a guy like Arakawa. The two of us are very compatible, that we've both had a lot of luck and a lot of success. I was very lucky in meeting a guy like Yamauchi, who I have tremendous admiration for. This has been a lot of fun."

"I'm not going to be here forever, that's for sure. I think there's a time for everything, there's a time for all seasons or whatever the expression is. I don't know how much longer I'll be in the video-game business or how much longer I'll be working in the daily at Nintendo. I'll be 59 [in February 1999]. Probably there'll be some changes in the next few years, but I really have loved everything I've done here. It's been so much fun. It's kind of like the guys who play for the Mariners who say, 'We should pay you to play baseball.' It's the same in the video-game business—it's that much fun. I'm sure I'll miss that part of it."

SELECTED BIBLIOGRAPHY

Adlum, Ed, "The Replay Years," *Replay*, November 1985: 121-172.

Brand, Stewart, *The Media Lab* (New York: Penguin, 1987).

Burstein, Daniel, *Yen! Japan's New Financial Empire and Its Threat to America* (New York: Ballantine, 1988).

Bylinsky, Gene, "The Marvels of Virtual Reality," *Fortune*, June 3, 1991: 146-150.

Carse, James P., *Finite and Infinite Games* (New York: Ballantine, 1987).

Choate, Pat, *Agents of Influence* (New York: Knopf, 1990).

Cohen, Scott, *Zap! The Rise and Fall of Atari* (New York: McGraw-Hill, 1984).

Corr, O. Casey, and Jimi Lott, "Nintendo Power: Bigger Than Mickey Mouse," *Pacific Magazine*, December 16, 1990: 10-14.

Crichton, Michael, *Rising Sun* (New York: Knopf, 1992).

Dertouzos, Michael L., Richard K. Lester, and Robert M. Solow, "Made in America: Regaining the Productive Edge," *The Report of the MIT Commissiono on Industrial Productivity* (Cambridge, Mass.: MIT Press, 1989).

Ditlea, Steve, "Inside Artificial Reality," *PC/Computing*, November 1989: 91-102.

Dvorak, John C., "Is the CD-ROM Rip-off About to End? Ask Nintendo," *PC/Computing*, September 1991: 66.

Fuller, Buckminster, *Education Automation* (Carbondale, Ill.: Southern Illinois University Press, 1961).

Garver, Lloyd, "No, You Can't Have Nintendo," *Newsweek*, June 11, 1990: 8.

Gibson, William, *Neuromancer* (New York: Berkley, 1984).

Greenfield, Patricia Marks, *Mind and Media*: *The Effects of Television, Video Games, and Computers* (Cambridge, Mass.: Harvard University Press, 1984).

Ishihara, Shintaro, and Akio Morita, *No to ieru Nihon [The Japan That Can Say No]* (Tokyo: Kobunsha, 1990).

"Japan Goes Hollywood," *Newsweek*, October 9, 1989: 62-72.

Japan Publications (ed.), *Hanafuda: The Flower Card Game* (Tokyo: Nichibo Shuppan-sho, 1970).

Katz, Donald R., "The New Generation Gap," *Esquire*, February 1990: 49-50.

Kidder, Tracy, *The Soul of a New Machine* (Boston: Little, Brown, 1981).

"Less Fun, Few Games? Toy Industry Analyst Is Bearish on Nintendo," *Barron's*, December 3, 1990: 16-63.

Levine, David, "Special Effects: Broadcast Arts Has Created a New Genre in the World of Television," *Continental Profiles*, April 1991: 23-40.

Levy, Richard C., and Ronald O. Weingartner, *Inside Santa's Workshop: How Toy Inventors Develop, Sell, and Cash In on Their Ideas* (New York: Henry Holdt, 1990).

Levy, Steven, Hackers: *Heroes of the Computer Revolution* (New York: Bantam Doubleday Dell, 1984).

Loftus, Geoffrey R., and Elizabeth F. Loftus, *Mind at Play: The Psychology of Video Games* (New York: Basic, 1983).

McLuhan, Marshall, *Understanding Media: The Extensions of Man* (New York: McGraw-Hill, 1964).

Moritz, Michael, *The Little Kingdom: The Private Story of Apple Computer* (New York: William Morrow, 1984), p. 13.

Nakata, Hiroyuki, *Success Legend in the Semiconductor Era. Nintendo's Great Strategy (The Day Mario Exceeded Toyota)* (Tokyo: JICC, 1990).

Okimoto, Daniel I., *Between MITI and the Market: Japanese Industrial Policy for High Technology* (Stanford: Stanford University Press, 1989).

Orey, Michael, "Godzilla in Toyland: Why Nobody Beats Nintendo of America at Its Own Game," *The American Lawyer*, April 1990: 62-69.

Papert, Seymour, *Mindstorms: Children, Computers and Powerful Ideas* (New York: Basic, 1980).

Perry, J. Nancy, "Will Sony Make It in Hollywood?" *Fortune*, September 9, 1991: 158-166.

Provenzo, Eugene F., Jr., *Video Kids: Making Sense of Nintendo* (Cambridge, Mass.: Harvard University Press, 1991).

Rayl, A.J.S., "The New, Improved Reality," *Los Angeles Times Magazine*, July 21, 1991: 17.

Reischauer, Edwin O., *The Japanese Today* (Cambridge, Mass.: Harvard University Press, 1977).

Rosenberg, Scott, "Condemned to Be Mario: The Video-game Plumber as Existential Hero," *Image*, September 1, 1991: 22-25.

Schlender, Brenton R., "The Future of the PC," *Fortune*, August 26, 1991: 40.

Schwartz, John, "The Next Revolution," *Newsweek*, April 6, 1992: 42-48.

Selnow, Gary W., "Playing Videogames: The Electronic Friend," *Journal of Communications*, Spring 1984: 140-148.

Sheff, David, "How Nintendo Did It," *Men's Life*, Fall 1990: 84.

———"Interview: Pat Choate," *Upside*, February 1991.

———"Interview: Trip Hawkins," *Upside,* August/September, 1990: 46.

———"Mario's Big Brother," *Rolling Stone*, January 9, 1992: 45.

————"Playboy Interview: Steve Jobs," *Playboy*, February 1985.

————"Playboy Interview: Robert Maxwell." *Playboy*, October 1991.

————"Reversal of Fortune," *San Francisco Focus*, May 1991: 54.

Skow, John, "Games That People Play," *Time*, January 18, 1982: 50-58.

Takahashi, Kenji, *Super Famicom: The Intrigue of Nintendo* (*Light and Shadow of an Enterprise That Surpassed Matsushita and Sony*) (Tokyo: Kobunsha, 1990).

Tatsuno, Sheridan M., *Created in Japan* (New York: Ballinger, 1990).

"Teaching Japan to Say No," *Time*, November 20, 1989: 81-82.

Tiley, Ed, *Tricks of the Nintendo Masters* (Carmel, Ind.: Hayden, 1990).

Turkle, Sherry, *The Second Self: Computers and the Human Spirit* (New York: Simon & Schuster, 1984).

Utsumi, Ichiro, *Secrets of Nintendo's Gulliverian Oligopoly* (Tokyo: Nihon Bungei-sha, 1989).

"Videogames: Fun or Serious Threat?" *U.S. News & World Report*, February 22, 1982: 7.

Warshofsky, Fred, *The Chip War* (New York: Scribner's, 1989).

Webb, Marcus, "Into the Unknown!: New Decade Brings Promises and Pitfalls for Hopeful Coin-op Tradesters," *Replay*, January 1990: 41-66.

INDEX

R&D 2 group of, 39-40, 41-42, 71
R&D 3 group of, 39-40, 42-44, 114
R&D 4 group of, 49, 59, 358
stock sales of, 20, 24, 34, 197
subcontractors for, 32, 79
Ultra toy series of, 22
values imparted by, 10
"Yamauchi's generals" at, 67
Nintendo Entertainment System (NES), 3, 5, 349-50, 358, 443, 444, 445, 460, 470
as computer terminal, 7, 29, 32-34, 267-68, 337, 394, 446
development of, 162-63
introduction of, 163-69
lock-out chips for, 161, 214, 247-50, 251, 256, 267, 286, 288
marketing of, 174-77
memory capability of, 33
retail price of, 29, 34
U.S. sales of, 169, 172, 221, 267, 367, 401
Nintendo Hands Free system, 210
Nintendo International, 413-14
Nintendo Network, 390-98
Nintendo of America (NOA):
advertising campaigns of, 187-91, 461-62
anti-Japanese resentment felt toward, 195, 203, 223-24, 261-62, 274-75, 279, 282
autonomy of, 453
battle over E3 and CES trade shows, 464-67
change in executive tier, 459
change in image, 461-62
as cultural force, 8-9, 188, 458
"Donkey Kong" as first success of, 109-11
dress code at, 198
founding of, 92-95
game counseling system of, 181-83
game evaluation process at, 184-85, 224-25
home video-game market entered by, 158-69
in investigations and lawsuits, 243-74, 285-91, 335-39
licensed spinoffs from, 191-192
licensee companies of, 214-35, 243-60, 269-71, 277-78, 365-66, 401, 402
management style at, 197-202
marketing strategies of, 193-96, 402, 408, 472

market share of, 349, 364, 402, 430-34
NCL's relationship with, 196-97
"Next-generation system (2000)," 444
number of people employed by, 410
"Play it Loud" ad campaign, 461-62
profitability of, 196, 221, 343
Redmond move of, 112-13
return policy of, 196
salaries and bonuses at, 200
sales team of, 97, 99-100, 102-3
Seattle move of, 105-6
use of FUD (fear, uncertainty and doubt), 450-51
Nintendo of Europe, 185, 269, 410, 414-16
Nintendo Player's Guide, 180-81
Nintendo Power, 178-80, 181, 184, 215, 220, 224, 270, 271, 290, 293, 339, 362, 365, 409
Nintendo PowerFest, 191
Nintendo Seal of Quality, 175
Nintendo World Championship, 191, 192
Nippon Keisai Shimbun (Japan Economic Journal), 74, 79, 391
NOA, see Nintendo of America
Nomura Securities Co., Ltd., 78, 392
NROM chips, 42

Odyssey, 108, 141, 174
Odyssey 2, 143
Ohga, Nhorio, 380
O'Keefe, Richard, 386
Olafsson, Olaf, 379, 381, 382
online banking, 454
Oprah, 206
Orbison, Roy, 193
organized crime, 15, 116, 137, 360-61
Origin, 455
Ota, Shigeru, 164, 185, 399, 410
otaku, 44
Ovitz, Mike, 192

Paauw, Donald, 247
pachinko, 57
"Pac-Man," 73, 100, 103, 149, 150, 159, 244, 443
Pajitnov, Alexey, 296-309, 313
background of, 296-98
in Japan, 345-47